SCIENCE

BELFAST PUBLIC LIBRARIES
539.04
ADVA
697511

Advances in
ATOMIC AND MOLECULAR PHYSICS

VOLUME 10

CONTRIBUTORS TO THIS VOLUME

LLOYD ARMSTRONG, Jr.

K. L. BELL

B. C. FAWCETT

SERGE FENEUILLE

WESLEY T. HUNTRESS, Jr.

A. E. KINGSTON

W. LANGE

J. LUTHER

W. C. PRICE

A. STEUDEL

ADVANCES IN
ATOMIC
AND MOLECULAR
PHYSICS

Edited by
D. R. Bates
DEPARTMENT OF APPLIED MATHEMATICS AND THEORETICAL PHYSICS
THE QUEEN'S UNIVERSITY OF BELFAST
BELFAST, NORTHERN IRELAND

Benjamin Bederson
DEPARTMENT OF PHYSICS
NEW YORK UNIVERSITY
NEW YORK, NEW YORK

VOLUME 10

 1974

ACADEMIC PRESS   New York   San Francisco   London
A Subsidiary of Harcourt Brace Jovanovich, Publishers

COPYRIGHT © 1974, BY ACADEMIC PRESS, INC.
ALL RIGHTS RESERVED.
NO PART OF THIS PUBLICATION MAY BE REPRODUCED OR
TRANSMITTED IN ANY FORM OR BY ANY MEANS, ELECTRONIC
OR MECHANICAL, INCLUDING PHOTOCOPY, RECORDING, OR ANY
INFORMATION STORAGE AND RETRIEVAL SYSTEM, WITHOUT
PERMISSION IN WRITING FROM THE PUBLISHER.

ACADEMIC PRESS, INC.
111 Fifth Avenue, New York, New York 10003

United Kingdom Edition published by
ACADEMIC PRESS, INC. (LONDON) LTD.
24/28 Oval Road, London NW1

LIBRARY OF CONGRESS CATALOG CARD NUMBER: 65-18423

ISBN 0-12-003810-2

PRINTED IN THE UNITED STATES OF AMERICA

Contents

LIST OF CONTRIBUTORS   vii
CONTENTS OF PREVIOUS VOLUMES   ix

Relativistic Effects in the Many-Electron Atom
Lloyd Armstrong, Jr. and Serge Feneuille

I. Introduction	1
II. Basic Concepts	3
III. Nonrelativistic Limits	10
IV. Effective Operators	21
V. Relativity Plus Nonrelativistic Perturbations	40
VI. Relativistic Radial Wave Functions	42
References	50

The First Born Approximation
K. L. Bell and A. E. Kingston

I. Introduction	53
II. Excitation of Atoms by Electron and Proton Impact	54
III. Ionization of Atoms by Electron and Proton Impact	87
IV. Atom–Atom and Ion–Atom Collisions	113
References	125

Photoelectron Spectroscopy
W. C. Price

I. Introduction	131
II. X-Ray Photoelectron Spectroscopy	134
III. Ultraviolet Photoelectron Spectroscopy	136
IV. Physical Aspects of Photoelectron Spectroscopy	155
V. Conclusion	169
References	170

Dye Lasers in Atomic Spectroscopy

W. Lange, J. Luther, and A. Steudel

I. Introduction	173
II. Properties of Dye Lasers	174
III. Applications of the High Spectral Density of Dye Lasers	177
IV. Applications of Tunable Dye Lasers with Extreme Narrow Bandwidth	197
References	217

Recent Progress in the Classification of the Spectra of Highly Ionized Atoms

B. C. Fawcett

I. Introduction	224
II. Laboratory Light Sources	225
III. Measuring and Experimental Techniques	232
IV. Theoretical Calculations	236
V. Line Classifications of Highly Ionized Systems	239
VI. Identification of Emission Lines of Highly Ionized Atoms in the Solar Spectrum	262
VII. Discussion	284
References	285

A Review of Jovian Ionospheric Chemistry

Wesley T. Huntress, Jr.

I. Introduction	295
II. Production of Ions by Photoionization	297
III. Ion-Neutral Reactions	300
IV. Terminal-Ion Loss Processes	309
V. Ion Loss Rates	322
VI. Ionization Processes of Photoelectrons	328
VII. Additional Photon Impact Processes	334
VIII. Concluding Remarks	335
References	338

SUBJECT INDEX	341

List of Contributors

Numbers in parentheses indicate the pages on which the authors' contributions begin.

LLOYD ARMSTRONG, Jr.,* Laboratoire Aimé Cotton, Centre National de la Recherche Scientifique, Campus d'Orsay, Essonne, France (1)

K. L. BELL, Department of Applied Mathematics and Theoretical Physics, The Queen's University, Belfast, Northern Ireland (53)

B. C. FAWCETT, Science Research Council, Astrophysics Research Division of the Appleton Laboratory, Culham Laboratory, near Abingdon, Berkshire, England (223)

SERGE FENEUILLE, Laboratoire Aimé Cotton, Centre National de la Recherche Scientifique, Campus d'Orsay, Essonne, France (1)

WESLEY T. HUNTRESS, Jr., Jet Propulsion Laboratory, California Institute of Technology, Pasadena, California (295)

A. E. KINGSTON, Department of Applied Mathematics and Theoretical Physics, The Queen's University, Belfast, Northern Ireland (53)

W. LANGE, Institut A für Experimentalphysik, Technische Universität Hannover, Hannover, Federal Republic of Germany (173)

J. LUTHER, Institut A für Experimentalphysik, Technische Universität Hannover, Hannover, Federal Republic of Germany (173)

W. C. PRICE, King's College, London, England (131)

A. STEUDEL, Institut A für Experimentalphysik, Technische Universität Hannover, Hannover, Federal Republic of Germany (173)

* Present address: Department of Physics, The Johns Hopkins University, Baltimore, Maryland.

Contents of Previous Volumes

Volume 1

Molecular Orbital Theory of the Spin Properties of Conjugated Molecules, *G. G. Hall and A. T. Amos*

Electron Affinities of Atoms and Molecules, *B. L. Moiseiwitsch*

Atomic Rearrangement Collisions, *B. H. Bransden*

The Production of Rotational and Vibrational Transitions in Encounters between Molecules, *K. Takayanagi*

The Study of Intermolecular Potentials with Molecular Beams at Thermal Energies, *H. Pauly and J. P. Toennies*

High Intensity and High Energy Molecular Beams, *J. B. Anderson, R. P. Andres, and J. B. Fenn*

AUTHOR INDEX—SUBJECT INDEX

Volume 2

The Calculation of van der Waals Interactions, *A. Dalgarno and W. D. Davison*

Thermal Diffusion in Gases, *E. A. Mason, R. J. Munn, and Francis J. Smith*

Spectroscopy in the Vacuum Ultraviolet, *W. R. S. Garton*

The Measurement of the Photoionization Cross Sections of the Atomic Gases, *James A. R. Samson*

The Theory of Electron-Atom Collisions, *R. Peterkop and V. Veldre*

Experimental Studies of Excitation in Collisions between Atomic and Ionic Systems, *F. J. de Heer*

Mass Spectrometry of Free Radicals, *S. N. Foner*

AUTHOR INDEX—SUBJECT INDEX

Volume 3

The Quantal Calculation of Photoionization Cross Sections, *A. L. Stewart*

Radiofrequency Spectroscopy of Stored Ions. I: Storage, *H. G. Dehmelt*

Optical Pumping Methods in Atomic Spectroscopy, *B. Budick*

Energy Transfer in Organic Molecular Crystals: A Survey of Experiments, *H. C. Wolf*

Atomic and Molecular Scattering from Solid Surfaces, *Robert E. Stickney*

Quantum Mechanics in Gas Crystal-Surface van der Waals Scattering, *F. Chanoch Beder*

Reactive Collisions between Gas and Surface Atoms, *Henry Wise and Bernard J. Wood*

AUTHOR INDEX—SUBJECT INDEX

Volume 4

H. S. W. Massey—A Sixtieth Birthday Tribute, *E. H. S. Burhop*
Electronic Eigenenergies of the Hydrogen Molecular Ion, *D. R. Bates and R. H. G. Reid*
Applications of Quantum Theory to the Viscosity of Dilute Gases, *R. A. Buckingham and E. Gal*
Positrons and Positronium in Gases, *P. A. Fraser*
Classical Theory of Atomic Scattering, *A. Burgess and I. C. Percival*
Born Expansions, *A. R. Holt and B. L. Moiseiwitsch*
Resonances in Electron Scattering by Atoms and Molecules, *P. G. Burke*
Relativistic Inner Shell Ionization, *C. B. O. Mohr*
Recent Measurements on Charge Transfer, *J. B. Hasted*
Measurements of Electron Excitation Functions, *D. W. O. Heddle and R. G. W. Kessing*
Some New Experimental Methods in Collision Physics, *R. F. Stebbings*
Atomic Collision Processes in Gaseous Nebulae, *M. J. Seaton*
Collisions in the Ionosphere, *A. Dalgarno*
The Direct Study of Ionization in Space, *R. L. F. Boyd*
AUTHOR INDEX—SUBJECT INDEX

Volume 5

Flowing Afterglow Measurements of Ion-Neutral Reactions, *E. E. Ferguson, F. C. Fehsenfeld, and A. L. Schmeltekopf*
Experiments with Merging Beams, *Roy H. Neynaber*
Radiofrequency Spectroscopy of Stored Ions II: Spectroscopy, *H. G. Dehmelt*
The Spectra of Molecular Solids, *O. Schnepp*
The Meaning of Collision Broadening of Spectral Lines: The Classical-Oscillator Analog, *A. Ben-Reuven*
The Calculation of Atomic Transition Probabilities, *R. J. S. Crossley*
Tables of One- and Two-Particle Coefficients of Fractional Parentage for Configurations $s^\lambda s'^\mu p^q$, *C. D. H. Chisholm, A. Dalgarno, and F. R. Innes*
Relativistic Z Dependent Corrections to Atomic Energy Levels, *Holly Thomis Doyle*
AUTHOR INDEX—SUBJECT INDEX

Volume 6

Dissociative Recombination, *J. N. Bardsley and M. A. Biondi*
Analysis of the Velocity Field in Plasma from the Doppler Broadening of Spectral Emission Lines, *A. S. Kaufman*
The Rotational Excitation of Molecules by Slow Electrons, *Kazuo Takayanagi and Yukikazu Itikawa*

The Diffusion of Atoms and Molecules, *E. A. Mason and T. R. Marrero*

Theory and Application of Sturmian Functions, *Manuel Rotenberg*

Use of Classical Mechanics in the Treatment of Collisions between Massive Systems, *D. R. Bates and A. E. Kingston*

AUTHOR INDEX—SUBJECT INDEX

Volume 7

Physics of the Hydrogen Maser, *C. Audoin, J. P. Schermann, and P. Grivet*

Molecular Wave Functions: Calculation and Use in Atomic and Molecular Processes, *J. C. Browne*

Localized Molecular Orbitals, *Harel Weinstein, Ruben Pauncz, and Maurice Cohen*

General Theory of Spin-Coupled Wave Functions for Atoms and Molecules, *J. Gerratt*

Diabatic States of Molecules—Quasistationary Electronic States, *Thomas F. O'Malley*

Selection Rules within Atomic Shells, *B. R. Judd*

Green's Function Technique in Atomic and Molecular Physics, *Gy. Csanak, H. S. Taylor, and Robert Yaris*

A Review of Pseudo-Potentials with Emphasis on Their Application to Liquid Metals, *Nathan Wiser and A. J. Greenfield*

AUTHOR INDEX—SUBJECT INDEX

Volume 8

Interstellar Molecules: Their Formation and Destruction, *D. McNally*

Monte Carlo Trajectory Calculations of Atomic and Molecular Excitation in Thermal Systems, *James C. Keck*

Nonrelativistic Off-Shell Two-Body Coulomb Amplitudes, *Joseph C. Y. Chen and Augustine C. Chen*

Photoionization with Molecular Beams, *R. B. Cairns, Halstead Harrison, and R. I. Schoen*

The Auger Effect, *E. H. S. Burhop and W. N. Asaad*

AUTHOR INDEX—SUBJECT INDEX

Volume 9

Correlation in Excited States of Atoms, *A. W. Weiss*

The Calculation of Electron–Atom Excitation Cross Sections, *M. R. H. Rudge*

Collision-Induced Transitions Between Rotational Levels, *Takeshi Oka*

The Differential Cross Section of Low Energy Electron–Atom Collisions, *D. Andrick*

Molecular Beam Electric Resonance Spectroscopy, *Jens C. Zorn and Thomas C. English*

Atomic and Molecular Processes in the Martian Atmosphere, *Michael B. McElroy*

AUTHOR INDEX—SUBJECT INDEX

RELATIVISTIC EFFECTS IN THE MANY-ELECTRON ATOM

LLOYD ARMSTRONG, JR.*† and SERGE FENEUILLE

Laboratoire Aimé Cotton
Centre National de la Recherche Scientifique
Campus d'Orsay
Essonne, France

I. Introduction	1
II. Basic Concepts	3
A. The Dirac Wave Equation	3
B. The Breit Interaction	5
C. Dirac Electron in a Central Field	7
III. Nonrelativistic Limits	10
A. The Central Field Hamiltonian H_C	11
B. Interaction with an External Field	13
C. The Perturbation Hamiltonian H_P	15
D. The Breit Perturbation H_B	16
E. Higher Order Terms	17
F. The Nonrelativistic Hamiltonian	20
IV. Effective Operators	21
A. General Concepts	21
B. Effective Operators for H_P and H_B	26
C. Effective Operators Involving Fields	30
V. Relativity Plus Nonrelativistic Perturbations	40
VI. Relativistic Radial Wave Functions	42
A. The Parametric Method	42
B. Hydrogenic Radial Wave Functions	43
C. The Pseudopotential Method	44
D. Self-consistent Field Calculations	46
E. Size and Mass of the Nucleus	48
F. Configuration Mixing and Correlations	49
References	50

I. Introduction

The study of relativistic effects in atomic structure theory is rather an old story, the first work on this subject having been done by Sommerfeld in 1916. Moreover, the basic ideas and much of the formalism that we still use

---

* On leave from The Johns Hopkins University during 1972–1973.
† Present address: Department of Physics, The Johns Hopkins University, Baltimore, Maryland.

at present were given by Dirac and Breit during the 1930's (Dirac, 1928; Breit, 1929). However, for a long time, relativistic investigations were confined to one- and sometimes two-electron atoms (Bethe and Salpeter, 1957), while the properties of many-electron atoms were studied in the Pauli approximation. In fact, the introduction of relativistic effects into the study of the many-electron atom was made possible mainly by the simplifications brought about through use of tensor operator techniques (Racah, 1942, 1943, 1949) and by the enormous power of electronic computers which permit rather precise self-consistent field calculations. Furthermore, the increasing precision of experimental data, particularly that of hyperfine structure, has progressively demonstrated the necessity of introducing relativistic effects into the study of atomic properties. More recently, speculations concerning the existence of super-heavy elements have produced growing interest in relativistic atomic structure calculations since, obviously, the physical and chemical properties of such elements can be predicted only if relativity is taken into account. Thus, one can easily understand that very important advances have recently been made in this field, because considerably improved calculations have become possible at a time when they are in great demand by experimental physicists and chemists.

However, it must be recalled that the problem of relativistic effects in atomic structure is far from being solved, even from a fundamental point of view. The Breit formulation of the interelectron interaction (Breit, 1929), the only formulation which can be conveniently used, is not Lorentz covariant and, in fact, is valid only for atoms with rather low nuclear charge Z. For large Z, it is possible in principle to use a modified form of the Breit interaction (Bethe and Salpeter, 1957), but this form has not actually been used. In any case, it neglects radiative corrections which may be significant at large Z. Thus, the problem of carrying out a rigorous, fully relativistic calculation for the many-electron atom remains impossible at this time.

There are already several excellent reviews of relativistic atomic structure calculations (Doyle, 1969; Grant, 1970; Ermolaev and Jones, 1973; Lindgren and Rosén, 1973), each emphasizing or highlighting a particular aspect of the problem (for example, the calculation of self-consistent-field wave functions). Our goal here is to give a more global review of the subject; we will admit, however, a propensity for the method of equivalent operators, which is described in some detail since it seems, despite its simplicity and utility, to be unfamiliar to many workers in the field.

We begin by reviewing in Section II the basic ideas which are the starting point of any relativistic calculation of atomic properties. Then, in Section III, nonrelativistic limits are discussed in order to exhibit the main differences between a relativistic and a nonrelativistic treatment of atomic structure. The equivalent operator formalism, initially introduced by Sandars and Beck (1965) for studies of hyperfine structure, is described in Section IV and is applied to fine and hyperfine structures, the Zeeman effect, and transition probabilities. In Section V, the problem of perturbation calculations in the relativistic scheme is briefly discussed. Finally, we consider in Section VI the calculation of relativistic radial wave functions.

II. Basic Concepts

A. The Dirac Wave Equation

If we consider the nucleus of an atom as a point charge Ze with infinite mass situated at the origin of the coordinate system, we are concerned only with the motion of the electrons. Then, the natural starting point of any relativistic study of the many-electron atom is the Dirac wave equation. There are many excellent discussions of Dirac theory (Dirac, 1930; Schiff, 1955; Messiah, 1959; Akhiezer and Berestetskii, 1965), but, for the convenience of the reader, it may be useful to briefly review some of the aspects of the theory.

First of all, most of the postulates at the origin of the Dirac theory are the same as in any quantum mechanical theory (Messiah, 1959). The additional requirements of the relativistic theory are (a) the equation of motion must be consistent with the principle of special relativity, and (b) the theory must be consistent with the correspondence principle in both the nonrelativistic and the nonquantum limits.

In order to satisfy the superposition principle, the wave equation is necessarily of the form:

$$H\psi = i\hbar(\partial\psi/\partial t) \tag{1}$$

where H is the energy operator. The relativistic covariance then implies that H is linear in the space derivatives, that is, in the components of the linear momentum p_k. Moreover, for free electrons, the correspondence principle requires that the energy E and the linear momentum $\mathbf{p}$ be related according to

$$E^2 = c^2(p^2 + m^2c^2) \tag{2}$$

Thus, the energy operator must be postulated to be of the form

$$H = c(\boldsymbol{\alpha} \cdot \mathbf{p}) + \beta mc^2 \tag{3}$$

where the components of the 4-vector $(\boldsymbol{\alpha}, \beta)$ must satisfy the relationship

$$\begin{aligned} \alpha_i \alpha_k + \alpha_k \alpha_i &= 2\delta_{ik} \quad (i, k = 1, 2, 3) \\ \alpha_i \beta + \beta \alpha_i &= 0 \\ \beta^2 &= 1 \end{aligned} \tag{4}$$

Furthermore, H must be Hermitian, which requires that the operators α and β also be Hermitian.

As Dirac has shown, a matrix representation for (α, β) can be found. In fact, in order to satisfy Eqs. (4), the minimum dimensionality of this matrix representation is 4, and for a 4 by 4 representation, there is an infinite choice. In this article, we shall use the standard representation in which α and β are, respectively, written in terms of the Pauli 2 by 2 σ matrices and the 2 by 2 unit matrix I. That is,

$$\alpha = \begin{pmatrix} 0 & \sigma \\ \sigma & 0 \end{pmatrix}, \quad \beta = \begin{pmatrix} I & 0 \\ 0 & -I \end{pmatrix} \tag{5}$$

The eigenfunction ψ is then described by a matrix of one column and four rows.

By multiplying Eq. (3) by β and by defining

$$\gamma_k = -i\beta\alpha_k$$
$$\gamma_4 = \beta \tag{6}$$
$$X_\mu = (\mathbf{r}, ict)$$

one obtains the covariant form of the Dirac wave equation

$$\left(\hbar \sum_\mu \gamma_\mu \frac{\partial}{\partial X_\mu} + mc\right)\psi = 0 \quad (\mu = 1, 2, 3, 4) \tag{7}$$

which is particularly convenient for studies of its invariance properties. For more details, the reader is referred to the books of Messiah (1959) and Bethe and Salpeter (1957).

The postulates which are at the origin of the Dirac theory for the electron still apply for electrons in electromagnetic fields if we replace E by $E + eV$ and $\mathbf{p}$ by

$$\pi = \mathbf{p} + (e/c)\mathbf{A} \tag{8}$$

and V and $\mathbf{A}$ are, respectively, the scalar and vector potentials of the field (the charge of the electron is $-e$). Then the Dirac equation for stationary states becomes

$$[c(\alpha \cdot \pi) + \beta mc^2 - (E + eV)]\psi = 0 \tag{9}$$

or, in covariant notation

$$\left[\sum_\mu \gamma_\mu\left(-i\hbar \frac{\partial}{\partial X_\mu} + \frac{e}{c} A_\mu\right) - imc\right]\psi = 0 \tag{10}$$

where $A_\mu = (\mathbf{A}, iV)$. However, in these equations, the potentials are taken as given quantities and are not quantized. This implies that the Dirac equation results from a single particle theory. Now, it is well known that the modifications produced in the Dirac theory by the quantization of the electromagnetic field can affect atomic calculations, and sometimes in a very dramatic way. However, these corrections due to field quantization are extremely difficult to evaluate for many-electron atoms, and thus are usually ignored. In cases where these "radiative corrections" are large enough to significantly affect the calculation (as in the case of the Zeeman effect), they are generally introduced in an *ad hoc* manner by taking their value for a free electron (e.g., anomalous moment of the electron) or for electrons in a Coulomb potential (e.g., Lamb shift).

Finally, let us note that, because Eq. (2) relates the energy squared to the electron mass and momentum, there are two values of E for a given $\mathbf{p}$, one of which is of the order of $+mc^2$ (assuming $|\mathbf{p}| \ll mc$), the other of the order of $-mc^2$. There are, correspondingly, different eigenfunctions of Eq. (1) for the different energies. In the presence of a weak attractive potential field, there are still two sets of eigenfunctions, one corresponding to positive energies; the other, to negative energies. The former set will contain both bound and continuum states, whereas the latter contains only continuum states. The negative energy solutions correspond, of course, to positron states. For our purposes, it will not be necessary to consider the complicated aspects of a complete positron-electron theory, and we can simply use the designations positive- and negative-energy electrons. These states will be of interest to us because, in any perturbation expansion based on a zeroth-order Dirac equation, intermediate state summations must include all eigenfunctions of the zeroth-order equation; that is, both positive- and negative-energy solutions.

B. The Breit Interaction

Obviously, in the theory of atomic structure we are concerned with interacting electrons, and the Dirac theory cannot be sufficient. However, at present there is no method known for describing in Hamiltonian form the interaction between relativistic electrons. The most commonly used approximation to this interaction is the Breit Hamiltonian (Breit, 1929)

$$H\psi = \left\{ \sum_{i=1}^{N} [c(\boldsymbol{\alpha}_i \cdot \boldsymbol{\pi}_i) + \beta_i mc^2 - eV(\mathbf{r}_i)] \right.$$
$$\left. + \sum_{i>j} \left[\frac{e^2}{r_{ij}} - \frac{e^2}{2r_{ij}} \left\{ (\boldsymbol{\alpha}_i \cdot \boldsymbol{\alpha}_j) + \frac{(\boldsymbol{\alpha}_i \cdot \mathbf{r}_{ij})(\boldsymbol{\alpha}_j \cdot \mathbf{r}_{ij})}{r_{ij}^2} \right\} \right] \right\} \psi = E\psi \quad (11)$$

where $r_{ij} = |\mathbf{r}_i - \mathbf{r}_j|$. The one-body part of this equation is clearly the sum of the Dirac equations for the N electrons of the atom. The two-body part arises from the exchange of a single photon between two electrons (Akhiezer and Berestetskii, 1965; Mann and Johnson, 1971). The origin of this term can be understood by referring to Eq. (10). We see that an external field A_μ produces a perturbation $(e/c)\sum_\mu \gamma_\mu A_\mu$ in the Dirac Hamiltonian. The matrix element of this perturbation, $(\psi'_1 | e \sum_\mu \gamma_\mu A_\mu | \psi_1)$ corresponds simply to the classical $(\mathbf{j} \cdot \mathbf{A})$ interaction between the current $\psi'_1 e\gamma_\mu(1)\psi_1$ of electron 1 and a field A_μ. In this case, the field is produced by the current of the second electron:

$$A_\mu(\mathbf{r}_1) = \int \frac{j_\mu(\mathbf{r}_2)}{r_{12}} e^{i\omega r_{12}/c} \, dr_2 \tag{12}$$

where $j_\mu(\mathbf{r}_2) = \psi'_2 e\gamma_\mu(2)\psi_2$ and $\hbar\omega = |E_2 - E'_2|$. The exponential appearing in A_μ is due to retardation. The matrix element of the interaction can then be written as

$$(\psi'_1\psi'_2 | (e^2/r_{12})[1 - (\boldsymbol{\alpha}_1 \cdot \boldsymbol{\alpha}_2)]e^{i\omega r_{12}/c} | \psi_1\psi_2) \tag{13}$$

where we have used $\sum_\mu \gamma_\mu(1)\gamma_\mu(2) = 1 - (\boldsymbol{\alpha}_1 \cdot \boldsymbol{\alpha}_2)$. Because energy must be conserved in this interaction, we also have $\hbar\omega = |E'_1 - E_1|$. When all states refer to electrons of positive energy, $\omega r_{12}/c \approx v/c \approx Z\alpha$ and the exponential may be expanded roughly in powers of $Z\alpha$. Keeping only the first term in this expansion (i.e., neglecting retardation), we find that the lowest order interaction between two Dirac electrons is given by (Gaunt, 1929)

$$B = (e^2/r_{12})[1 - (\boldsymbol{\alpha}_1 \cdot \boldsymbol{\alpha}_2)] \tag{14}$$

i.e., the usual Coulomb interaction plus a correction. This correction is, as we shall see later, of order $(Z\alpha)^2$, the Coulomb term. To be consistent, therefore, we should keep retardation terms of the same order. In doing this, one finds an additional interaction (Breit, 1929, 1930)

$$B' = \frac{e^2}{2r_{12}}\left[(\boldsymbol{\alpha}_1 \cdot \boldsymbol{\alpha}_2) - \frac{(\boldsymbol{\alpha}_1 \cdot \mathbf{r}_{12})(\boldsymbol{\alpha}_2 \cdot \mathbf{r}_{12})}{r_{12}^2}\right] \tag{15}$$

The sum of B and B', generalized to many-electron form, makes up the two-body part of H. Finally, let us note that, had ψ_2 described an electron of positive energy and ψ'_2 an electron of negative energy, then $\hbar\omega \approx 2mc^2$; in this case, an expansion of the retardation exponential would clearly not have been justified.

Thus we see that the Breit Hamiltonian considers only one-photon exchange processes between positive energy electrons and, even there, keeps only the leading terms in the retardation. As might be expected from such an approximation, H is not Lorentz invariant. A Lorentz invariant theory is, in principle, available—the covariant Bethe–Salpeter wave equation (Bethe and Salpeter, 1957)—but, at least at present, its application to many-electron atoms seems hopeless.

The fact that the Breit Hamiltonian describes only interactions between positive energy electrons has an important consequence: the treatment of the Breit operator

$$H_B = B + B' - \frac{e^2}{r_{ij}} = \sum_{i<j} -\frac{e^2}{2r_{ij}} \left[(\boldsymbol{\alpha}_i \cdot \boldsymbol{\alpha}_j) + \frac{(\boldsymbol{\alpha}_i \cdot \mathbf{r}_{ij})(\boldsymbol{\alpha}_j \cdot \mathbf{r}_{ij})}{r_{ij}^2} \right] \quad (16)$$

must generally be restricted to first-order perturbation theory. That is, in higher order perturbation expressions, intermediate sums would have to include negative energy states, and in such cases, H_B usually does not describe the interaction correctly. There are many methods which in principle can be used to evaluate higher order corrections, but in fact, they are all too complicated to be used in the many-electron atom. For details, the reader is referred to Bethe and Salpeter (1957) and Akhiezer and Berestetskii (1965).

C. Dirac Electron in a Central Field

Although the Breit equation is already an approximation, it cannot be solved exactly. Therefore it is necessary to work with approximation methods, which may be more-or-less sophisticated according to the problem under study. In any case, most of these methods start from some sort of central field model in which the electrons of the atom are regarded as moving independently in a central field, with the residual interactions being treated as perturbations. That is, we write H of Eq. (11) as

$$H = H_C + H_P + H_B$$

where

$$H_C = \sum_i H_D(i) \quad (17)$$

$$H_D(i) = c(\boldsymbol{\alpha}_i \cdot \mathbf{p}_i) + \beta_i mc^2 - eU(r_i)$$

and

$$H_P = \sum_i \left[eU(r_i) - \frac{Ze^2}{r_i} \right] + \sum_{i<j} \frac{e^2}{r_{ij}}$$

(we assume, for the moment, that $\mathbf{A} = 0$). Eigenfunctions of H_C can easily be found (at least numerically), and thus this term is used as the zeroth-order Hamiltonian; H_P and H_B are then treated as perturbations.

In this section, we shall be concerned only with the positive energy eigenfunctions of H_C or, more precisely, of H_D. Thus, we are faced with a one-particle problem provided that the motion of the nucleus can be eliminated.

This can be done in a trivial way by taking the mass of the nucleus to be infinite. A better approximation can be obtained by replacing the electron mass by the reduced mass; of course, reduced mass is not unambiguously defined in relativistic problems, but the error introduced by this ambiguity is of the same order of magnitude as the radiative corrections. It can therefore be neglected in the study of most of the properties of atomic energy levels. In the following, the electron coordinates will be defined with respect to the central field source, which will be regarded as motionless.

In the nonrelativistic central field model, the three components of the orbital angular momentum operator defined by $\hbar \mathbf{l} = \mathbf{r} \times \mathbf{p}$ and the three components of $\boldsymbol{\sigma}$ commute with the Hamiltonian. In the relativistic case, this is no longer true since

$$[H_D, \mathbf{l}] = -ic\{\boldsymbol{\alpha} \times \mathbf{p}\}$$
$$[H_D, \boldsymbol{\sigma}] = 2ic\{\boldsymbol{\alpha} \times \mathbf{p}\} \tag{18}$$

However, it is clear that the three components of the total angular momentum defined by $\mathbf{j} = \mathbf{l} + \frac{1}{2}\boldsymbol{\sigma} = \mathbf{l} + \mathbf{s}$ commute with H_D. Therefore, the eigenfunctions of H_D can be labeled by the quantum numbers j and m_j, which are defined by

$$j^2 |\gamma j m_j\rangle = j(j+1) |\gamma j m_j\rangle$$
$$j_z |\gamma j m_j\rangle = m |\gamma j m_j\rangle \tag{19}$$

Using the properties of the $\boldsymbol{\alpha}$ and β matrices [Eq. (5)], $(H_D - E) = 0$ can be rewritten in the form

$$(E + eU - mc^2)\xi - c(\boldsymbol{\sigma} \cdot \mathbf{p})\eta = 0$$
$$-c(\boldsymbol{\sigma} \cdot \mathbf{p})\xi + (E + eU + mc^2)\eta = 0 \tag{20}$$

where ξ and η are matrices with one column and two rows. Now, it is clear that in the nonrelativistic limit $(E \to mc^2)$, η goes to zero and ξ goes to a solution of the one-electron Schrödinger equation

$$H_{nr} |\psi\rangle = \left\{\frac{1}{2m}p^2 - eU(r)\right\} |\psi\rangle = W |\psi\rangle \tag{21}$$

The functions $|\psi\rangle$ which are simultaneously eigenfunctions of j^2 and j_z can be written as

$$|\psi\rangle = (1/r)R_l |ljm_j\rangle \tag{22}$$

where

$$|ljm_j\rangle = \sum_{m_s, m_l} \begin{pmatrix} \frac{1}{2} & l & j \\ m_s & m_l & -m_j \end{pmatrix} (-1)^{1/2 - l - m_j} [j]^{1/2} Y_{lm_l}(\theta, \varphi) \chi_{m_s}^{(1/2)} \tag{23}$$

The quantity $\chi^{(1/2)}_{m_s}$ is a Pauli spin matrix defined by

$$\chi^{(1/2)}_{1/2} = \begin{pmatrix} 1 \\ 0 \end{pmatrix}, \quad \chi^{(1/2)}_{-1/2} = \begin{pmatrix} 0 \\ 1 \end{pmatrix}$$

and $Y_{lm_l}(\theta, \varphi)$ is a spherical harmonic. Note that the spin is placed to the left of l in the $3-j$ symbol of Eq. (23). Some other commonly used definitions may, therefore, differ from ours by a phase. We have also used a rounded ket $|\,)$ to denote a relativistic wave function, and an angular ket $|\,\rangle$ for a non-relativistic wave function; we shall adopt this as a convention.

Equation (22) implies that ξ can be written as

$$\xi = [F(r)/r]\,|ljm_j\rangle \tag{24}$$

Using angular momentum algebra, it can be shown that

$$(\boldsymbol{\sigma} \cdot \mathbf{p})\xi = -i\hbar \left\{ \frac{1}{r}\frac{d}{dr}r + \frac{\varkappa}{r} \right\} \frac{F(r)}{r} \,|\bar{l}jm_j\rangle \tag{25}$$

where $\bar{l} = 2j - l$

$$\varkappa = \tfrac{1}{2}[\bar{l}(\bar{l}+1) - l(l+1)] = (-1)^{l+j-1/2}(j+\tfrac{1}{2}). \tag{26}$$

Equation (24), in conjunction with Eqs. (20) and (25), then requires that η be written in the form

$$\eta = i[G(r)/r]\,|\bar{l}jm_j\rangle \tag{27}$$

Thus, the eigenfunctions of H_D can be written as

$$|\psi) = \begin{pmatrix} [F(r)/r]\,|ljm_j\rangle \\ [iG(r)/r]\,|\bar{l}jm_j\rangle \end{pmatrix} \tag{28}$$

where the radial functions $F(r)$ and $G(r)$ satisfy the first-order coupled differential equations

$$\begin{aligned} \left(\frac{d}{dr} + \frac{\varkappa}{r}\right) G(r) &= \frac{-1}{\hbar c}[-E + mc^2 - eU(r)]F(r) \\ \left(\frac{d}{dr} - \frac{\varkappa}{r}\right) F(r) &= \frac{-1}{\hbar c}[E + mc^2 + eU(r)]G(r) \end{aligned} \tag{29}$$

and are normalized according to

$$\int_0^\infty (F^2 + G^2)\,dr = 1 \tag{30}$$

Equation (29) can be solved exactly for an attractive Coulomb potential; the solutions will be examined in Section VI. Let us remark, however, that the discrete positive eigenvalues in this case depend directly on the principal quantum number n, which, as in the nonrelativistic case, is a positive integer greater than l. Thus, for a Coulomb potential, the eigenfunctions of H_D can be labeled $|nljm_j\rangle$. For the many-electron atom, U still is attractive but remains Coulombic only at the origin ($U = Ze/r$) and at infinity [$U = (Z - N + 1)e/r$]. In this case, Eqs. (29) can only be solved numerically, but it is still possible to define a principal quantum number n and to label the eigenfunctions as $|nljm_j\rangle$. Obviously, the eigenvalue in the general case is no longer a simple function of n; n is, however, related to the number of nodes of the functions F and G. More precisely, the number of nodes (excluding the origin and infinity) of F and G are $(n - l - 1)$ and $(n - |\varkappa|)$, respectively. This can be shown by means of basically the same techniques as used to relate n to the number of nodes in the nonrelativistic case (Rose, 1961). We shall not consider continuum solutions to the eigenvalue equation for H_D, since we shall have no need for them; a description of the continuum functions for the Coulomb potential can be found in Rose (1961).

III. Nonrelativistic Limits

In the previous section, we discussed in a very general fashion the calculation of atomic properties using relativistic Hamiltonians. However, as we saw, these calculations cannot be, even in principle, completely accurate due to the approximations required in the derivation of the Hamiltonian operators themselves. Thus, the central field calculation described in Section II,C necessarily contains errors of order $(Z\alpha)^4 W$ because of errors in the Breit Hamiltonian, even though the interaction between the electron and the central field is correct to all orders in $Z\alpha$.

This innate limitation in the accuracy of the relativistic calculations suggests that in many cases it may be preferable to start with the many-electron Schrödinger equation

$$H_S|\psi\rangle = \left\{\sum_i \left(\frac{p_i^2}{2m} - \frac{Ze^2}{r_i}\right) + \sum_{i<j} \frac{e^2}{r_{ij}}\right\}|\psi\rangle = E_S|\psi\rangle \quad (31)$$

and take account of relativistic effects by adding perturbation operators of order $(Z\alpha)^2 E_S$; these operators can, of course, be obtained by expanding in powers of $Z\alpha$ the relativistic Hamiltonian operators discussed in the second section. In this section, we consider such an expansion.

We will use a development which is based on the realization that, in actual calculations, the Hamiltonian of Eq. (31) is too complicated to be solved, and one must use some variant of the central field model to obtain the nonrelativistic counterpart of H_C [Eq. (17)]

$$H_{SC}|\psi_C\rangle = \sum_i \left(\frac{p_i^2}{2m} - eU(r_i)\right)|\psi_C\rangle = E_C|\psi_C\rangle. \tag{32}$$

Our results are then expressed in terms of operators which are to be treated as perturbations to the Hamiltonian of Eq. (32).

An alternative (and much more elegant) discussion of nonrelativistic limits could be carried out using the Foldy–Wouthuysen transformation (Foldy and Wouthuysen, 1950; Messiah, 1959). We have not followed this approach, since we feel that in many cases it obscures the relationship between the relativistic operators and their nonrelativistic limits. We also do not consider the question of errors introduced in higher order terms by use of the approximate Hamiltonian of Eq. (32) as a zeroth-order approximation; this has been discussed quite thoroughly and ably by Ermolaev and Jones (1973).

A. The Central Field Hamiltonian H_C

We first consider the equation

$$(\psi|H_C|\psi) = E \tag{33}$$

where $|\psi\rangle$ is a positive energy state. By carrying out an expansion of the matrix element on the left-hand side of Eq. (33), we will find an operator which, when evaluated between the states $|\psi_C\rangle$, produces an energy eigenvalue equal to E through order $(Z\alpha)^2 E_C$. Because we are at this point considering a Hamiltonian which is a sum of one-electron operators, we can simplify our problem somewhat by considering only a single electron moving in the central field; the complete wave function is a product (antisymmetrized of course) of N one-electron wave functions of this type.

We begin by rewriting the second of Eqs. (20) in the form

$$\eta = (1/2Mc)(\boldsymbol{\sigma} \cdot \mathbf{p})\xi \tag{34}$$

where $2Mc^2 = E + eU + mc^2$. Then, the first of Eqs. (20) can be written as an equation in ξ only:

$$\{(\boldsymbol{\sigma} \cdot \mathbf{p})(1/2M)(\boldsymbol{\sigma} \cdot \mathbf{p}) - eU\}\xi = (E - mc^2)\xi \tag{35}$$

In the limit of nonrelativistic electron energies and weak central fields, we can approximate M by m, and rewrite Eq. (35) as

$$[(p^2/2m) - eU]\xi = W\xi \tag{36}$$

In this limit, we can replace ξ by $|\psi\rangle$, thus obtaining Eq. (21); Eq. (36) is therefore the one-electron form of Eq. (32). We denote this limit as the Pauli limit, since consideration of external fields in this approximation leads to the Pauli Hamiltonian (see Section III,B). In this approximation, one generally neglects matrix elements of an operator Op evaluated between the small components η of the wave function. The only exception occurs when Op is simply mc^2, in which case one must use the higher order approximation discussed below.

In the Pauli limit the wave function ξ is normalized to (but not including) order $(Z\alpha)^2$, and W is defined to order $(Z\alpha)^2$. However, we will be interested in calculating perturbations which are of order $(Z\alpha)^2 W$; thus, we must define our wave functions more precisely than is done in the Pauli approximation. This is easily done (Akhiezer and Berestetskii, 1965). First, we expand the operator $1/2Mc$ of Eq. (34)

$$\frac{1}{2Mc} \cong \frac{1}{2mc}\left\{1 - \frac{E - mc^2 + eU}{2mc^2} + \cdots\right\}$$

$$\cong \frac{1}{2mc}\left\{1 - \frac{W + eU}{2mc^2} + \cdots\right\}$$

Thus

$$\eta \cong \frac{1}{2mc}\left\{1 - \frac{W + eU}{2mc^2}\right\}(\boldsymbol{\sigma} \cdot \mathbf{p})\xi \qquad (37)$$

We must also preserve normalization through order $(Z\alpha)^2$; using Eq. (37) we have

$$1 = (\psi|\psi) = \langle\psi|\psi\rangle = \int (|\eta|^2 + |\xi|^2)d\mathbf{r} = \int \xi^*\left(1 + \frac{p^2}{4m^2c^2}\right)\xi \, d\tau.$$

This equation will be satisfied if

$$|\psi\rangle = [1 + (p^2/8m^2c^2)]\xi$$

or

$$\xi = [1 - (p^2/8m^2c^2)]|\psi\rangle \qquad (38)$$

We shall call the limit given by Eqs. (37) and (38) the second Pauli approximation.

Returning now to the one-electron form of Eq. (33), we write

$$(\psi|H_{\text{D}}|\psi) = (\psi|c(\boldsymbol{\alpha}\cdot\mathbf{p}) + \beta mc^2 - eU|\psi)$$

which becomes, after use of Eqs. (37) and (38), with particular care being paid to the βmc^2 term

$$(\psi | H_D | \psi) = W + mc^2 + \langle \psi | \left\{ -\frac{p^4}{8m^3c^2} + \frac{e}{8m^2c^2}(p^2 U + U p^2) \right.$$

$$\left. -\frac{e}{4m^2c^2}(\boldsymbol{\sigma} \cdot \mathbf{p}) U (\boldsymbol{\sigma} \cdot \mathbf{p}) \right\} | \psi \rangle$$

where we have kept all terms through order $(Z\alpha)^2 W$. Rewriting and combining the last two terms on the right above, we find, finally

$$(\psi | H_D | \psi) = W + mc^2 + \langle \psi | \left\{ -\frac{p^4}{8m^3c^2} - \frac{e\hbar^2}{8m^2c^2} \nabla^2 U \right.$$

$$\left. + \frac{e\hbar}{4m^2c^2}(\boldsymbol{\sigma} \cdot \{\boldsymbol{\mathscr{E}} \times \mathbf{p}\}) \right\} | \psi \rangle \qquad (39)$$

where $\boldsymbol{\mathscr{E}} = -\nabla U$. The first of the correction terms in Eq. (39) is simply a relativistic correction to the kinetic energy of the electron, that is, the third term in the expansion of $[(mc^2)^2 + (pc)^2]^{1/2}$; the second is the Darwin term (Darwin, 1928). The third term can be put into a much more familiar form if we write

$$\boldsymbol{\mathscr{E}} = -\nabla U = -(\mathbf{r}/r)(dU/dr)$$

or

$$\frac{e\hbar}{4m^2c^2}(\boldsymbol{\sigma} \cdot \{\boldsymbol{\mathscr{E}} \times \mathbf{p}\}) = -\frac{e\hbar^2}{2m^2c^2} \frac{1}{r} \frac{dU}{dr} (\mathbf{s} \cdot \mathbf{l}) \qquad (40)$$

in other words, the third term is the spin–orbit interaction in a central field (Condon and Shortley, 1935). This demonstrates the well-known result that the Dirac Hamiltonian contains the spin–orbit interaction. The Dirac Hamiltonian in a central field will, naturally enough, contain the spin–orbit interaction in a central field.

B. Interaction with an External Field

Many of the effects of interest in atomic physics involve an interaction of the electrons with an external vector field; this interaction is introduced relativistically through a term $e \sum_i (\boldsymbol{\alpha}_i \cdot \mathbf{A}_i)$ in the Hamiltonian. Again, since this is a one-electron operator, we can consider the effect of the field on a e electron. In the first Pauli limit, the matrix element of interest reduces simply

$$\langle\psi|e(\boldsymbol{\alpha}\cdot\mathbf{A})|\psi\rangle = \langle\xi|e(\boldsymbol{\sigma}\cdot\mathbf{A})|\eta\rangle + \langle\eta|e(\boldsymbol{\sigma}\cdot\mathbf{A})|\xi\rangle$$

$$= \frac{e}{2mc}\langle\xi|(\boldsymbol{\sigma}\cdot\mathbf{A})(\boldsymbol{\sigma}\cdot\mathbf{p}) + (\boldsymbol{\sigma}\cdot\mathbf{p})(\boldsymbol{\sigma}\cdot\mathbf{A})|\xi\rangle$$

$$= \frac{e}{2mc}\langle\psi|(\mathbf{A}\cdot\mathbf{p}) + (\mathbf{p}\cdot\mathbf{A}) + \hbar(\boldsymbol{\sigma}\cdot\mathcal{H})|\psi\rangle \quad (41)$$

where $\mathcal{H} = \nabla \times \mathbf{A}$.

The term involving $\boldsymbol{\sigma}$ above demonstrates the well-known result that a nonrelativistic electron can be considered to have a spin magnetic moment of

$$\boldsymbol{\mu} = -(e\hbar/2mc)\boldsymbol{\sigma} = -\mu_0\boldsymbol{\sigma} = -2\mu_0\,\mathbf{s} \quad (42)$$

Further, writing $\mathbf{A} = \frac{1}{2}(\mathcal{H} \times \mathbf{r})$, the first two operators above can be put into the form

$$(e/2mc)\{(\mathbf{A}\cdot\mathbf{p}) + (\mathbf{p}\cdot\mathbf{A})\} = (e\hbar/2mc)(\mathbf{l}\cdot\mathcal{H}) = \mu_0(\mathbf{l}\cdot\mathcal{H}) \quad (43)$$

showing that the atomic electron also has an orbital magnetic moment of $-\mu_0\,\mathbf{l}$.

We can also consider the second Pauli limit of this interaction; using in a straightforward fashion Eqs. (37) and (38) above, we find that this will result in the added interaction

$$-(e/4m^2c^3)\langle\psi|\{[(p^2/2m), (\boldsymbol{\sigma}\cdot\mathbf{A})(\boldsymbol{\sigma}\cdot\mathbf{p}) + (\boldsymbol{\sigma}\cdot\mathbf{p})(\boldsymbol{\sigma}\cdot\mathbf{A})]_+$$
$$+ (\boldsymbol{\sigma}\cdot\{\mathbf{A}\times\nabla(-eU)\}) + \tfrac{1}{2}[H_{SC}, \{\hbar(\boldsymbol{\sigma}\cdot\mathcal{H}) - 2i(\boldsymbol{\sigma}\cdot\{\mathbf{A}\times\mathbf{p}\})\}]\}|\psi\rangle$$
$$(44)$$

where $[,]_+$ is an anticommutator. The interaction of Eq. (44) is also valid for off-diagonal matrix elements and agrees with the results obtained by Drake (1971) using a time-dependent Foldy–Wouthuysen transformation; the commutator term vanishes, of course, for diagonal matrix elements. The anticommutator term can be written, for diagonal matrix elements, as

$$\mu_0([\mathbf{l} + 2\mathbf{s}]\mathcal{H})(-T/mc^2)$$

where T is the kinetic energy of the electron. This term, which clearly arises from the relativistic change of mass of the electron, is called the Breit Margenau correction (Breit, 1928; Margenau, 1940). We shall consider higher order interactions with $\mathbf{A}$ in Section III,E.

C. The Perturbation Hamiltonian H_P

The perturbation term H_P is composed of one-body interactions

$$\sum_i \left(-\frac{Ze^2}{r_i} + eU(r_i) \right)$$

and two-body interactions

$$\left(\sum_{i<j} \frac{e^2}{r_{ij}} \right)$$

The term $eU(r)$ was, of course, discussed in Section III,A. The Pauli limit of $-\sum_i (Ze^2/r_i)$ can also be obtained directly from the results of Section III,A by replacing $-eU(r)$ by $-(Ze^2/r)$. Thus, we find immediately that, in the second Pauli limit

$$\langle \psi | -\frac{Ze^2}{r} | \psi \rangle = \langle \psi | \left[-\frac{Ze^2}{r} + \frac{Ze^2\hbar^2}{2m^2c^2} \pi\delta(r) + \frac{Ze^2\hbar^2}{2m^2c^2} \frac{1}{r^3} (\mathbf{s} \cdot \mathbf{l}) \right] | \psi \rangle \quad (45)$$

Here, the Darwin term (the second term on the right, above) can clearly affect only electrons with a nonvanishing probability density at $r = 0$, i.e., electrons in s states.

In order to study the second Pauli limit of the two-body interactions, we can use a two-electron wave function of the type

$$|\Psi\rangle = |\psi_1\psi_2\rangle$$

where $|\psi_1\rangle$ is a relativistic central field wave function for electron 1, etc. The limiting procedure is itself not dependent on the coupling between the electrons; our results are then equally valid for a more complicated form of $|\Psi\rangle$ involving coupled antisymmetric electron states.

Proceeding as above, we write

$$\langle \Psi | \frac{e^2}{r_{12}} | \Psi \rangle = \langle \xi_1\xi_2 | \frac{e^2}{r_{12}} | \xi_1\xi_2 \rangle + \langle \eta_1\eta_2 | \frac{e^2}{r_{12}} | \eta_1\eta_2 \rangle + \langle \xi_1\eta_2 | \frac{e^2}{r_{12}} | \xi_1\eta_2 \rangle$$

$$+ \langle \eta_1\xi_2 | \frac{e^2}{r_{12}} | \eta_1\xi_2 \rangle$$

$$= \langle \xi_1\xi_2 | \frac{e^2}{r_{12}} | \xi_1\xi_2 \rangle + \frac{1}{4m^2c^2} \langle \xi_1\xi_2 |$$

$$\times \left[(\boldsymbol{\sigma}_1 \cdot \mathbf{p}_1) \frac{e^2}{r_{12}} (\boldsymbol{\sigma}_1 \cdot \mathbf{p}_1) + (\boldsymbol{\sigma}_2 \cdot \mathbf{p}_2) \frac{e^2}{r_{12}} (\boldsymbol{\sigma}_2 \cdot \mathbf{p}_2) \right] | \xi_1\xi_2 \rangle$$

After use of Eqs. (37) and (38) and some straightforward but tedious manipulations, this becomes (Akhiezer and Berestetskii, 1965)

$$\langle \psi_1 \psi_2 | \frac{e^2}{r_{12}} + \frac{\hbar^2}{8m^2c^2} (\nabla_1^2 + \nabla_2^2)\left(\frac{e^2}{r_{12}}\right)$$
$$+ \frac{\hbar}{4m^2c^2}\left[\left(\boldsymbol{\sigma}_1 \cdot \left\{\nabla_1\left(\frac{e^2}{r_{12}}\right) \times \mathbf{p}_1\right\}\right) + \left(\boldsymbol{\sigma}_2 \cdot \left\{\nabla_2\left(\frac{e^2}{r_{12}}\right) \times \mathbf{p}_2\right\}\right)\right] |\psi_1 \psi_2\rangle \tag{46}$$

The second term above is a Darwin-like term; it can be rewritten as

$$\frac{\hbar^2}{8m^2c^2} (\nabla_1^2 + \nabla_2^2)\left(\frac{e^2}{r_{12}}\right) = -\frac{e^2\hbar^2}{m^2c^2} \pi\delta(r_{12}) \tag{47}$$

The third term is a combination of the spin–other-orbit interaction and the spin–orbit interaction obtained using the correct interelectron potential. A further contribution to the spin–other-orbit interaction comes from the reduction of the Breit interaction, as we shall show in the next section.

D. The Breit Perturbation H_B

The reduction of the Breit interaction to its Pauli limit proceeds in the same manner as followed above. The calculation is somewhat simplified in that, since H_B is itself of order $(Z\alpha)^2 W$, the first Pauli limit is sufficiently accurate. It is complicated, however, in that a number of integrations by part must be carried out during the procedure, and care must be taken near the points $r_1 = r_2$. The Pauli limit of the Breit equation is found to be (Akhiezer and Berestetskii, 1965)

$$(\psi_1\psi_2 | H_B | \psi_1\psi_2) = \langle \psi_1\psi_2 |$$
$$\left\{ \frac{\hbar}{2m^2c^2}\left[\left(\boldsymbol{\sigma}_2 \cdot \left\{\nabla_1\left(\frac{e^2}{r_{12}}\right) \times \mathbf{p}_1\right\}\right)\right.\right.$$
$$\left.+ \left(\boldsymbol{\sigma}_1 \cdot \left\{\nabla_2\left(\frac{e^2}{r_{12}}\right) \times \mathbf{p}_2\right\}\right)\right]$$
$$- \frac{e^2}{2m^2c^2}\left[\frac{(\mathbf{p}_1 \cdot \mathbf{p}_2)}{r_{12}} + \frac{(\mathbf{r}_{12}(\mathbf{r}_{12} \cdot \mathbf{p}_1) \cdot \mathbf{p}_2)}{r_{12}^3}\right]$$
$$+ \frac{e^2\hbar^2}{4m^2c^2}\left[\frac{(\boldsymbol{\sigma}_1 \cdot \boldsymbol{\sigma}_2)}{r_{12}^3} - \frac{3(\boldsymbol{\sigma}_1 \cdot \mathbf{r}_{12})(\boldsymbol{\sigma}_2 \cdot \mathbf{r}_{12})}{r_{12}^5} - \frac{8\pi}{3}(\boldsymbol{\sigma}_1 \cdot \boldsymbol{\sigma}_2)\delta(r_{12})\right]\right\} |\psi_1\psi_2\rangle \tag{48}$$

The first term in square brackets above is the remainder of the spin–other-orbit interaction; the second is the orbit–orbit interaction; and the last is the spin–spin and spin–spin-contact interactions.

E. Higher Order Terms

The reader may have remarked that certain familiar nonrelativistic terms, such as the diamagnetic interaction, have not appeared in the results of the previous sections. In particular, the nonrelativistic Hamiltonian as derived thus far does not reflect the fact that an external vector potential $\mathbf{A}$ is incorporated into a nonrelativistic Hamiltonian by the replacement of $\mathbf{p}$ by π.

The terms which are missing at this point arise, interestingly enough, from perturbation theory. For example, the diamagnetic term is obtained (Sternheim, 1962; Feneuille, 1973) from the reduction to the Pauli limit of the second-order expression

$$\Delta = \sum_{\psi'} (\psi | e(\boldsymbol{\alpha} \cdot \mathbf{A}) | \psi')(\psi' | e(\boldsymbol{\alpha} \cdot \mathbf{A}) | \psi)/(E_\psi - E_{\psi'}) \tag{49}$$

The sum over ψ' in this equation includes both positive and negative energy states of the electron. However, because of the nature of the operators and states involved, Δ^-, the part of Δ arising from the sum over negative energy states, can be evaluated explicitly in the Pauli limit.

When considering only the case where ψ' describes a negative energy electron, one can replace, in the Pauli limit, the denominator $E_\psi - E_{\psi'}$ by $2mc^2$. Then, we have

$$\Delta^- = \frac{e^2}{2mc^2} \sum_{\psi^-} (\psi | (\boldsymbol{\alpha} \cdot \mathbf{A}) | \psi^-)(\psi^- | (\boldsymbol{\alpha} \cdot \mathbf{A}) | \psi) \tag{50}$$

To the accuracy of the Pauli approximation, one can define a projection operator

$$\Lambda_- = \tfrac{1}{2}\{1 - \beta - [(\boldsymbol{\alpha} \cdot \mathbf{p})/mc]\} \tag{51}$$

which projects out positive energy states $|\psi^+)$ and does not effect negative energy states $|\psi^-)$. Thus, we can replace $|\psi^-)$ by $\Lambda_-|\psi)$ in Eq. (50), and then extend the sum over all states, both positive and negative, without changing the value of the sum to the accuracy of the Pauli limit. Closure then permits the removal of the sum over intermediate states, and we have

$$\begin{aligned}
\Delta^- &= \frac{e^2}{2mc^2} (\psi | (\boldsymbol{\alpha} \cdot \mathbf{A})(\boldsymbol{\alpha} \cdot \mathbf{A}) | \psi) \\
&= \frac{e^2}{2mc^2} (\psi | (\boldsymbol{\sigma} \cdot \mathbf{A})(\boldsymbol{\sigma} \cdot \mathbf{A}) | \psi) \\
&= \frac{e^2}{2mc^2} \langle \psi | A^2 | \psi \rangle
\end{aligned} \tag{52}$$

where, at each step, we have kept only terms consistent with the first Pauli limit. We see that Δ^- is just the usual diamagnetic term. We note that, because $(\boldsymbol{\alpha} \cdot \mathbf{A})$ is an odd operator, the projection operator Λ_- is not actually required in the evaluation of Δ^-; that is, it can be shown that extending the sum in Eq. (50) to include positive energy states does not affect the value of Δ^- in the Pauli limit (Feneuille, 1973). We have introduced Λ_- here for convenience and clarity.

The natural extension of this technique is to consider second-order perturbations of the sort

$$W^{(2)} = \sum_{\Psi'} \left\{ (\Psi | \sum_i e(\boldsymbol{\alpha}_i \cdot \mathbf{A}_i) | \Psi')(\Psi' | H_\mathrm{P} + H_\mathrm{B} | \Psi)/(E_\Psi - E_{\Psi'}) + \mathrm{C.C.} \right\} \tag{53}$$

However, as we noted in Section II, the Breit interaction is derived in such a fashion that it describes only interactions between positive energy electrons, whereas the sum over $|\Psi'\rangle$ in Eq. (53) must include states having one negative energy electron. Thus it would seem that Eq. (53) is not a correct expression for the second-order energy change in the electrons. Nevertheless, Drake (1972), working directly with the equations of quantum electrodynamics, showed that in fact, Eq. (53) is correct. The restriction that H_B connect only electron states of positive energy arises essentially from energy conservation considerations; in the second-order term of Eq. (53), $|\Psi'\rangle$ need not be a state which conserves energy.

As in the discussion of the diamagnetic term, we concentrate first on $W_-^{(2)}$, the negative energy contribution to $W^{(2)}$. Once again, keeping only terms consistent with the Pauli limit, we can write

$$W_-^{(2)} = \sum_{\Psi'} \left\{ (\Psi | e \sum_i (\boldsymbol{\alpha}_i \cdot \mathbf{A}_i) \Lambda_-^i | \Psi')(\Psi' | H_\mathrm{P} + H_\mathrm{B} | \Psi)/(E_\Psi - E_{\Psi'}) + \mathrm{C.C.} \right\} \tag{54}$$

From this point, one proceeds exactly as above, replacing $E_\Psi - E_{\Psi'}$ by $2mc^2$, using closure, and replacing the relativistic wave functions $|\Psi\rangle$ by their Pauli limits. The result of this calculation can be simply stated: replace $\mathbf{p}$ by $2\boldsymbol{\pi}$ in the last term of Eq. (39) [changing also the overall sign to reflect the fact that H_P contains $eU(r)$ rather than $-eU(r)$] and in Eq. (46), and $\mathbf{p}$ by $\boldsymbol{\pi}$ in Eq. (48); finally, keep only the terms linear in $\mathbf{A}$.

We find that we can also explicitly evaluate a part of $W_+^{(2)}$, the positive energy contribution to $W^{(2)}$, in the second Pauli limit. That is, we consider

$$W_+^{(2)} = \sum_{\Psi^+} \left\{ (\Psi | e \sum_i (\boldsymbol{\alpha}_i \cdot \mathbf{A}_i) | \Psi^+)(\Psi^+ | H_\mathrm{P} + H_\mathrm{B} | \Psi)/(E_\Psi - E_{\Psi'}) + \mathrm{C.C.} \right\} \tag{55}$$

and replace the matrix elements by their first (H_B), or second (H_P, $e\boldsymbol{\alpha} \cdot \mathbf{A}$) Pauli limits using the results of Sections III,B, C, and D. The perturbation involving the commutator term in the Zeeman effect can then easily be evaluated; the resulting expression is described by replacing $\mathbf{p}$ by $-\boldsymbol{\pi}$ in the last term of Eq. (39) (again, with an overall sign change) and in Eq. (46), and keeping only the terms linear in $\mathbf{A}$. Thus, combining $W_-^{(2)}$ and the explicitly summable part of $W_+^{(2)}$, we see that $\mathbf{p}$ can be replaced by $\boldsymbol{\pi}$ in the Hamiltonian, at least through terms linear in $\mathbf{A}$. That is, the effect of the external field is reproduced in the Pauli limit by the Hamiltonian interactions of Section III,B plus the Hamiltonian term

$$H_A = \sum_i \left\{ \frac{e\hbar}{4m^2c^3} \left(\boldsymbol{\sigma}_i \cdot \left\{ \boldsymbol{\nabla}_i \left(eU(r_i) - \frac{Ze^2}{r_i} + \sum_{i \neq j} \frac{e^2}{r_{ij}} \right) \times \mathbf{A}_i \right\} \right) \right\}$$
$$+ \sum_{i \neq j} \left\{ \frac{e\hbar}{2m^2c^3} \left(\boldsymbol{\sigma}_i \cdot \left\{ \boldsymbol{\nabla}_j \left(\frac{e^2}{r_{ij}} \right) \times \mathbf{A}_j \right\} \right) \right.$$
$$\left. - \frac{e^2}{2m^2c^3} \left[\frac{(\mathbf{A}_i \cdot \mathbf{p}_j)}{r_{ij}} + \frac{(\mathbf{r}_{ij} \cdot \mathbf{A}_i)(\mathbf{r}_{ij} \cdot \mathbf{p}_j)}{r_{ij}^3} \right] \right\}. \tag{56}$$

The reader should recall, however, that the commutator term of the Zeeman effect is to not be used in the evaluation of second-order terms if H_A is used in first order.

The terms of H_A are, in conjunction with the results of Section III,B, very important in accurate calculations involving the Zeeman effect (Abragam and Van Vleck, 1953; Perl and Hughes, 1953; Perl, 1953; Judd and Lindgren, 1961), and transition probabilities (Drake, 1972); their effect on hyperfine structure has been discussed by Feneuille and Armstrong (1973).

Finally, a few words seem to be called for at this point to explain why we have been forced to use second-order perturbation theory to find, for example, the diamagnetic term, while other works (e.g., Messiah, 1959; Bethe and Salpeter, 1957) find these terms in first order. The explanation is actually quite straightforward. Had we assumed that $|\psi\rangle$ was a solution to the equation

$$\left\{ \frac{(\boldsymbol{\sigma} \cdot \boldsymbol{\pi})^2}{2m} - eU(r) \right\} |\psi\rangle = W' |\psi\rangle \tag{57}$$

and that η was related to $|\psi\rangle$ by

$$\eta = (1/2mc)(\boldsymbol{\sigma} \cdot \boldsymbol{\pi}) |\psi\rangle$$

in the first Pauli limit, we would have obtained the traditional first-order expansions containing the diamagnetic term, etc. However, because Eq. (57) does not, in fact, describe the usual zeroth-order equation with which one works, we did not use it despite the obvious simplifications it could have produced.

F. The Nonrelativistic Hamiltonian

We can now gather together the results of Section III, generalized to the many-electron case. We have found that the energy can be calculated in the nonrelativistic scheme through order $(Z\alpha)^2 E_C$ by use of Hamiltonian $(\mathbf{A} = 0)$

$$H_{nr} = H_C + H_P + H_R + H_{SO} + H_{SOO} + H_{OO} + H_{SS} \tag{58}$$

where

$$H_C = \sum_i \left[\frac{p_i^2}{2m} - eU(r_i) \right]$$

$$H_P = \sum_i \left[eU(r_i) - \frac{Ze^2}{r_i} \right] + \sum_{i<j} \frac{e^2}{r_{ij}}$$

$$H_R = \sum_i \left[-\frac{p_i^4}{8m^3 c^2} + \frac{\hbar^2}{8m^2 c^2} \nabla_i^2 \left(-\frac{Ze^2}{r_i} + \sum_{i \neq j} \frac{e^2}{r_{ij}} \right) \right]$$

$$H_{SO} = \sum_i \frac{\hbar}{4m^2 c^2} \left(\boldsymbol{\sigma}_i \cdot \left\{ \boldsymbol{\nabla}_i \left(-\frac{Ze^2}{r_i} + \sum_{i \neq j}' \frac{e^2}{r_{ij}} \right) \times \mathbf{p}_i \right\} \right)$$

$$H_{SOO} = \sum_i \frac{\hbar}{4m^2 c^2} \left(\boldsymbol{\sigma}_i \cdot \left\{ \boldsymbol{\nabla}_i \left(\sum_{i \neq j}'' \frac{e^2}{r_{ij}} \right) \times \mathbf{p}_i \right\} \right)$$

$$+ \sum_{i \neq j} \frac{\hbar}{2m^2 c^2} \left(\boldsymbol{\sigma}_i \cdot \left\{ \boldsymbol{\nabla}_j \left(\frac{e^2}{r_{ij}} \right) \times \mathbf{p}_j \right\} \right)$$

$$H_{OO} = -\frac{e^2}{2m^2 c^2} \sum_{i<j} \left\{ \frac{(\mathbf{p}_i \cdot \mathbf{p}_j)}{r_{ij}} + \frac{(\mathbf{r}_{ij}(\mathbf{r}_{ij} \cdot \mathbf{p}_i) \cdot \mathbf{p}_j)}{r_{ij}^3} \right\}$$

$$H_{SS} = \frac{e^2 \hbar^2}{4m^2 c^2} \sum_{i<j} \left\{ \frac{(\boldsymbol{\sigma}_i \cdot \boldsymbol{\sigma}_j)}{r_{ij}^3} - \frac{3(\boldsymbol{\sigma}_i \cdot \mathbf{r}_{ij})(\boldsymbol{\sigma}_j \cdot \mathbf{r}_{ij})}{r_{ij}^5} \right.$$

$$\left. - \frac{8\pi}{3} (\boldsymbol{\sigma}_i \cdot \boldsymbol{\sigma}_j) \delta(r_{ij}) \right\}$$

In our derivation of this expression, we have assumed that H_C would be used as a zeroth-order Hamiltonian, with the remaining terms being treated as perturbation. The term $\sum' e^2/r_{ij}$ implies that only the contribution from $r_i > r_j$ in the first term in the expansion of $1/r_{ij}$ is to be kept; $\sum'' e^2/r_{ij}$ is defined as $\sum e^2/r_{ij} - \sum' e^2/r_{ij}$. The implications of all of these terms have been discussed above. We will find it convenient to define the Breit–Pauli Hamiltonian by

$$H_{BP} = H_{SOO} + H_{OO} + H_{SS} \tag{59}$$

It can be seen that the separation of H_{SO} (spin–orbit) from H_{SOO} (spin–other-orbit) is somewhat arbitrary, as the two terms are very similar. Blume and Watson (1962, 1963) have pointed out that this separation is not, in fact, really justified, as the two effects are physically quite intermingled. Such a separation is, however, quite convenient for discussions of the approximations contained in the central field model.

IV. Effective Operators

A. GENERAL CONCEPTS

Despite the difficulties discussed in the previous two sections concerning the definition of a proper relativistic Hamiltonian, it is often desirable to work directly with relativistic central field wave functions rather than to use the techniques of the previous section. In the relativistic central field calculation, certain interactions are, of course, included to all orders of perturbation theory. If these interactions are larger than the effects ignored by the relativistic central field, then the results obtained from the central field calculation will be greatly superior to those obtained with the techniques of the previous chapter. This is the case, for example, if the spin–orbit interaction is large compared to H_P and H_B. In this section, we discuss a very powerful method for carrying out calculations based on a relativistic central field—that of effective operators.

The principle of effective operators is easily defined (Sandars and Beck, 1965). If $|\psi)$ is a relativistic state which in the nonrelativistic limit goes into the state $|\psi\rangle$, for any relativistic operator H one can define the effective operator H^e by the equation

$$(\psi | H | \psi) = \langle \psi | H^e | \psi \rangle \tag{60}$$

This approach is important for several reasons. First, from the calculational point of view, it enables one to avoid using complicated relativistic $j - j$ coupled states by providing operators which can be evaluated between nonrelativistic LS coupled states. This is very useful because most of the powerful algebraic techniques developed during the past 30 years have been for LS-coupled states (Racah, 1942, 1943, 1949; Judd, 1963; Nielson and Koster, 1963). Second, almost all experimental data in atomic physics have been and continue to be analyzed in terms of nonrelativistic descriptions of the states and operators. This technique puts the relativistic operators into a form in which they have the same transformation properties as do the nonrelativistic operators, thus enabling parameters obtained from fitting of experimental data to be directly interpreted using either nonrelativistic or relativistic theory. Finally, it is in the effective operator form that one can most clearly see how familiar nonrelativistic interactions are changed by relativistic effects. A much more complete discussion of these points has been given by Armstrong (1971).

We begin by defining effective operators for some very general one- and two-electron operators; in Sections IV,B and C, these techniques shall be applied to specific interactions. The first step in all of the calculations is to describe the relativistic interactions in terms of spherical tensors in the $R(3)$ space defined by $\mathbf{J} = \sum_i \mathbf{j}_i$, where $\mathbf{j}_i$ is the total angular momentum of the ith electron. The theory of spherical tensors has been fully developed in other works (see, for example, Judd, 1963) and will not be repeated here.

1. One-Electron Operators

We consider first a one-electron tensor operator of the type

$$V_Q^{(K)} = \sum_i v_Q^{(K)}(i) \tag{61}$$

The Wigner–Eckart theorem (Judd, 1963) indicates that the equivalent operator for $V_Q^{(K)}$, which we denote by $R_Q^{(K)}$, can be expressed in the form

$$R_Q^{(K)} = \sum A_{\varkappa k}^K(nl, n'l') W_Q^{(\varkappa k)K}(nl, n'l') \tag{62}$$

where the sum is over $\varkappa$, k, nl, and $n'l'$. The operator $W^{(\varkappa k)K}$ is a nonrelativistic tensor operator having rank $\varkappa$ in the space defined by $\mathbf{S} = \sum_i \mathbf{s}_i$, k in the space defined by $\mathbf{L} = \sum_i \mathbf{l}_i$. These two tensors are then coupled using usual angular momentum coupling rules to a tensor of rank K in the space of $\mathbf{J} = \mathbf{L} + \mathbf{S}$. The magnitude of $\mathbf{W}^{(\varkappa k)}$ is defined through the equations (Feneuille, 1967)

$$W_{\pi q}^{(\varkappa k)}(nl, n'l') = \sum_i w_{\pi q}^{(\varkappa k)}(nl, n'l')_i$$

$$\langle n_a \tfrac{1}{2} l_a \| w^{(\varkappa k)}(nl, n'l') \| n_b \tfrac{1}{2} l_b \rangle = [\varkappa, k]^{1/2} \delta(n_a l_a, nl) \delta(n_b l_b, n'l')$$

where $[a, b, \ldots,] = (2a + 1)(2b + 1) \ldots$.

Inserting Eqs. (61) and (62) into Eq. (60), we have

$$(\psi | V_Q^{(K)} | \psi) = \langle \psi | R_Q^{(K)} | \psi \rangle \tag{63}$$

Once again, since we are considering a one-electron operator, we can assume that $|\psi\rangle$ is a one-electron wave function of the type $|nljm_j\rangle$, etc. Then Eq. (63) can be written

$$(nljm_j | v_Q^{(K)} | n'l'j'm'_j) = \langle nljm_j | \sum_{\varkappa, k} A_{\varkappa k}^K(nl, n'l') w_Q^{(\varkappa k)K}(nl, n'l') | n'l'j'm'_j \rangle$$

or

$$\int (nlj\|v^{(K)}\|n'l'j')r^2\,dr$$

$$= \sum_{\varkappa,k} A^K_{\varkappa k}(nl, n'l')\langle nlj\|w^{(\varkappa k)K}(nl, n'l')\|n'l'j'\rangle$$

$$= \sum_{\varkappa,k} A^K_{\varkappa k}(nl, n'l')[j, K, j', \varkappa, k]^{1/2} \begin{Bmatrix} \tfrac{1}{2} & \tfrac{1}{2} & \varkappa \\ l & l' & k \\ j & j' & K \end{Bmatrix}$$

In the notation we shall use, relativistic reduced matrix elements are reduced only with respect to angular variables and do not imply integration over radial variables. Thus, for example, $(nlj\|v^{(K)}\|n'l'j')$ is a function of r. We can easily solve for the coefficients $A^K_{\varkappa k}$ by multiplying both sides of this equation by

$$\begin{Bmatrix} \varkappa & \tfrac{1}{2} & \tfrac{1}{2} \\ l & l' & k \\ j & j' & K \end{Bmatrix} [j,j']^{1/2}$$

and summing over j and j'. Orthogonality rules for $9-j$ symbols (Brink and Satchler, 1962; Judd, 1963) then give

$$A^K_{\varkappa k}(nl, n'l') = \sum_{j,j'} [j, j', \varkappa, k]^{1/2}[K]^{-1/2} \begin{Bmatrix} \tfrac{1}{2} & \tfrac{1}{2} & \varkappa \\ l & l' & k \\ j & j' & K \end{Bmatrix}$$

$$\int (nlj\|v^{(K)}\|n'l'j')r^2\,dr \quad (64)$$

We shall consider specific applications of Eq. (62) in later sections. At this point, however, we can point out one of the important aspects of this approach. We see that a general relativistic one-body operator $V^{(K)}_Q$ has been split into two parts: a part $A^K_{\varkappa k}$ which contains integrals over relativistic radial functions, and an angular part to be evaluated between nonrelativistic states, which is expressed in terms of spherical tensors $W^{(\varkappa k)K}_Q$. Thus, all of the powerful algebraic techniques developed for calculations in nonrelativistic atomic physics (see, e.g., Judd, 1963) can be brought directly to bear on the evaluation of matrix elements.

2. Two-Electron Operators

In this section we consider a class of two-electron operators, composed of those operators which are scalar in the total **J** space. These operators can be written in the form

$$S = \sum_{i<j} (\mathbf{v}_i^{(K)} \cdot \mathbf{v}_j^{(K)}) \tag{65}$$

In this case, the Wigner–Eckart theorem requires that the equivalent operator T also be given by a scalar product of tensor operators

$$T = \sum B_{\varkappa'k'}^{\varkappa k}(n_1 l_1, n_2 l_2; n_3 l_3, n_4 l_4 : K)$$
$$\times \sum_{i<j} [\mathbf{w}_i^{(\varkappa k)K}(n_1 l_1, n_3 l_3) \cdot \mathbf{w}_j^{(\varkappa' k')K}(n_2 l_2, n_4 l_4)] \tag{66}$$

where the first sum is over $\varkappa, \varkappa', k, k', n_1 l_1, n_2 l_2, n_3 l_3$, and $n_4 l_4$. Since S and T are two-body operators, Eq. (60) takes the form

$$(n_1 l_1 j_1 m_{j1}, n_2 l_2 j_2 m_{j2} | S | n_3 l_3 j_3 m_{j3}, n_4 l_4 j_4 m_{j4})$$
$$= \langle n_1 l_1 j_1 m_{j1}, n_2 l_2 j_2 m_{j2} | T | n_3 l_3 j_3 m_{j3}, n_4 l_4 j_4 m_{j4} \rangle$$

Proceeding exactly as above, one finds the unknown coefficients B

$$B_{\varkappa'k'}^{\varkappa k}(n_1 l_1, n_2 l_2; n_3 l_3, n_4 l_4 : K)$$

$$= \sum [\varkappa, k, \varkappa', k', j_1, j_2, j_3, j_4]^{1/2} [K]^{-1} \begin{Bmatrix} \tfrac{1}{2} & \tfrac{1}{2} & \varkappa \\ l_1 & l_3 & k \\ j_1 & j_3 & K \end{Bmatrix}$$

$$\times \begin{Bmatrix} \tfrac{1}{2} & \tfrac{1}{2} & \varkappa' \\ l_2 & l_4 & k' \\ j_2 & j_4 & K \end{Bmatrix} \iint (n_1 l_1 j_1 \| v^{(K)} \| n_3 l_3 j_3)_1$$

$$\times (n_2 l_2 j_2 \| v^{(K)} \| n_4 l_4 j_4)_2 r_1^2 r_2^2 \, dr_1 \, dr_2 \tag{67}$$

where the sum is over all j values, and the subscripts 1 and 2 to the reduced matrix elements imply that the radial elements therein are functions of the variables r_1 and r_2, respectively. We shall consider specific applications of Eq. (66) in later sections.

3. Relativistic Reduced Matrix Elements

The results of the previous two sections depend on reduced matrix elements of the operators $\mathbf{v}^{(k)}$; we will find two particular spherical tensors appearing often in the specific calculations which follow, and it is convenient to define those reduced matrix elements here. We do not present details of the calculations since they can easily be found in the literature (Armstrong, 1966, 1968, 1971; Feneuille, 1971a).

The simplest type of operator $\mathbf{v}^{(k)}$ that will be of interest is $\mathbf{C}^{(k)}$, where $C_q^{(k)}$ is related to the spherical harmonic Y_{kq} by $C_q^{(k)} = (4\pi/2k+1)^{1/2} Y_{kq}$. Straightforward application of tensor techniques leads to

$$(n_1 l_1 j_1 \| C^{(k)} \| n_2 l_2 j_2) = (-1)^{j_1 - 1/2} \begin{pmatrix} j_1 & k & j_2 \\ -\tfrac{1}{2} & 0 & \tfrac{1}{2} \end{pmatrix}$$
$$\times [j_1, j_2]^{1/2} [\Delta(l_1, l_2, k)(F_1 F_2/r^2) + \Delta(\overline{l_1}, \overline{l_2}, k)(G_1 G_2/r^2)] \quad (68)$$

where $\Delta(l_1, l_2, k)$ is equal to 1 if l_1, l_2, and k satisfy the triangular condition and their sum is even; $\Delta(l_1, l_2, k)$ is equal to 0 otherwise. In order to simplify our notation, we have written $F_{n_1 l_1 j_1}$ as F_1, etc.

The second reduced matrix element of interest is given by

$$(n_1 l_1 j_1 \| \{\alpha C^{(k)}\}^{(K)} \| n_2 l_2 j_2) = i(G_2 F_1/r^2)(n_1 l_1 j_1 \| \{\sigma C^{(k)}\}^{(K)} \| n_2 \overline{l_2} j_2)$$
$$- i(G_1 F_2/r^2)(n_1 \overline{l_1} j_1 \| \{\sigma C^{(k)}\}^{(K)} \| n_2 l_2 j_2)$$

Using again standard tensorial techniques and simplifying $9 - j$ symbols, we obtain

$$(n_1 l_1 j_1 \| \{\alpha C^{(k)}\}^{(K)} \| n_2 l_2 j_2) = i[K, j_1, j_2]^{1/2}$$
$$\times \left\{ 2^{1/2}(-1)^{l_1 + k + 1} \begin{pmatrix} 1 & k & K \\ -1 & 0 & 1 \end{pmatrix} \begin{pmatrix} j_1 & j_2 & K \\ -\tfrac{1}{2} & -\tfrac{1}{2} & 1 \end{pmatrix} \right.$$
$$\times [\Delta(l_1, \overline{l_2}, k)(F_1 G_2/r^2) + \Delta(\overline{l_1}, l_2, k)(G_1 F_2/r^2)]$$
$$+ (-1)^{j_2 + 1/2} \begin{pmatrix} 1 & k & K \\ 0 & 0 & 0 \end{pmatrix} \begin{pmatrix} j_1 & j_2 & K \\ -\tfrac{1}{2} & \tfrac{1}{2} & 0 \end{pmatrix}$$
$$\left. \times [\Delta(l_1, \overline{l_2}, k)(F_1 G_2/r^2) - \Delta(\overline{l_1}, l_2, k)(G_1 F_2/r^2)] \right\}. \quad (69)$$

4. Nonrelativistic Limits

Nonrelativistic limits are particularly easy to obtain when using the effective operator approach. The angular part of the operator is already in nonrelativistic form; only the coefficients A [Eq. (64)] or B [Eq. (67)] need be reduced. The Pauli limits of F and G are obtained in an obvious fashion from Eq. (29), that is, one finds that

$$G_{nlj} = -\frac{\alpha a_0}{2}\left(\frac{d}{dr} - \frac{\varkappa}{r}\right) F_{nlj}; \qquad F_{nlj} = R_{nl} \quad (70)$$

The reduction of A and B then involves replacing F and G by their Pauli limits as given in Eq. (70), carrying out radial integrations where possible, and summing over $n - j$ symbols.

B. Effective Operators for H_P and H_B

In this section, we will consider the effective operator form of the perturbations H_P and H_B. These operators are of interest in evaluating the energy of the atomic state and in interpretating parameters obtained from experimental data. We shall confine our calculation to a configuration $(nl)^N$ (Armstrong, 1966); results applicable to more general configurations can be found in the literature (Armstrong, 1968; Armstrong and Feneuille, 1968).

1. Effective Operator for the Central Field

For this one-body operator, we utilize the equations of Section IV,A,1. The operator $U(r_i)$ is spherically symmetric and thus a tensor of rank $K = 0$. Simplifying the $9 - j$ symbols in Eq. (64), we have

$$\sum_i eU(r_i) = \sum \left[[k]^{1/2}(-1)^{l+j+1/2+k} \begin{Bmatrix} \tfrac{1}{2} & l & j \\ l & \tfrac{1}{2} & k \end{Bmatrix} [j] \right. \\
\left. \times \int (F_{nlj}^2 + G_{nlj}^2) eU(r)\,dr \right] W_0^{(kk)0}(nl, nl) \qquad (71)$$

where the sum is over j and k. The $6 - j$ symbol limits the possible values of k to 0 and 1. For $k = 0$, this term will simply add a constant to the energy, independent of the coupling of the electrons. The $k = 1$ term can be rewritten, using the relationship

$$w_i^{(11)0}(nl, nl) = -\{\tfrac{1}{2}l(l+1)(2l+1)\}^{-1/2}(\mathbf{s}_i \cdot \mathbf{l}_i)$$

as

$$-H_{\text{SOCD}} = -\sum_i a_{\text{SOCD}}(i)(\mathbf{s}_i \cdot \mathbf{l}_i) \qquad (72)$$

where

$$a_{\text{SOCD}}(i) = \sum (-1)^{l+j+1/2}[j] \begin{Bmatrix} \tfrac{1}{2} & l & j \\ l & \tfrac{1}{2} & 1 \end{Bmatrix} \left[\frac{6}{l(l+1)(2l+1)} \right]^{1/2} \\
\times \int (F_{nlj}^2 + G_{nlj}^2)(-eU(r_i))\,dr_i$$

The quantity a_{SOCD} is the relativistic form of the spin–orbit coupling constant in a central field. That is to say, this part of the interaction of the relativistic electron with a central field has transformation properties which are identical with those of the usual nonrelativistic spin–orbit interaction.

The implication of this is discussed below. The minus sign in Eq. (72) obviously appears because the central field contribution in H_P is subtracted.

The nonrelativistic limit of a_{SOCD} is easily obtained using the technique described above. We find that the Pauli limit of Eq. (72) is

$$-H_{SOC} = -\sum a_{SOC}(i)(\mathbf{s}_i \cdot \mathbf{l}_i)$$

with

$$a_{SOC}(i) = 2\left(\frac{\mu_0}{e}\right)^2 \int R_{nl}^2(r_i) \frac{1}{r_i} \frac{d}{dr_i}[-eU(r_i)]\, dr_i, \tag{73}$$

that is, the usual nonrelativistic spin–orbit interaction in a central field.

When attempting to interpret experimental spectroscopic data, one usually introduces a spin–orbit interaction $a_i(\mathbf{s}_i \cdot \mathbf{l}_i)$ and determines the value of a_i which best fits the data. One can then try to explain this experimentally determined value of a_i in terms of a central field calculation. The relativistic theory predicts that the experimental a_i will be given by Eq. (72); the nonrelativistic theory, by Eq. (73).

2. Effective Operator Form of the Coulomb Central Field

This operator is of the same tensorial form as $\sum eU(r_i)$. The effective operator is thus the same as that of Eq. (71) with $eU(r)$ replaced by $-Ze^2/r$. The $k = 1$ term in this case is simply the part of the relativistic spin–orbit coupling constant produced by the field of the nucleus. We consider this term further in the next section.

3. Effective Operator for the Coulomb Interaction

The first step in the equivalent operator formulation of e^2/r_{ij} is to express this operator in terms of spherical tensors. This is easily done using the traditional expansion of $1/r_{ij}$

$$\sum_{i<j} \frac{e^2}{r_{ij}} = e^2 \sum_{i<j, K} \frac{r_<^K}{r_>^{K+1}} (\mathbf{C}_i^{(K)} \cdot \mathbf{C}_j^{(K)}) \tag{74}$$

where $r_<$ is the smaller of r_i, r_j, etc. The results of Section IV,A,2 can now be used directly. Thus, for this operator

$$B_{\varkappa'k'}^{\varkappa k}(nl:K) = \sum [\varkappa, k, \varkappa', k']^{1/2} [j_1, j_2, j_3, j_4][K]^{-1}(-1)^{j_1+j_2+1}$$

$$\times \begin{pmatrix} j_1 & K & j_3 \\ -\tfrac{1}{2} & 0 & \tfrac{1}{2} \end{pmatrix} \begin{pmatrix} j_2 & K & j_4 \\ -\tfrac{1}{2} & 0 & \tfrac{1}{2} \end{pmatrix} \begin{Bmatrix} \tfrac{1}{2} & \tfrac{1}{2} & \varkappa \\ l & l & k \\ j_1 & j_3 & K \end{Bmatrix} \begin{Bmatrix} \tfrac{1}{2} & \tfrac{1}{2} & \varkappa' \\ l & l & k' \\ j_2 & j_4 & K \end{Bmatrix} \Delta(l, l, K) \mathscr{F}^K \tag{75}$$

In obtaining Eq. (75), we have used Eq. (68) to calculate the reduced matrix element of $C^{(k)}$; $\mathscr{F}^k$ is a "relativistic Slater integral" defined by

$$\mathscr{F}^k = e^2 \iint (F_1 F_3 + G_1 G_3)_i (F_2 F_4 + G_2 G_4)_j (r_<^K / r_>^{K+1}) \, dr_i \, dr_j \quad (76)$$

Consideration of the symmetry properties of the $n - j$ symbols appearing in Eq. (75) shows that $B_{\varkappa \varkappa'}^{k k'}$ will vanish unless $\varkappa + k$ and $\varkappa' + k'$ are both even; K must also be even.

We can, once again, describe certain terms in this expression as relativistic forms of certain familiar nonrelativistic operators. Thus, in the same sense as used in the previous section, we can say that the terms proportional to $[\mathbf{w}_i^{(0k)k} \cdot \mathbf{w}_j^{(0k)k}]$ (k even) are the relativistic Coulomb interaction terms; the two terms $[\mathbf{w}_i^{(00)0} \cdot \mathbf{w}_j^{(11)0}]$ and $[\mathbf{w}_i^{(11)0} \cdot \mathbf{w}_j^{(00)0}]$ describe the relativistic spin–orbit interaction involving the correct interelectron interaction rather than a central field; the terms proportional to $[\mathbf{w}_i^{(0k)k} \cdot \mathbf{w}_j^{(1k\pm 1)k}]$ (k even) are the relativistic form of a part of the spin–other-orbit interaction [the part contained in the Hamiltonian of Eq. (46)]; finally, there are terms proportional to $[\mathbf{w}_i^{(1k\pm 1)k} \cdot \mathbf{w}_j^{(1k\pm 1)k}]$ which disappear in the Pauli limit.

It is interesting to regard in detail the spin–orbit portion of Eq. (75) so that we may complete the discussion begun in the previous two sections. We find that

$$B_{00}^{11}(nl:0) = \sum (-1)^{l+j-1/2} \begin{Bmatrix} \frac{1}{2} & l & j \\ l & \frac{1}{2} & 1 \end{Bmatrix} \{3/2(2l+1)\}^{1/2} [j,j']$$

$$\times \int (F_j^2 + G_j^2)_1 (F_{j'}^2 + G_{j'}^2)_2 \frac{1}{r_>} \, dr_1 \, dr_2$$

This result can be rewritten in a more familiar form by using the relation

$$\sum_{i>j=1} [\mathbf{w}_i^{(11)0}(nl, nl) \cdot \mathbf{w}_j^{(00)0}(nl, nl)] = (4l+2)^{-1/2} \left(\frac{N-1}{2}\right) \sum_i \mathbf{w}_i^{(11)0}(nl, nl)$$

In addition, we can combine this term with the result of the previous section to obtain an equation for the relativistic spin–orbit coupling constant produced by the field of the nucleus $(-Ze^2/r)$ and the correct interelectron interaction (e^2/r_{ij}):

$$H_{\text{SOD}} = \sum a_{\text{SOD}}(i)(\mathbf{s}_i \cdot \mathbf{l}_i) \quad (77)$$

where

$$a_{\text{SOD}}(i) = \sum (-1)^{l+j+1/2} \begin{Bmatrix} \frac{1}{2} & l & j \\ l & \frac{1}{2} & 1 \end{Bmatrix} [j]$$

$$\left[\frac{6}{l(l+1)(2l+1)}\right]^{1/2} \int (F_j^2 + G_j^2)_i V(r_i) \, dr_i$$

with

$$V(r) = -\frac{Ze^2}{r} + (N-1)e^2 \sum_{j'} \frac{[j']}{2[l]} \int (F_{j'}^2 + G_{j'}^2)_k \frac{1}{r_>} dr_k$$

The nonrelativistic limit of Eq. (77) is easily obtained; it is

$$H_{SO} = \sum a_{SO}(i)(\mathbf{s}_i \cdot \mathbf{l}_i) \tag{78}$$

where

$$a_{SO} = 2\left(\frac{\mu_0}{e}\right)^2 \int R_{nl}^2(r_i) \frac{1}{r_i} \frac{d}{dr} V(r_i) dr_i,$$

$$V(r) = -\frac{Ze^2}{r} + (N-1)e^2 \int R_{nl}^2(r_k) \frac{1}{r_>} dr_k$$

is the usual nonrelativistic spin–orbit coupling constant calculated in the correct potential field. As in Section IV,B,1, if one attempts to explain a measured spin–orbit constant a using the relativistic theory with correct interelectron interactions, Eq. (77) should be used; for a nonrelativistic calculation, Eq. (78) should be used. The reader should recall, however, the difficulty in separating spin–orbit and spin–other-orbit effects (Section III,F).

The Dirac equation in a central field contains, of course, the central field spin–orbit interaction. Thus the sum of Eqs. (73) and (77) is a perturbation operator which corrects for the approximate nature of the central field spin–orbit interaction which is contained in H_C.

4. *Effective Operator Form of* H_B

The effective operator form for H_B is somewhat difficult to obtain because of problems involved in expressing H_B in spherical tensor notation. The tensor form of H_B can, nevertheless, be obtained after straightforward but tedious use of algebraic techniques (Armstrong, 1968, 1971). We shall not give any details here concerning the effective operator form of H_B because it is too complicated to be directly of much use. It is very important, however, in the derivation of the Breit–Pauli operators (that is, the spin–spin, spin–other-orbit, and orbit–orbit interactions) written in spherical tensor notation for use with mixed configurations. In the effective operator notation, in which there is a separation of angular and radial operators, one finds that the reduction to nonrelativistic limits of the effective operator form of H_B is reasonably simple and straightforward. Thus, it was in this fashion that the tensor operator expressions of the Breit–Pauli operators valid for mixed configurations were first obtained (Armstrong and Feneuille, 1968); these operator expressions are, on the other hand, very difficult to obtain starting from the operators of Section III,D.

C. Effective Operators Involving Fields

All of the operators considered in this section describe interactions between the atomic electrons and fields—in one case, an external magnetic field (Zeeman effect, Section IV,C,1); in another, the multipole fields of the nucleus (hyperfine structure, Section IV,C,2); and finally, the field producing (and produced by) an electronic transition (Section IV,C,3). We will find that the effective operators obtained in this section are much less complicated than those obtained in the previous section. In Sections IV,C,1 and 2, we shall consider only operators diagonal with respect to configuration.

1. Zeeman Effect

The interaction between atomic electrons and an external magnetic field has been considered by a number of authors (see, for example, Abragam and Van Vleck, 1953; Judd and Lindgren, 1961), generally using the techniques described in Section III. In this section we wish to consider the same interaction using the techniques of equivalent operators.

As we have noted, it is not the magnetic field which appears in the relativistic Hamiltonian, but rather the vector potential $\mathbf{A}$. We write then

$$\mathbf{A}(\mathbf{r}) = -\tfrac{1}{2}\{\mathbf{r} \times \mathscr{H}\} \tag{79}$$

and consider the effect of the perturbation operator

$$H_Z = e \sum_i [\boldsymbol{\alpha}_i \cdot \mathbf{A}(\mathbf{r}_i)] \tag{80}$$

Combining Eqs. (79) and (80), and using standard tensorial techniques, we see that H_Z can be expressed in terms of tensor operators as

$$H_Z = ie2^{-1/2} \sum_i (r_i \{\boldsymbol{\alpha}_i \ \mathbf{C}_i^{(1)}\}^{(1)} \cdot \mathscr{H}) \tag{81}$$

Because Eq. (81) is of the form of a one-electron operator, we use the result of Section IV,A,1 to obtain the effective operator form of H_Z, which we shall denote by H_Z^e. We find

$$H_Z^e = \mu_0 \bigg(\sum_{nl} \Big\{ (2[l])^{1/2} a_{nl}(10) \mathbf{W}^{(10)1}(nl, nl)$$
$$+ \tfrac{1}{3}[6l(l+1)(2l+1)]^{1/2} a_{nl}(01) \mathbf{W}^{(01)1}(nl, nl)$$
$$+ 2 \left[\frac{l(l+1)(2l+1)}{(2l+3)(2l-1)} \right]^{1/2} a_{nl}(12) \mathbf{W}^{(12)1}(nl, nl) \Big\} \cdot \mathscr{H} \bigg) \tag{82}$$

where

$$a_{nl}(10) = \frac{2e}{3\mu_0} \frac{1}{[l]^2} \{(l+1)^2 D_{++} + 2l(l+1)D_{+-} + l^2 D_{--}\}$$

$$a_{nl}(01) = -\frac{2e}{\mu_0} \frac{1}{[l]^2} \{-(l+1)D_{++} + D_{+-} + lD_{--}\}$$

$$a_{nl}(12) = -\frac{e}{3\mu_0} \frac{1}{[l]^2}$$
$$\times \{2(l+1)(2l-1)D_{++} - (2l-1)(2l+3)D_{+-} + 2l(2l+3)D_{--}\}$$

and

$$D_{jj'} = \frac{1}{2} \int r[F_j G_{j'} + F_{j'} G_j] \, dr$$

We have used the subscripts $\pm$ to represent the values $j = l \pm \frac{1}{2}$. For the case in which H_Z^e is used only within the configuration $(nl)^N$, Eq. (82) can be written as

$$H_Z^e = \mu_0 \left(\left[2a(10)\mathbf{S} + a(01)\mathbf{L} - 2(10)^{1/2} a(12) \sum_i (\mathbf{s}_i \mathbf{C}_i^{(2)})^{(1)} \right] \cdot \mathcal{H} \right) \tag{83}$$

One can also calculate the relativistic form of the Landé-g factor. The Wigner–Eckart theorem implies that for matrix elements diagonal in J, H_Z^e can be written in the form

$$H_Z^e = g_J \mu_0 (\mathbf{J} \cdot \mathcal{H}) \tag{84}$$

(The reader should be warned that another common definition of g_J differs by a sign from ours.) The value of g_J is given quite generally by

$$g_J = \frac{1}{\mu_0} \frac{\langle \psi | H_Z^e | \psi \rangle}{\langle \psi | (\mathbf{J} \cdot \mathcal{H}) | \psi \rangle}$$

For the special case in which $|\psi\rangle$ is of the form $|\psi\rangle = |l^N \gamma SLJM_J\rangle$ g_J can be evaluated rather simply using Eq. (83)

$$g_J = \left\{ 2a(10) \frac{J(J+1) + S(S+1) - L(L+1)}{2J(J+1)} \right.$$
$$+ a(01) \frac{J(J+1) + L(L+1) - S(S+1)}{2J(J+1)}$$
$$\left. - 2(10)^{1/2} a(12) \frac{\langle l^N \gamma SLJ \| \sum_i (\mathbf{s}_i \mathbf{C}_i^{(2)})^{(1)} \| l^N \gamma SLJ \rangle}{[J(J+1)(2J+1)]^{1/2}} \right\} \tag{85}$$

Finally, we can easily find the nonrelativistic limits of the equations above using Eq. (70). In the Pauli limit

$$D_{++} = (\mu_0/e)(l + \tfrac{3}{2})$$
$$D_{--} = -(\mu_0/e)(l - \tfrac{1}{2})$$
$$D_{+-} = \mu_0/e$$

which gives $a_{nl}(10) = a_{nl}(01) = 1$, and $a_{nl}(12) = 0$, leading to the usual expressions for the Zeeman effect.

We see that in the nonrelativistic limit, the Zeeman effect does not depend on any radial parameters. One might expect, therefore, that the dependence of the relativistic Zeeman effect on the radial wave functions would be quite weak; that is, that the $D_{jj'}$ would not vary greatly from their nonrelativistic values when evaluated using any realistic radial wave functions. In fact, calculations have generally shown that the relativistic effects are small in the Zeeman effect, generally of the order of the radiative corrections. Thus, one can usually obtain quite accurate results using the Pauli-limit equations of Sections III,B and E. The radiative corrections can be easily introduced in this limit, one has only to replace the 2 in Eq. (42) by the value

$$2[1 + (\alpha/2\pi) - 0.328(\alpha^2/\pi^2) + \cdots] = 2[1 + 0.001\ 159\ 644(7)]$$

(Schwinger, 1949; Karplus and Kroll, 1950; Sommerfield, 1957; Wesley and Rich, 1970).

2. Hyperfine Structure

It was recognized quite early that relativistic effects are important in hyperfine structure (Fermi, 1930; Breit, 1931; Racah, 1931a,b). The calculation of these effects is, however, rather complicated, and for a long time calculations were more or less restricted to one-electron atoms (Schwartz, 1955). It was, in fact, only with the introduction of the effective operator technique by Sandars and Beck (1965) that relativistic hyperfine calculations in the many-electron atom became generally feasible.

Hyperfine structure arises from the interaction of the atomic electrons with the various moments of the nucleus. It can be shown on quite general grounds that this interaction can be written in the form (Schwartz, 1955)

$$H_{\text{hyp}} = \sum_k (\mathbf{T}_e^{(k)} \cdot \mathbf{T}_n^{(k)}) \tag{86}$$

where $\mathbf{T}_e^{(k)}$ is a tensor of rank k in the electronic space, and $\mathbf{T}_n^{(k)}$ a tensor of rank k in the nuclear space. We shall not discuss the derivation of the general form of the nuclear and electronic tensor operators, as these results can be obtained elsewhere (Schwartz, 1955; Armstrong, 1971). We shall limit our discussion here to the $k = 1$ (magnetic dipole) and $k = 2$ (electric quadrupole) terms, since these are the most important interactions experimentally.

The discussion of the magnetic dipole interaction follows quite closely the discussion of the Zeeman effect in the preceding section. In this case, the magnetic field is produced by the magnetic dipole moment $\mathbf{\mu} = \mu_I \mathbf{I}/I$ of the nucleus. We then write

$$\mathbf{A}(\mathbf{r}) = \frac{1}{r^3}\{\mathbf{\mu} \times \mathbf{r}\} = \frac{\mu_I}{I}\frac{1}{r^3}\{\mathbf{I} \times \mathbf{r}\}$$

and obtain the Hamiltonian

$$H_M = \frac{e\mu_I}{I}\sum_i\left(\mathbf{\alpha}_i\frac{\{\mathbf{I} \times \mathbf{r}_{ij}\}}{r_i^3}\right) = i\frac{\mu_I e}{I}2^{1/2}\sum_i\left(\frac{\{\mathbf{\alpha}_i\,\mathbf{C}_i^{(1)}\}^{(1)}}{r_i^2}\cdot\mathbf{I}\right) \qquad (87)$$

The equivalent operator for H_M is found using the results of Section III,A,1. We have

$$H_M^e = \frac{\mu_I}{I}\left(\sum_{nl}\left\{2\frac{\mu_0}{3}[2(2l + 1)]^{1/2}\left\langle\frac{\delta(r)}{r^2}\right\rangle_{10}\mathbf{W}^{(10)1}(nl, nl)\right.\right.$$
$$+ 2\frac{\mu_0}{3}[6l(l + 1)(2l + 1)]^{1/2}\left\langle\frac{1}{r^3}\right\rangle_{01}\mathbf{W}^{(01)1}(nl, nl)$$
$$\left.\left.+ 2\mu_0\left|\frac{l(l + 1)(2l + 1)}{(2l + 3)(2l - 1)}\right|^{1/2}\left\langle\frac{1}{r^3}\right\rangle_{12}\mathbf{W}^{(12)1}(nl, nl)\right\}\cdot\mathbf{I}\right)$$

where

$$\left\langle\frac{\delta(r)}{r^2}\right\rangle_{10} = \frac{e}{\mu_0}\frac{1}{[l]^2}\{(l + 1)^2 P_{++} + 2l(l + 1)P_{+-} + l^2 P_{--}\}$$

$$\left\langle\frac{1}{r^3}\right\rangle_{01} = -\frac{e}{\mu_0}\frac{1}{[l]^2}\{-(l + 1)P_{++} + P_{+-} + lP_{--}\}$$

$$\left\langle\frac{1}{r^3}\right\rangle_{12} = -\frac{e}{3\mu_0}\frac{1}{[l]^2}\{2(l + 1)(2l - 1)P_{++}$$
$$- (2l - 1)(2l + 3)P_{+-} + 2l(2l + 3)P_{--}\} \qquad (88)$$

with

$$P_{jj'} = \int (F_j G_{j'} + G_j F_{j'})\frac{1}{r^2}dr$$

Once again, if we restrict our attention to a configuration $(nl)^N$, we find that

$$H_M^e = \frac{\mu_0 \mu_I}{I}\left(\left\{\frac{4}{3}\left\langle\frac{\delta(r)}{r^2}\right\rangle_{10} \mathbf{S} + 2\left\langle\frac{1}{r^3}\right\rangle_{01} \mathbf{L} - 2(10)^{1/2}\right.\right.$$

$$\left.\left.\left\langle\frac{1}{r^3}\right\rangle_{12} \sum_i (\mathbf{s}_i \mathbf{C}_i^{(2)})^{(1)}\right\} \cdot \mathbf{I}\right) \quad (89)$$

The Wigner–Eckart theorem implies that, within a manifold of states arising from a given $\mathbf{I}$ and $\mathbf{J}$, H_M^e can be written as

$$H_M^e = A(\mathbf{I} \cdot \mathbf{J}) \quad (90)$$

The quantity A is directly obtained from Eq. (90) as

$$A = (\mu_I/I)[J(J+1)(2J+1)]^{-1/2}\langle J\|M^{(1)}\|J\rangle \quad (91)$$

where $M^{(1)}$ is the quantity in curly brackets in Eq. (89).

The nonrelativistic limits of Eqs. (88) and (89) can easily be found. One obtains in this limit

$$\left\langle\frac{1}{r^3}\right\rangle_{01} = \left\langle\frac{1}{r^3}\right\rangle_{12} = \left\langle\frac{1}{r^3}\right\rangle; \quad \left\langle\frac{\delta(r)}{r^2}\right\rangle_{10} = \left\langle\frac{\delta(r)}{r^2}\right\rangle$$

where $\langle 1/r^3\rangle$ is the expectation value of $1/r^3$ and $\langle\delta(r)/r^2\rangle$, the expectation value of $\delta(r)/r^2$. That is, in the nonrelativistic limit, both $\langle 1/r^3\rangle_{01}$ and $\langle 1/r^3\rangle_{12}$ have the same value, whereas when relativity is taken into account, the values may be quite different. Parametric studies of hyperfine structure have indeed often shown that the effective radial parameters in the two cases are different (e.g., Harvey, 1965). This cannot immediately be taken as proof of the presence of relativistic effects, however, since core polarization can also cause deviations between the two parameters (Sandars and Beck, 1965; Armstrong, 1971). We also see, that in the Pauli limit, $\langle\delta(r)/r^2\rangle_{10}$ becomes a delta function at the origin; this will have nonzero expectation value only for s electrons. Relativistically, however, this quantity may not be zero for electrons of any angular momentum. Once again, however, core polarization may produce the same type of effect, thus confusing the interpretation of experimental data.

The origin of the quadrupole interaction can easily be seen by considering the form of the Coulomb interaction between the individual nucleons in the nucleus and the individual electrons. This interaction is simply

$$-\sum_{\gamma,i}\frac{e^2}{r_{\gamma i}}$$

where the sum over γ runs over all protons in the nucleus and the sum over i, over all orbital electrons. Expanding this term, assuming that $r_\gamma < r_i$ for all

γ, i, one obtains

$$-\sum_{\gamma,i} \frac{e^2}{r_{\gamma i}} = -\sum_{\gamma,i} \sum_k e^2 \frac{(r_\gamma)^k}{(r_i)^{k+1}} (\mathbf{C}_\gamma^{(k)} \cdot \mathbf{C}_i^{(k)})$$

For $k = 0$, this becomes simply $-\sum_i Ze^2/r_i$, i.e., the usual interaction of a point nucleus with the electrons. The $k = 1$ term vanishes because of parity considerations; the $k = 2$ term is the desired quadrupole interaction

$$H_Q = -e^2 \sum_{\gamma,i} \frac{r_\gamma^2}{r_i^3} (\mathbf{C}_\gamma^{(2)} \cdot \mathbf{C}_i^{(2)}) \tag{92}$$

An effective operator for the electronic part of this interaction is easily found using once again the results of Section IV,A,1. We obtain

$$H_Q^e = \left(\sum_{nl} \left\{ e \left[\frac{l(l+1)(2l+1)}{6} \right]^{1/2} \left\langle \frac{1}{r^3} \right\rangle_{11} W^{(11)2}(nl, nl) \right. \right.$$
$$- e \left[\frac{(l-1)l(2l+1)(l+1)(l+2)}{56(2l+3)(2l-1)} \right]^{1/2} \left\langle \frac{1}{r^3} \right\rangle_{13} W^{(13)2}(nl, nl)$$
$$+ e \left[\frac{2l(2l+1)(l+1)}{5(2l-1)(2l+3)} \right]^{1/2} \left\langle \frac{1}{r^3} \right\rangle_{02} W^{(02)2}(nl, nl) \right\}$$
$$\left. \cdot \sum_\gamma e r_\gamma^2 \mathbf{C}_\gamma^{(2)} \right) \tag{93}$$

where

$$\left\langle \frac{1}{r^3} \right\rangle_{11} = -\frac{2(6)^{1/2}}{5[l]^2} [-(l+2)T_{++} + 3T_{+-} + (l-1)T_{--}]$$

$$\left\langle \frac{1}{r^3} \right\rangle_{13} = \frac{4(21)^{1/2}}{5[l]^2} [(2l-1)T_{++} + 4T_{+-} - (2l+3)T_{--}]$$

$$\left\langle \frac{1}{r^3} \right\rangle_{02} = \frac{1}{[l]^2} [(2l-1)(l+2)T_{++} + 6T_{+-} + (l-1)(2l+3)T_{--}]$$

with

$$T_{jj'} = \int (F_j F_{j'} + G_j G_{j'}) \frac{1}{r^3} dr$$

For matrix elements in $(nl)^N$, we can rewrite Eq. (93) as

$$H_Q^e = e \left(\sum_i \left\{ \left\langle \frac{1}{r^3} \right\rangle_{11} (\mathbf{s}_i \mathbf{l}_i)^{(2)} + \left\langle \frac{1}{r^3} \right\rangle_{13} (\mathbf{s}_i \{\mathbf{C}_i^{(4)} \mathbf{l}_i\}^{(3)})^{(2)} \right. \right.$$
$$\left. - \left\langle \frac{1}{r^3} \right\rangle_{02} \mathbf{C}_i^{(2)} \right\} \cdot \sum_\gamma e r_\gamma^2 \mathbf{C}_\gamma^{(2)} \right) \tag{94}$$

The nuclear quadrupole moment is defined as

$$Q = 2/e \langle II | \sum_\gamma er_\gamma^2 C_{\gamma 0}^{(2)} | II \rangle \tag{95}$$

It is customary to also define the quantity q_J by

$$q_J = 2/e \langle JJ | Q_0^{(2)} | JJ \rangle \tag{96}$$

where $Q^{(2)}$ is the operator in curly brackets in Eq. (93). Then, for states in the manifold arising from a given I and J, we have

$$H_Q^e = b \frac{3(\mathbf{I} \cdot \mathbf{J})^2 + \tfrac{3}{2}(\mathbf{I} \cdot \mathbf{J}) - I(I+1)J(J+1)}{2IJ(2I-1)(2J-1)} \tag{97}$$

where

$$b = e^2 q_J Q$$

In the Pauli limit, Eqs. (93) and (94) simplify greatly, since in this limit $\langle 1/r^3 \rangle_{11} = \langle 1/r^3 \rangle_{13} = 0$; $\langle 1/r^3 \rangle_{02} = \langle 1/r^3 \rangle$. It is rather striking that $\langle 1/r^3 \rangle_{02}$ has the same nonrelativistic limit as $\langle 1/r^3 \rangle_{01}$ and $\langle 1/r^3 \rangle_{12}$, considering how differently the three parameters look in their relativistic form. It should be remarked, however, that core polarization effects are very strong in the quadrupole interaction (Sternheimer, 1950, 1971), thus making comparisons between the radial parameters of H_Q^e and H_M^e very difficult.

3. Transition Probabilities

The influence of relativistic effects on transition probabilities has been ably discussed by Drake (1971, 1972). He used in his investigation Pauli-limit equations such as those discussed in Section III, including the second-order terms of Section III,E. In this section, we wish to consider the fully relativistic calculation of transition probabilities based on the central field model and the use of effective operators. This approach has been used by Feneuille (1971a) to investigate certain types of transitions.

In first-order perturbation theory, the probability per unit time of an atom undergoing a transition from a state $|\psi\rangle$ to a state $|\psi'\rangle$ through emission of absorption of a photon of energy $\hbar\omega = \hbar c |\mathbf{k}| = |E - E'|$, direction of propagation $\hat{\mathbf{k}} = \mathbf{k}/|\mathbf{k}|$ (defined within a solid angle $d\Omega$), and polarization $\hat{\mathbf{e}}$ is given by

$$P(\omega, \mathbf{k}, \hat{\mathbf{e}}) \, d\Omega = (e^2 \omega / 2\pi \hbar c) |\langle \psi' | \sum_i (\boldsymbol{\alpha}_i \cdot \hat{\mathbf{e}}) \exp[\pm i(\mathbf{k}_i \cdot \mathbf{r}_i)] | \psi \rangle|^2 N \, d\Omega \tag{98}$$

Here, N is the number of photons of type $(\omega, \mathbf{k}, \hat{\mathbf{e}})$ present after the emission or before the absorption; the minus sign in the exponential factor holds for emission, the plus for absorption. We shall consider only the former occurrence; the results are trivially extended by replacing **k** by $-\mathbf{k}$.

The first step in obtaining the equivalent operator form of the transition operator

$$T = \sum_i (\boldsymbol{\alpha}_i \cdot \hat{\mathbf{e}}) \exp[i(\mathbf{k}_i \cdot \mathbf{r}_i)] \tag{99}$$

is, as usual, to express it in terms of spherical tensors. This is easily done using vector spherical harmonics $\mathbf{Y}_{JLM}$ (Brink and Satchler, 1962; Akhiezer and Berestetskii, 1965), where

$$\mathbf{Y}_{JLM} = \sum_{Qq} Y_{LQ} \varepsilon_q \begin{pmatrix} L & 1 & J \\ Q & q & -M \end{pmatrix} (-1)^{L+1-M} [J]^{1/2} \tag{100}$$

Here, Y_{LQ} is the usual spherical harmonic, and ε is a tensor of rank 1 composed of the unit vectors $\hat{\varepsilon}_x$, $\hat{\varepsilon}_y$, and $\hat{\varepsilon}_z$

$$\varepsilon_0 = \hat{\varepsilon}_z$$

$$\varepsilon_{\pm 1} = \mp 2^{-1/2}(\hat{\varepsilon}_x \pm i\hat{\varepsilon}_y)$$

Then, we can expand $\hat{\mathbf{e}} e^{i(\mathbf{k}\cdot\mathbf{r})}$ as

$$\hat{\mathbf{e}} e^{i(\mathbf{k}\cdot\mathbf{r})} = \sum_{JKQ} [\hat{\mathbf{e}} \cdot \mathbf{Y}^*_{JKQ}(\hat{\mathbf{k}})] g_K(kr) \mathbf{Y}_{JKQ}(\hat{\mathbf{r}}) \tag{101}$$

where

$$g_K(kr) = (2\pi)^{3/2} i^K J_{K+1/2}(kr)/(kr)^{1/2}$$

$k = |\mathbf{k}|$, and $J_{K+1/2}(kr)$ is a Bessel function. Straightforward use of tensor techniques leads to the relationship

$$(\boldsymbol{\alpha} \cdot \mathbf{Y}_{JKM}) = \left(\frac{2K+1}{4\pi}\right)^{1/2} (\alpha C^{(K)})^{(J)}_M (-1)^{J+K+1} \tag{102}$$

Finally, combining Eqs. (101) and (102) with the results of Section IV,A,1, we find that T can be expressed in the effective operator form

$$T = \sum_{\substack{JKQ \\ \varkappa\beta}} [\hat{\mathbf{e}} \cdot \mathbf{Y}^*_{JKQ}(\hat{\mathbf{k}})] A^{JK}_{\varkappa\beta}(nl, n'l') W^{(\varkappa\beta)J}_Q(nl, n'l') \tag{103}$$

where

$$A^{JK}_{\varkappa\beta}(nl, n'l') = \sum_{j,j'} \frac{[\varkappa, \beta, j, j', K]^{1/2}}{[J]^{1/2}(4\pi)^{1/2}} \begin{Bmatrix} \tfrac{1}{2} & \tfrac{1}{2} & \varkappa \\ l & l' & \beta \\ j & j' & J \end{Bmatrix}$$

$$\times (-1)^{K+J+1} \int (nlj\|\{\alpha C^{(K)}\}^{(J)}\|n'l'j') g_K(kr) r^2 \, dr$$

The reduced matrix elements appearing in Eq. (103) can, of course, be evaluated using the results of Section IV,A,3.

It is interesting to consider the nonrelativistic limit of Eq. (103) in order to better identify the terms in the expansion. Considerable care must be exercised, however, because $kr \sim Z\alpha$; thus, $g_K(kr)$ is effectively given by an expansion in $(Z\alpha)^2$. If only the first term in the expansion of $g_K(kr)$ is retained, therefore, the wave function should be treated in the first Pauli limit. If, on the other hand, the first two terms in $g_K(kr)$ are retained, the second Pauli limit should be used for the wave function, etc. We shall consider below only the former case.

The dominant term above is clearly the $K = 0, J = 1$ term, as it is the only one which contains $g_0(kr) \sim 4\pi$. Keeping only this term and the lowest order expression for $g_0(kr)$ is equivalent to setting $e^{i(\mathbf{k} \cdot \mathbf{r})} = 1$ in Eq. (98). Then, in this approximation for $g_0(kr)$, we have

$$A^{10}_{\varkappa\beta} = \sum_{j, j'} ([\varkappa, \beta, j, j']/3)^{1/2}(4\pi)^{1/2} \begin{Bmatrix} \tfrac{1}{2} & \tfrac{1}{2} & \varkappa \\ l & l' & \beta \\ j & j' & 1 \end{Bmatrix}$$

$$\times \int (nlj\|\alpha\|n'l'j')r^2\,dr \quad (104)$$

In the Pauli limit, the reduced matrix element above becomes

$$(nlj\|\alpha\|n'l'j') = (1/mc)\langle nlj\|p\|n'l'j'\rangle$$

$$= (-1)^{l'+j-1/2}(1/mc)[j,j']^{1/2}\begin{Bmatrix} j & 1 & j' \\ l' & \tfrac{1}{2} & l \end{Bmatrix}$$

$$\times \langle nl\|p\|n'l'\rangle$$

where **p** is the linear momentum.

Inserting this value in Eq. (104) and carrying out the sums over j and j', we obtain

$$A^{10}_{\varkappa\beta} = \delta(\varkappa, 0)\delta(\beta, 1)(8\pi/3)^{1/2}\int \langle nl\|\frac{p}{mc}\|n'l'\rangle r^2\,dr$$

In other words, $A^{10}_{\varkappa\beta} w^{(\varkappa\beta)1}_Q (nl, n'l')$ is the spherical tensor form of the operator $(4\pi)^{1/2} p_Q/mc$. Finally, expanding the "polarization" part of Eq. (103)

$$(\hat{\mathbf{e}} \cdot Y^*_{10Q}(\hat{\mathbf{k}})) = (\hat{\mathbf{e}} \cdot \varepsilon^*_Q)(4\pi)^{-1/2}$$

and combining it with the value of $A^{10}_{\varkappa\beta}$, we see that in the Pauli limit, the dominant term in T is simply

$$(\hat{\mathbf{e}} \cdot \mathbf{p}/mc) \quad (105)$$

i.e., the electric dipole interaction.

The next largest term in T has $K = 1$, with $g_1(kr)$ approximated, again, by its leading term $g_1(kr) \sim 4\pi ikr/3$. There are, in this case, three possible values of J in Eq. (103): $J = 0, 1, 2$. The $J = 0$ term is easily disposed of by noting that $\mathbf{Y}_{010}(\hat{\mathbf{k}}) \sim \hat{\mathbf{k}}$. Thus $[\hat{\mathbf{e}} \cdot \mathbf{Y}_{010}(\hat{\mathbf{k}})] \sim (\hat{\mathbf{e}} \cdot \hat{\mathbf{k}}) = 0$, due to the transverse nature of the field.

We consider next the $J = 1$ term; then, using only the leading term in $g_1(kr)$, we have

$$A_{\varkappa\beta}^{11} = -i \sum_{jj'} [\varkappa, \beta, j, j']^{1/2} (4\pi/9)^{1/2} \begin{Bmatrix} \frac{1}{2} & \frac{1}{2} & \varkappa \\ l & l' & \beta \\ j & j' & 1 \end{Bmatrix}$$

$$\times \int (nlj \| (\alpha C^{(1)})^{(1)} \| n'l'j') kr^3 \, dr \qquad (106)$$

Working directly with Eqs. (69) and (70) in order to find the nonrelativistic limit of the reduced matrix element above, we find the Pauli limit form of $A_{\varkappa\beta}^{11}$ to be

$$A_{\varkappa\beta}^{11} = -(8\pi/9)^{1/2} k\mu_0/e [\langle \tfrac{1}{2}l \| l \| \tfrac{1}{2}l' \rangle 2^{1/2} \delta(\varkappa, 0) \delta(\beta, 1)$$
$$+ \langle \tfrac{1}{2}l \| \sigma \| \tfrac{1}{2}l' \rangle [l]^{1/2} \delta(\varkappa, 1) \delta(\beta, 0)]$$

Thus, in this limit, the effective operator $A_{\varkappa\beta}^{11} w_Q^{(\varkappa\beta)1}(nl, n'l')$ is just the tensor operator form of

$$-(8\pi/3)^{1/2} (k\mu_0/e)(L_Q + 2S_Q) \qquad (107)$$

and this term corresponds to the familiar magnetic dipole transition operator.

Treating the $J = 2$ term in the same limit, we have

$$A_{\varkappa\beta}^{21} = \sum_{jj'} \frac{1}{5^{1/2}} [\varkappa, \beta, j, j']^{1/2} (4\pi/3)^{1/2} ik \begin{Bmatrix} \frac{1}{2} & \frac{1}{2} & \varkappa \\ l & l' & \beta \\ j & j' & 2 \end{Bmatrix}$$

$$\times \int (nlj \| (\alpha C^{(1)})^{(2)} \| n'l'j') r^3 \, dr \qquad (108)$$

In order to find the Pauli limit of the reduced matrix element above, we first consider the matrix element

$$(nljm_j | (\boldsymbol{\alpha} \cdot \mathbf{Y}_{21Q}) r | n'l'j'm_j')$$

Using standard algebraic techniques (Brink and Satchler, 1962) this can be rewritten as

$$(1/10)^{1/2}(nljm_j|(\boldsymbol{\alpha}\cdot\mathbf{V}(r^2Y_{2Q}))|n'l'j'm'_j)$$

$$=10^{-1/2}\frac{i}{\hbar}(nljm_j|[(\boldsymbol{\alpha}\cdot\mathbf{p})r^2Y_{2Q}-r^2Y_{2Q}(\boldsymbol{\alpha}\cdot\mathbf{p})]|n'l'j'm'_j)$$

Because $-eU(r)$ commutes with r^2Y_{2Q}, $(\boldsymbol{\alpha}\cdot\mathbf{p})$ can be replaced by H_D/c [Eq. (17)], and our last form can be again rewritten as

$$(10)^{-1/2}\frac{i}{\hbar c}(nljm_j|[H_D r^2Y_{2Q}-r^2Y_{2Q}H_D]|n'l'j'm'_j)$$

$$=(10)^{-1/2}ik(nljm_j|r^2Y_{2Q}|n'l'j'm'_j)$$

Combining this result with Eq. (102), we find that, in the Pauli limit

$$\int (nlj\|(\alpha C^{(1)})^{(2)}\|n'l'j')r^2\,dr = ik(6)^{-1/2}\int \langle nlj\|r^2 C^{(2)}\|n'l'j'\rangle r^2\,dr$$

Evaluating the nonrelativistic reduced matrix element above using standard techniques (Judd, 1963) and carrying out the sums over j and j', we find finally the Pauli limit of $A^{21}_{\varkappa\beta}$:

$$A^{21}_{\varkappa\beta}=\frac{k^2}{15}(20\pi)^{1/2}\delta(\varkappa,0)\delta(\beta,2)\int\langle nl\|r^2 C^{(2)}\|n'l'\rangle r^2\,dr$$

We see, therefore, that the nonrelativistic limit of the effective operator $A^{21}_{\varkappa\beta}w^{(\varkappa\beta)2}_Q$ is simply the tensor form of the operator

$$(k^2/3)(2\pi)^{1/2}r^2 C^{(2)}_Q \tag{109}$$

that is, the usual electric quadrupole transition operator.

As can be seen from the nonrelativistic limits discussed above, the transitions arising from the terms proportional to $g_J(kr)\mathbf{Y}_{JJM}$ involve magnetic multipole operators. These are called the magnetic 2^J-pole transitions. Transitions arising from terms proportional to $g_k(kr)\mathbf{Y}_{JKM}$, where $J\neq K$, on the other hand, involve electric multipole operators. These are called the electric 2^J-pole transitions.

V. Relativity Plus Nonrelativistic Perturbations

Because the central field wave functions of the previous section are not obtained using the correct interelectronic interactions, the results of Section IV will not be, of course, exact. One is tempted to suggest that these results could be improved by employing perturbation theory and considering higher order terms involving the operator of interest and $H_P + H_B$. That is, one can consider the calculation

$$(O_p) = (\Psi|O_p|\Psi) + \left\{\frac{\sum(\Psi|O_p|\Psi')(\Psi'|H_P + H_B|\Psi)}{(E - E')} + \text{C.C.}\right\} + \text{higher order} \tag{110}$$

where (O_p) is the matrix element of the operator O_p evaluated between the exact relativistic wave functions of the system. Ignoring for the moment difficulties involved in using H_B in perturbation expansions, one quickly finds that, due to the complexity of the relativistic operators and wave functions, such a program would be very difficult to carry out. In nonrelativistic quantum mechanics, however, perturbation calculations of this sort can be and have been carried out to very high order (e.g., Kelly, 1971). Thus, one would like to find some approximation which allows the use of nonrelativistic wave functions and operators for the calculation of the perturbation expansion. The lowest order terms in such an approach can easily be obtained from the results of Section III.

We shall consider the case in which

$$O_p = e \sum_i (\boldsymbol{\alpha}_i \cdot \mathbf{A}_i)$$

which is physically of very great interest. The matrix element $(\Psi|O_p|\Psi)$ can be evaluated using the techniques discussed in Section IV,C. Next, the second-order relativistic perturbation

$$\left[\sum \frac{(\Psi|\sum_i e(\boldsymbol{\alpha}_i \cdot \mathbf{A}_i)|\Psi')(\Psi'|H_P + H_B|\Psi)}{E - E'} + \text{C.C.}\right] \tag{111}$$

can be evaluated in the Pauli limit using the results of Section III,E. In this limit, we find that the term above becomes

$$\langle\Psi|\sum_i (e/mc)\{(\mathbf{p}_i \cdot \mathbf{A}_i) + (\mathbf{A}_i \cdot \mathbf{p}_i) + \hbar(\boldsymbol{\sigma}_i \cdot \mathscr{H}_i)\}$$

$$\sum \frac{|\Psi'\rangle\langle\Psi'|H_P + H_{BP} + H_{SO} - H_{SOC}|\Psi\rangle}{E - E'}$$

$$+ \text{C.C.} + \langle\Psi|H_A|\Psi\rangle \tag{112}$$

The second-order perturbation term in this expression comes, in effect, directly from the replacement of the relativistic wave functions and operators by their nonrelativistic limits. The contribution which involves H_P in this second-order expression is a core polarization-correlation term of the type which has often been calculated (Kelly, 1971); the contribution from the combination of operators $H_{SO} - H_{SOC}$ seeks to correct the errors introduced into $(\Psi|O_p|\Psi)$ through use of wave functions containing spin–orbit effects calculated in an approximate central field. Finally, the term involving H_{BP} attempts to correct the first-order matrix element for neglect of the Breit interaction in the relativistic central field. The first-order term arises, as shown in Section III,E, partially from the sum over negative energy states in the original relativistic perturbation. Generalizing this result, we have that the second- and higher order terms in Eq. (110) can be evaluated by replacing the operators and wave functions by their nonrelativistic limits, with the first correction term to this approach being given by the operator H_A. Thus if the expectation value of H_A is small compared to the second-order term of Eq. (112), one is probably quite justified in evaluating the higher order corrections to $(\Psi | O_p | \Psi)$ in the nonrelativistic framework. If the expectation value of H_A is not small, the accuracy of this approach may be rather suspect; it is, nevertheless, probable that H_A will be the largest correction term.

This problem can also be studied in an equivalent manner by comparing Foldy–Wouthuysen transformations of H_C [Eq. (17)] and of H [Eq. (11)]. A calculation of this type has been carried out by Feneuille and Armstrong (1973) for the case in which A is the magnetic dipole hyperfine field.

VI. Relativistic Radial Wave Functions

A. THE PARAMETRIC METHOD

The difficult problem of calculating atomic radial wave functions can be, in some applications, avoided by considering the radial integrals appearing in energy formulas as parameters to be fitted with experimental data. This parametric method is particularly useful in the analysis of complex spectra (Feneuille, 1971b); it can provide fundamental information concerning the atomic wave function such as, for example, the intermediate coupling characteristics of the states. The generalization of this method to a relativistic scheme is not difficult in principle, especially if one uses the equivalent operator formalism described in Section IV. That is, as we have already noted, in the equivalent operator notation, relativistic operators are given the transformation properties of nonrelativistic operators. Thus the parameterization procedure remains virtually unchanged when going to the relativistic scheme—only the interpretation of the parameters thus obtained need be revised [see Armstrong (1971) for a more complete discussion of this question]. However, the number of parameters which must be introduced to account for all of the relativistic operators is usually very large and, very often, it becomes larger than the number of experimental data. In fact, to our knowledge no parametric study based on the Breit Hamiltonian H [Eq. (11)] has yet been carried out; there are, however, a few studies which include some magnetic terms such as the spin–spin or spin–other-orbit interactions (Judd et al., 1968; Goldschmit, 1973).

Interpretation of parametric data is further complicated by the fact that the angular dependence of purely relativistic terms is the same as that of effective operators which take into account far-configuration interaction effects, and thus the relativistic corrections cannot be simply distinguished from configuration interaction effects. Let us recall, as an example, that the famous Trees parametric correction term $\alpha L(L + 1)$ could arise either from the orbit–orbit interaction or from far-configuration interaction (Wybourne, 1965). This mixing between correlations and relativistic corrections occurs also in hyperfine structure problems, as we noted in Section IV,C,2. Here, however, the number of parameters involved is small enough that a careful analysis of data may enable one to separate relativistic and configuration interaction effects. A well-known example of such a parametric treatment is provided by the hyperfine structure of the $(4f^6)^7F_J$ of the ground term of SmI (reviewed by Armstrong, 1971). Another detailed example is to be found in the study performed by Liberman (1971) on the hyperfine structure of 6s, 5d, 6p, and 7p levels of ^{129}Xe.

B. Hydrogenic Radial Wave Functions

A very rapid way to obtain a crude idea of the magnitude of relativistic corrections is to use screened hydrogenic radial wave functions. Then, any integral can be calculated algebraically, at least in principle. In fact, for a long time hydrogenic relativistic radial integrals have been calculated by an approximate method initially introduced by Racah (1931b) and Casimir (1936) in the study of hyperfine structure. The key point of this method is to replace E by mc^2 in Eqs. (29). The wave functions F and G can then be expressed in terms of Bessel functions and general formulas for one-body integrals can be obtained. An exact treatment for diagonal radial integrals has been proposed recently by Crubellier and Feneuille (1971). Here, the starting point is the use of solutions of the generalized Kepler problem to describe hydrogenic relativistic radial wave functions. The factorization method of Infeld and Hull (1951) and group theory are then used to provide some simple results for radial matrix elements of r^k and r^{-k} which are diagonal in energy. One obtains, for example

$$\int_0^\infty F_{nlj} G_{nlj} r^{-2} \, dr = \frac{-a^2 \varepsilon \alpha Z(2\varepsilon \varkappa + 1)}{\gamma v^3 (4\gamma^2 - 1)} \tag{113}$$

where

$$a = mc\varepsilon\alpha Z/\hbar, \qquad v = n - |\varkappa| + \gamma,$$
$$\gamma = (\varkappa^2 - \alpha^2 Z^2)^{1/2}, \qquad \varepsilon = E_{nlj}/mc^2.$$

However, the hydrogenic approximation is very poor for external atomic orbitals and results derived in this manner are usually quite unsatisfactory. One possible way to improve the integrals is to evaluate in the hydrogenic scheme only the ratio of the desired relativistic integral to its nonrelativistic limit, and finally to calculate this nonrelativistic limit by more sophisticated techniques (e.g., Hartree–Fock). In the theory of hyperfine structure, these ratios are called Casimir correction factors (Casimir, 1936); they can be obtained from tables given by Kopfermann (1958). Obviously, the problem of choosing the effective nuclear charge for use in a many-electron atom remains; it is traditional to take $Z_{\text{eff}} = Z$ and $Z_{\text{eff}} = Z - 4$ for s and p electrons, respectively. In any case, recent relativistic calculations using more sophisticated techniques show that, in most cases studied, the Casimir factors are significantly in error, especially for d and f electrons (Rosén and Lindgren, 1972).

Layzer (1959) has developed a Z expansion technique which uses hydrogenic functions to calculate quite accurately nonrelativistic atomic properties. This technique has also been extended to the relativistic case (Layzer and Bahcall, 1962); this approach has been discussed in detail elsewhere (Doyle, 1969; Ermolaev and Jones, 1973) and will not be considered further here.

C. The Pseudopotential Method

A better approximation for relativistic atomic wave functions can be obtained using a non-Coulombic central potential such as discussed in Section II. However, only the "boundary" conditions for this potential are defined by physical conditions—that is (neglecting the finite size of the nucleus), $U(r) = Ze/r$ at the origin, and $U(r) = (Z + 1 - N)e/r$ for large values of r. Obviously, what we want is to choose the best potential, but the definition of "best" potential implies a quality criterion. This criterion is usually expressed in terms of some functional which must be minimized to obtain the optimum solution.

Such a functional is provided, according to the variation principle, by the total energy of the ground level of the atom. Other choices are possible, however. For example, one can calculate the energy of a certain number of experimentally known atomic levels, changing the form of the potential so as to minimize the root mean square deviation between the observed and the calculated energies of the levels. Obviously, this quality criterion is semiempirical, but it is well adapted to the study of a spectrum as a whole, and thus is usually called "the spectroscopic criterium." Let us remark that, in general, all quality criteria are based on the energy; it is not at all obvious that the potential which leads to the best energies (which is roughly a measure of the expectation value of $1/r$) will also lead to accurate results in the calculation of hyperfine structure (which is approximately the expectation value of $1/r^3$) or transition probabilities (roughly the expectation value of r in the dipole approximation), etc.

Having adopted the idea of a potential chosen according to some type of energy criterion, the practical question arises as to the form to be given to the potential. One possible solution is, of course, a totally numerical potential, but this choice is not very convenient for actual computation. A very practical choice from a computational view is an analytic form depending on a few parameters which can be varied to optimize the solutions. A possibility for such a form is the scaled Thomas–Fermi potential (Stewart and Rotenberg, 1965) which has been extensively used in nonrelativistic calculations, but not often in relativistic ones. Another analytic form, initially introduced by Klapisch (1971) for nonrelativistic calculations, has, however, been successfully extended by Luc-Koenig (1972) for use in relativistic calculations. This potential is derived on the assumption that the charge distribution of an electron in a subshell nl is given by

$$D_{nl} = -e \frac{\alpha_{nl}^{(2l+3)}}{(2l+2)!} [r^{l+1} \exp(-\tfrac{1}{2}\alpha_{nl} r)]^2 \tag{114}$$

where α_{nl} is a parameter to be varied.

Corresponding to such a charge distribution is the potential

$$u_{nl}(r) = \left[-\frac{e}{r}\left\{ 1 - \exp(-\alpha_{nl} r) \sum_{j=0}^{2l+1} \frac{1}{j!}\left(1 - \frac{j}{2l+2}\right)(\alpha_{nl} r)^j \right\} \right]$$

$$\equiv \left[-\frac{e}{r}\{1 - f_{nl}(r)\} \right] \tag{115}$$

In order to obtain $U(r)$, one sums the $u_{nl}(r)$ over the various orbitals, excluding only the most external. Thus the general form of $U(r)$ is

$$U(r) = -\frac{e}{r}\left\{ \sum_{nl} q_{nl} f_{nl}(\alpha_{nl} r) + 1 + Z - N \right\} \tag{116}$$

where q_{nl} is the number of electrons in the subshell nl. If there are many electrons in the atom, the number of parameters in this model is rather large, but it is possible to reduce the independent parameters by using the empirical relation (Klapisch, 1971):

$$\alpha_{nl} = \alpha_n \left\{ \frac{l+1}{1 - 0.03l(l+1)} \right\}$$

In this way, for example, the number of parameters to be introduced for Cs I is five. This potential optimized, using either the total energy or the spectroscopic criterium, has been applied to the spectra of Cs I, Xe I (Koenig, 1972; Luc-Koenig, 1972), Sb I, Br I, and I I (Luc-Koenig et al., 1973) and Ar XVII (Feneuille and Koenig, 1972). In spite of the simplicity of the model, the results obtained are usually in good agreement with experimental data. A similar potential has been used by Matese and Johnson (1965) in computing relativistic photoeffect cross sections, and by Darewych et al. (1971) in calculating atomic energy levels.

Clearly, this form for $U(r)$ is well adapted to the study of discrete spectra in which only monoelectronic excitation from the most external orbital occur; this is, of course, the case for alkaline and noble gases. In the more general case, it has been shown in the nonrelativistic approximation that it is still possible to use the same form for $U(r)$ provided that one introduces a limited number of nearly degenerate configuration interactions (Aymar, 1973).

D. Self-consistent Field Calculations

The potential introduced in the previous section has no well-defined physical meaning; it is only a mathematical intermediary used to compute radial wave functions and its determination is essentially semiempirical. An alternative to such an approach is to demand that the potential be self-consistent. This requirement leads, in a nonrelativistic treatment, to the famous Hartree–Fock equations.

The first attempt to generalize the Hartree–Fock equations to the relativistic case was carried out by Swirles (1935, 1936) for a closed-shell configuration. However, the formalism used was rather intractable and for a long time only the Hartree approximation (neglecting the effects of exchange) was used in actual calculations. The earliest of such Hartree calculations were done by Williams in 1940 for Cu II. Similar work was carried out by Cohen (1960) on Fe I, W I, Pt I, Hg I, and Hg III, and by Mayers (1957) for Hg I.

After the reformulation of the problem by Grant (1961, 1965) in terms of coupled tensor operators, and the introduction of very powerful computers, Dirac–Hartree–Fock-type calculations have became possible, and during the last few years, many studies have been reported.

Because an excellent review of self-consistent field relativistic equations has been given by Grant (1970), we shall only indicate the method here. In brief, the calculation proceeds in much the same manner as that of a nonrelativistic Hartree–Fock calculation (although greatly complicated as a result of the coupled relativistic equations), with the electrostatic interaction energies formulated in terms of the relativistic analogs of the familiar F^k and G^k Slater integrals (Section IV,B,3), or in terms of the relativistic equivalent of Hartree's Y function

$$Y_k(1,2) = r^{-k} \int_0^r s^k (G_1 G_2 + F_1 F_2) \, ds + r^{k+1} \int_r^\infty (G_1 G_2 + F_1 F_2) s^{-(k+1)} \, ds \quad (117)$$

where s is the variable of integration. Assuming that the atomic wave function is a single Slater determinant, the angular factors can be expressed in terms of Grant's Γ coefficients (1961, 1965) which can easily be calculated from tables of $3-j$ symbols (Rotenberg et al., 1959)

$$\Gamma_{j_1 k j_2} = 2 \begin{pmatrix} j_1 & k & j_2 \\ \frac{1}{2} & 0 & -\frac{1}{2} \end{pmatrix}^2 \quad (118)$$

For more complicated wave functions, in $L-S$ coupling for example, the angular factors can be obtained by using the equivalent operator formalism described in the previous section.

In any case, although the use of Racahs tensor operator techniques (Judd, 1963) has enormously simplified the problem, the Dirac-Hartree-Fock formulation remains rather complicated, and many authors have preferred to treat the effect of exchange in an approximate manner. More precisely, they take exchange into account by adding the direct terms a local potential derived originally by statistical (Thomas Fermi) considerations. The following form, originally proposed by Slater (1951)

$$V_{\text{exc}} = -\tfrac{3}{2}[3\rho(r)/4\pi^2 r^2]^{1/3} \quad (119)$$

where $\rho(r)$ is the radial electron density, has been corrected by Gáspár (1954), Kohn and Sham (1965), and Cowan et al. (1966), who suppressed the factor 3/2 in Slater's formula. Other semiempirical modifications have been proposed by various authors. In particular, Rosén and Lindgren (1968) proposed a parametrized exchange potential of the form

$$V_{\text{exc}} = -(C/r)[81 r^n \rho^m(r)/32\pi^2]^{1/3}$$

where C, n, and m are adjustable parameters, equal to unity in the Hartree-Fock-Slater formulation. In all cases, these Slater-type exchange potentials do not vary as $1/r$ when r goes to infinity. Thus, it is necessary to define a value r_L such that, for $r > r_L$, the potential is given by

$$V(r) = -(Z - N + 1)e^2/r$$

Naturally, care must be taken to assure the continuity of the exchange potential at $r = r_L$. This is the so-called "Latter tail correction" (Latter,

1955). Let us also remark that, within these exchange approximations, the relativistic analog of the virial theorem (Kim, 1967a,b) is not exactly satisfied. This point has recently been discussed in detail by Lindgren and Rosén (1973).

This Hartree–Fock–Slater approximation (modified or unmodified), has been used extensively in relativistic calculations. The earliest work of this type was done by Liberman et al (1965) and provides essentially a comparison with previous calculations of Williams (1940), Mayers (1957), and Cohen (1960). Since this first paper, attention has centered on calculations of binding energies (Liberman et al., 1965, 1971; Cowan et al., 1966; Rosén and Lindgren, 1968), X-ray scattering factors (Waber and Cromer, 1965; Cowan et al., 1966; Cromer and Liberman, 1970), and hyperfine structure (Rosén, 1969, 1972; Rosén and Lindgren, 1972; Rosén and Nyqvist, 1972). Considerable work has also been done on superheavy elements (Waber et al., 1969; Lu et al., 1971; Fricke et al., 1971; Fricke and Waber, 1971; Penneman et al., 1971). However, here the coupling parameter $Z\alpha$ is no longer small with respect to one, strong quantum electrodynamical effects could appear; thus the value of these results may be open to question.

Because of the added complexity of the method, Dirac–Hartree–Fock results obtained by such calculations are less numerous, but certainly in principle at least, more precise. Binding energies (Coulthard, 1967; Smith and Johnson, 1967; Kim, 1967a; Mann, 1969; Mann and Waber, 1970, 1973; Desclaux et al., 1971b; Fricke et al., 1972), energies of excited states (Desclaux et al., 1971a), and hyperfine constants (Desclaux, 1972) have now been obtained for various elements, in particular for the alkali atoms where the calculations are greatly simplified.

In any case, many programs have now been published, and numerous relativistic calculations of all types will certainly soon appear in the literature.

E. Size and Mass of the Nucleus

In previous sections, the nucleus was considered to be a point charge and thus the nuclear potential was taken as purely coulombic. However, for penetrating orbitals and large values of the nuclear charge, the finite size of the nucleus can no longer be neglected; in particular, it is well known that for $\alpha Z > 1$, $j = \frac{1}{2}$ states appear only because of this finite nuclear size. The modification of penetrating radial wave functions by the nuclear size has been described in detail by various authors (e.g., Rose, 1961). In general, the model assumed for the nuclear charge density is generally either a uniformly charged sphere (Smith and Johnson, 1967) or a two-parameter Fermi distribution (Mann and Waber, 1973). In either case, the main effect of the finite size of the nucleus is to simplify the series expansions for the wave functions at small r, since in this case only integer powers of r occur; moreover, F/r and G/r no longer diverge at the origin for $j = \frac{1}{2}$ states. An idea of the possible importance of size effects can be seen in the result of Desclaux (1972), who found a 15% reduction of the magnetic hyperfine structure constant for the ground state of francium upon (partial) introduction of finite nuclear size effects. Even more impressive is the statement of Fricke *et al.* (1972), who conclude after precise calculation of atomic electron binding energies in fermium, that "energies of electronic transitions can now be used to estimate the nuclear radius." The optimistic tone of this statement may, however, be muted somewhat if one considers the approximations required to make a relativistic self-consistent field calculation.

Generally, the mass of the nucleus is considered to be infinite, which allows one to ignore nuclear motion. However, it is well known that the finite nuclear mass affects atomic spectra through the so-called mass isotope shift (normal and specific). In the many-electron atom, relativistic corrections to this effect have been investigated only in the Pauli limit (Stone, 1961, 1963; Bauche, 1969).

F. CONFIGURATION MIXING AND CORRELATIONS

Electronic correlation effects, which are already difficult to calculate in the nonrelativistic scheme, become practically incalculable in the relativistic scheme. Even the role of configuration mixing in the quasidegenerate case, which corresponds in a relativistic treatment to the usual intermediate coupling problem, has not yet been examined in self-consistent field calculations. Thus, the only investigations concern either pure $j - j$ coupling states or mean energies of configurations. However, a multiconfiguration Dirac–Hartree–Fock program now exists (Desclaux, 1972); thus such intermediate coupling calculations will soon be feasible. In any case, the calculations will be tremendously complicated. It may be that pseudopotential methods in combination with the equivalent operator formalism are more adapted, because of their relative simplicity, to investigate this difficult problem.

Far-configuration interactions have been introduced in very few relativistic studies and there, only partially. In the study of the hyperfine structure of the alkali atoms and of some p^N configurations, Desclaux (1972) used an unrestricted version of his Dirac–Hartree–Fock program, but it is difficult to understand exactly what excitations are taken into account by this procedure. In a study of the fine structure splittings of p^N configurations, some quasidegenerate configuration mixing (in the nonrelativistic sense) have also been introduced by Desclaux (1972). However, in most relativistic studies, relativistic corrections and correlation effects are considered to be additive; thus the latter can be calculated in a nonrelativistic scheme using one of the many methods which are now available. We have already discussed this problem in Section V.

As we have seen, many important advances in atomic physics in the last decade were the result of the parallel development of two distinct fields—nonrelativistic correlation calculations and relativistic techniques. At present only tentative attempts have been made to combine these two disciplines. We can expect, therefore, that an active and productive area of research in the coming decade will concern the development of accurate techniques for calculating relativistic correlation effects.

REFERENCES

Abragam, A., and Van Vleck, J. H. (1953). *Phys. Rev.* **92**, 1448.
Akhiezer, A. I., and Berestetskii, V. B. (1965). "Quantum Electrodynamics." Wiley (Interscience), New York.
Armstrong, L., Jr. (1966). *J. Math. Phys.* **7**, 1891.
Armstrong, L., Jr. (1968). *J. Math. Phys.* **9**, 1083.
Armstrong, L., Jr. (1971). "Theory of the Hyperfine Structure of Free Atoms." Wiley (Interscience), New York.
Armstrong, L., Jr., and Feneuille, S. (1968). *Phys. Rev.* **173**, 58.
Aymar, M. (1973). *Nucl. Instrum. & Methods* **110**, 211.
Bauche, J. (1969). Ph.D. Thesis Centre d'Orsay, Paris XI.
Bethe, H. A., and Salpeter, E. E. (1957). "Quantum Mechanics of One- and Two-Electron Atoms." Springer-Verlag, Berlin and New York.
Blume, M., and Watson, R. E. (1962). *Proc. Roy. Soc., Ser. A* **270**, 127.
Blume, M., and Watson, R. E. (1963). *Proc. Roy. Soc., Ser. A* **271**, 565.
Breit, G. (1928). *Nature (London)* **122**, 649.
Breit, G. (1929). *Phys. Rev.* **34**, 553.
Breit, G. (1930). *Phys. Rev.* **36**, 383.
Breit, G. (1931). *Phys. Rev.* **38**, 463.
Brink, D. M., and Satchler, G. R. (1962). "Angular Momentum," 1st ed. Oxford Univ. Press, London and New York (2nd ed., 1968).
Casimir H. B. G. (1936). "On the Interaction between Atomic Nuclei and Electrons." Teyler's Tweede Genootshap, Haarlem (reprinted by Freeman, San Francisco, California, 1963).
Cohen, S. (1960). *Phys. Rev.* **118**, 489.
Condon, E. U., and Shortley G. H. (1935). "The Theory of Atomic Spectra," 1st ed. Cambridge Univ. Press, London and New York (8th ed., 1967).
Coulthard, M. A. (1967). *Proc. Phys. Soc., London* **91**, 44.
Cowan, R. D., and Mann, J. B. (1971). *In* "Atomic Physics 2" (G. K. Woodgate and P. G. H. Sandars, eds.), p. 215. Plenum, New York.
Cowan, R. D., Larson, A. C., Liberman, D., and Waber, J. T. (1966). *Phys. Rev.* **144**, 5.
Cromer, D. T., and Liberman, D. A. (1970). *J. Chem. Phys.* **53**, 1891.
Crubellier, A., and Feneuille, S. (1971). *J. Phys. (Paris)* **32**, 405.
Darewych, J. W., Green, A. E. S., and Sellin, D. L. (1971). *Phys. Rev. A* **3**, 502.
Darwin, C. G. (1928). *Proc. Roy. Soc., Ser. A* **118**, 654.
Desclaux, J. P. (1972). *Int. J. Quantum Chem.* **6**, 25.
Desclaux, J. P., Moser, C. M., and Verhaegen G. (1971a). *Proc. Phys. Soc., London (At. Mol. Phys.)* [2], *B* **4**, 296.

Desclaux, J. P., Mayers, D. F., and O'Brien F. (1971b). *Proc. Phys. Soc., London (At. Mol. Phys.)* B **4**, 631.
Dirac, P. A. M. (1928). *Proc. Roy. Soc., Ser. A* **117**, 610; **118**, 351.
Dirac, P. A. M. (1930). "Quantum Mechanics," 1st ed. Oxford Univ. Press, London and New York (4th ed., 1958).
Doyle, H. T. (1969). *In* "Advances in Atomic and Molecular Physics" (D. R. Bates, ed.), Vol. 5, p. 337. Academic Press, New York.
Drake, G. W. F. (1971). *Phys. Rev. A* **3**, 908.
Drake, G. W. F. (1972). *Phys. Rev. A* **5**, 1979.
Ermolaev, A. M., and Jones, M. (1973). *Proc. Phys. Soc., London (At. Mol. Phys.)* B **6**, 1.
Feneuille, S. (1967). *J. Phys. (Paris)* **28**, 61.
Feneuille, S. (1971a). *Physica* **53**, 143.
Feneuille, S. (1971b). *In* "Atomic Physics 2" (G. K. Woodgate and P. G. H. Sandars, eds.), p. 201. Plenum, New York.
Feneuille, S. (1973). *J. Phys. (Paris)* **37**, 1.
Feneuille, S., and Armstrong, L., Jr. (1973). *Phys. Rev. A* **8**, 3.
Feneuille, S., and Koenig, E. (1972). *C. R. Acad. Sci., Ser. B* **274**, 46.
Fermi, E. (1930). *Z. Phys.* **60**, 320.
Foldy, L., and Wouthuysen, S. A. (1950). *Phys. Rev.* **78**, 29.
Fricke, B., and Waber, J. T. (1971). *Actinides Rev.* **1**, 433.
Fricke, B., Greiner, W., and Waber, J. T. (1971). *Theor. Chim. Acta* **21**, 235.
Fricke, B., Desclaux, J. P., and Waber, J. T. (1972). *Phys. Rev. Lett.* **28**, 714.
Gáspár, R. (1954). *Acta Phys.* **3**, 263.
Gaunt, J. A. (1929). *Proc. Roy. Soc., Ser. A* **122**, 513.
Goldschmit, Z. B. (1973). *In* "Atomic Physics 3" (S. J. Smith and G. K. Walters, eds.), p. 221. Plenum, New York.
Grant, I. P. (1961). *Proc. Roy. Soc., Ser. A* **262**, 555.
Grant, I. P. (1965). *Proc. Roy. Soc., Ser A* **286**, 523.
Grant, I. P. (1970). *Advan. Phys.* **19**, 747.
Harvey, J. S. M. (1965). *Proc. Roy. Soc., Ser. A* **285**, 581.
Infeld, L., and Hull, T. E. (1951). *Rev. Mod. Phys.* **23**, 21.
Judd, B. R. (1963). "Operator Techniques in Atomic Spectroscopy." McGraw-Hill, New York.
Judd, B. R., and Lindgren, I. (1961). *Phys. Rev.* **122**, 1802.
Judd, B. R., Crosswhite, H. M., and Crosswhite, H. (1968). *Phys. Rev.* **169**, 130.
Karplus, R., and Kroll, N. (1950). *Phys. Rev.* **77**, 536.
Kelly, H. P. (1971). *In* "Atomic Physics 2" (G. K. Woodgate and P. G. H. Sandars, eds.), p. 227. Plenum, New York.
Kim, Y. K. (1967a). *Phys. Rev.* **154**, 17.
Kim, Y. K. (1967b). *Phys. Rev.* **159**, 190.
Klapisch, M. (1971). *Comput. Phys. Commun.* **2**, 239.
Koenig, E. (1972). *Physica* **62**, 393.
Kohn, W., and Sham, J. (1965). *Phys. Rev. A* **140**, 1133.
Kopfermann, H. (1958). "Nuclear Moments." Academic Press, New York.
Latter, D. (1955). *Phys. Rev.* **99**, 510.
Layzer, D. (1959). *Ann. Phys. (New York)* **8**, 271.
Layzer, D., and Bahcall, J. N. (1962). *Ann. Phys. (New York)* **17**, 177.
Liberman, D. A., Waber, J. T., and Cromer, D. T. (1965). *Phys. Rev. A* **137**, 27.
Liberman, D. A., Cromer, D. T., and Waber, J. T. (1971). *Comput. Phys. Commun.* **2**, 107.
Liberman, S. (1971). *J. Phys. (Paris)* **32**, 867.
Lindgren, I., and Rosén, A. (1974). *Case Stud. At. Phys.* (to be published).
Lu, C. C., Carlson, T. A., Malik, F. B., Tucker, T. C., and Mestor, C. W. (1971). *At. Data* **3**, 1.

Luc-Koenig, E. (1972). *J. Phys.* (*Paris*) **33**, 847.
Luc-Koenig, E., Morillon, C., and Vergès, J. (1973). *Physica* **70**, 175.
Mann, J. B. (1969). *J. Chem. Phys.* **51**, 841.
Mann, J. B., and Johnson, W. R. (1971). *Phys. Rev. A* **4**, 41.
Mann, J. B., and Waber, J. T. (1970). *J. Chem. Phys.* **53**, 2397.
Mann, J. B., and Waber, J. T. (1973). *At. Data* **5** (in press).
Margenau, H. (1940). *Phys. Rev.* **57**, 383.
Matese, J. J., and Johnson, W. R. (1965). *Phys. Rev. A* **140**, 1.
Mayers, D. F. (1957). *Proc. Roy. Soc., Ser. A* **241**, 93.
Messiah, A. (1959). "Mécanique Quantique," 1st ed. Dunod, Paris (4th ed., 1972).
Nielson, C. W., and Koster, G. F. (1963). "Spectroscopic Coefficients for the p^N, d^N and f^N Configurations." MIT Press, Cambridge, Massachusetts.
Penneman, R. A., Mann, J. B., and Jrgensen, C. K. (1971). *Chem. Phys. Lett.* **8**, 321.
Perl, W. (1953). *Phys. Rev.* **91**, 852.
Perl, W., and Hughes, V. (1953). *Phys. Rev.* **91**, 842.
Racah, G. (1931a). *Z. Phys.* **71**, 431.
Racah, G. (1931b). *Nuovo Cimento* **8**, 178.
Racah, G. (1942). *Phys. Rev.* **62**, 438.
Racah, G. (1943). *Phys. Rev.* **63**, 367.
Racah, G. (1949). *Phys. Rev.* **76**, 1352.
Rose, M. E. (1961). "Relativistic Electron Theory." Wiley (Interscience), New York.
Rosén, A. (1969). *Proc. Phys. Soc., London* (*At. Mol. Phys.*) *B* **2**, 1257.
Rosén, A. (1972). *Phys. Scripta* **6**, 37.
Rosén, A., and Lindgren, I. (1968). *Phys. Rev.* **176**, 114.
Rosén, A., and Lindgren, I. (1972). *Phys. Scripta* **6**, 109.
Rosén, A., and Nyqvist, H. (1972). *Phys. Scripta* **6**, 24.
Rotenberg, M., Bivins, R., Metropolis, N., and Wooten, J. K., Jr. (1959). "The $3-j$ and $6-j$ Symbols." MIT Press, Cambridge, Massachusetts.
Sandars, P. G. H., and Beck, J. (1965). *Proc. Roy. Soc., Ser. A* **289**, 97.
Schiff, L. I. (1955). "Quantum Mechanics," 1st ed. McGraw-Hill, New York (3rd ed., 1968).
Schwartz, C. (1955). *Phys. Rev.* **97**, 380.
Schwinger, J. (1949). *Phys. Rev.* **75**, 1912.
Slater, J. C. (1951). *Phys. Rev.* **81**, 385.
Smith, F. C., and Johnson, W. R. (1967). *Phys. Rev.* **160**, 136.
Sommerfeld, A. (1916). *Ann. Phys.* (*Leipzig*) [4] **51**, 1.
Sommerfield, L. M. (1957). *Phys. Rev.* **107**, 328.
Sternheim, M. M. (1962). *Phys. Rev.* **128**, 676.
Sternheimer, R. M. (1950). *Phys. Rev.* **80**, 102.
Sternheimer, R. M. (1971). *Phys. Rev. A* **3**, 837.
Stewart, J. C., and Rotenberg, M. (1965). *Phys. Rev.* **140**, A1508.
Stone, A. P. (1961). *Proc. Phys. Soc., London* **77**, 786.
Stone, A. P. (1963). *Proc. Phys. Soc., London* **81**, 868.
Swirles, B. (1935). *Proc. Roy. Soc., Ser. A* **152**, 625.
Swirles, B. (1936). *Proc. Roy. Soc., Ser. A* **151**, 680.
Waber, J. T., and Cromer, D. T. (1965). *J. Chem. Phys.* **42**, 4116.
Waber, J. T., Cromer, D. T., and Liberman, D. (1969). *J. Chem. Phys.* **51**, 664.
Wesley, J. C., and Rich, A. (1970). *Phys. Rev. Lett.* **24**, 1320.
Williams, A. O. (1940). *Phys. Rev.* **58**, 723.
Wybourne, B. G. (1965). "Spectroscopic Properties of Rare Earths." Wiley (Interscience), New York.

THE FIRST BORN APPROXIMATION

K. L. BELL and A. E. KINGSTON

Department of Applied Mathematics and Theoretical Physics
The Queen's University
Belfast, Northern Ireland

I. Introduction 53
II. Excitation of Atoms by Electron and Proton Impact 54
   A. Electron Impact Excitation 57
   B. Proton Impact Excitation 78
III. Ionization of Atoms by Electron and Proton Impact 87
   A. Electron Ionization 89
   B. Proton Ionization 105
IV. Atom–Atom and Ion–Atom Collisions 113
   A. Excitation 114
   B. Electron Loss and Ionization 119
   C. Metastable Atom Destruction 124
   References 125

I. Introduction

It is important, for many applications, to have reliable data on the magnitudes of the cross sections for excitation and ionization of atoms by electrons, protons, and other atoms. Shortly after the introduction of quantum mechanics, Born (1926) suggested an approximate method for calculating these cross sections. His approximation depends upon the assumption that the incident system interacts only slightly with the target atom so that its wave function may be closely approximated by the plane wave, which would be the correct function in the absence of all interaction. Born does not appear to have used the approximation which bears his name; his approximation was first utilized by Bethe (1930). One may anticipate that this approximation should be valid when the speed of the incident system is great in comparison with that of the electrons in the target atom. However, one immediately asks the question "Above what impact energy does the Born approximation give reliable cross sections?" No theoretical answer has yet been given and one must rely on a comparison of theory and experiment.

The first major review article devoted mainly to a consideration of the first Born approximation is that of Bates et al. (1950). At that time, however, experimental data was extremely limited and it was not possible to make definite statements about the validity of the Born approximation. It is the purpose of the present article to review calculations using the Born approximation and attempt, by comparison of the theoretical results with experimental data, to derive lower limits to the incident system energy above which the Born approximation may be considered to give reliable cross sections. The article is concerned only with atomic targets and with excitation and ionization processes. Elastic scattering and processes involving molecular species are not considered.

In the following three sections, which are devoted to excitation by electron and proton impact, ionization by electron and proton impact, and excitation and ionization by atomic impact, each section attempts an extensive review of all appropriate Born approximation calculations and then compares these results with available experimental data. (Throughout this review we shall often use Born approximation, by which we mean the first Born approximation. Higher order Born approximations will be specifically designated as such.) Finally, we draw attention to the review article of Inokuti (1971) which is concerned mainly with the Born matrix elements and with the asymptotic behavior of the Born cross sections.

Throughout the article atomic units will be used unless otherwise indicated.

II. Excitation of Atoms by Electron and Proton Impact

The Born approximation for excitation of atoms by electron and proton impact was first considered by Bethe (1930). In the wave theory treatment, the total cross section is readily derived (cf. Mott and Massey, 1965). Thus for excitation of a target atom (or ion), of nuclear charge Z_A and having N_A electrons, from initial state i to final state f by an incident nucleus of charge Z, the total cross section is given by

$$Q(i \to f) = \frac{8\pi Z^2}{V^2} \int_{K_1}^{K_2} \frac{dK}{K^3} \varepsilon_{if}(K) \qquad (1)$$

with

$$\varepsilon_{if}(K) = \left| \int d\mathbf{r}\, \phi_f^*(\mathbf{r}_1, \ldots, \mathbf{r}_{N_A}) \left[\sum_{t=1}^{N_A} \exp(iKz_t) \right] \phi_i(\mathbf{r}_1, \ldots, \mathbf{r}_{N_A}) \right|^2 \qquad (2)$$

where ϕ_i and ϕ_f denote the unperturbed atomic wave functions of the initial and final states, respectively, V is the incident relative velocity, $K_1 = |K_i - K_f|$, $K_2 = K_i + K_f$. $K_i = M_R V$, where M_R is the reduced mass of the colliding systems and K_f is determined by the energy conservation equation, $K_f^2 = K_i^2 + 2M_R \Delta E_{if}$, ΔE_{if} being the energy difference between the initial and final atomic states. Equation (2) is generally referred to as the length formulation of the Born matrix element. However, Bates et al. (1950) showed that by using the properties of the atomic wave functions, Eq. (2) may be transformed to the velocity formula

$$\varepsilon_{if}(K) = \left| \frac{1}{2\Delta E_{if}} \int d\mathbf{r}\, \phi_f^* \left\{ K^2 \left[\sum_{t=1}^{N_A} \exp(iKz_t) \right] \phi_i - 2iK \left[\sum_{t=1}^{N_A} \exp(iKz_t) \frac{\partial \phi_i}{\partial z_t} \right] \right\} \right|^2 \tag{3}$$

Equations (2) and (3) both give the exact result if the wave functions are exact. However, apart from atomic hydrogen, approximate wave functions must be employed and the difference between the results of using Eqs. (2) and (3) in (1) can be interpreted as a measure of the inaccuracy in the Born cross section due to the approximate wave functions.

It should also be emphasized that the derivation of Eq. (1) assumes orthogonality of ϕ_i and ϕ_f. If the wave functions are not orthogonal, considerable errors can arise in the Born approximation cross section. In general, approximate wave functions will not be orthogonal but the final state wave function ϕ_f is usually made orthogonal to the initial state wave function ϕ_i by replacing ϕ_f by $(1 - \gamma^2)^{-1/2}(\phi_f - \gamma\phi_i)$, where $\gamma = \langle \phi_f | \phi_i \rangle$ is the overlap integral. This method of orthogonalizing the wave functions can itself lead to serious errors if the overlap integral is not small. For example, Bell et al. (1969a) in a study of $1^1S - 3^1S$ excitation of helium, employed a six-parameter correlated ground state wave function with a modified hydrogenic excited state representation. With these functions $\gamma = 0.11$ and the matrix elements (2) and (3) were found to differ by as much as 50%, while the corresponding length and velocity electron impact cross sections were found to be in error by factors of 2 and 3 when compared with a highly accurate first Born approximation calculation.

It should also be pointed out that for proton impact excitation, (cf. Bates, 1962), the first Born approximation may also be used in an impact parameter formulation. The first Born approximation cross section for excitation as derived using the wave and impact parameter treatments have been shown to be equivalent (Frame, 1931; Arthurs, 1961; Moiseiwitsch, 1966; Crothers and Holt, 1966; Vinogradov, 1967; McCarroll and Salin, 1966). Most calculations have, however, been performed using the wave theory equations.

At high impact velocities, the first Born approximation cross sections for electron and proton impact are closely related. From Eq. (1) Bates and Griffing (1953) have shown that for proton impact velocities V_p such that

$V_p^2 \gg (2\Delta E_{if}/M_R)$, the electron impact cross section at the electron velocity $V_e = \gamma V_p$ is given by

$$Q_e(V_e) = \gamma^{-2}\left\{Q_p(V_p) - \frac{V^2}{V_p^2} Q_p(V)\right\} \quad (4)$$

where

$$V = \Delta E_{if}/2V_p \qquad \gamma = 1 + \Delta E_{if}/2V_p^2$$

and $Q_{e,p}$ denote the electron and proton cross sections. Clearly the electron–atom cross sections are smaller than the corresponding proton–atom cross sections, but as $V_p \to \infty$, the electron and proton impact cross sections become equal for the same relative velocity. A further result of importance at high impact energies is that of Bethe (1930), who showed that the first Born approximation cross section for electron and proton impact has the asymptotic form

$$Q \to (A/V^2)\ln V + B/V^2 + \cdots \quad (5)$$

where A is related to the oscillator strength for optically allowed transitions and is zero for dipole forbidden transitions.

It is known that at sufficiently high impact energies the first Born approximation provides cross sections which are exact, but at lower energies electron exchange and coupling mechanisms become important and the Born approximation is not valid. It is still not possible from a theoretical analysis to determine the energy at which the Born approximation breaks down. Hence to determine the range of validity of the Born approximation we must resort to a comparison of the Born cross sections with measurements or a comparison of the Born cross sections with more sophisticated theoretical treatments. Before doing this, however, attention should be drawn to the work of Starostin (1967). By using dispersion and unitarity relations, he has attempted to obtain criteria for the validity of Born's approximation for elastic and inelastic collisions of electrons and highly excited atoms. For transitions $n \to n'$ $n, n' \gg 1$, $|n - n'| \sim 1$, he finds that the Born approximation is valid if

$$nE \gg \ln(4E/J_n)$$

where J_n is undetermined (but for estimates could be taken as the ionization potential of the nth level) and where E is the electron impact energy (in a.u.). This criterion is very imprecise but does show that with increasing n one may expect the Born approximation to be accurate to lower impact energies.

In the following subsections, we shall review first Born approximation calculations for electron and proton excitation of atoms, and when possible attempt to derive a limit to the energy range for which the approximation may be considered satisfactory.

THE FIRST BORN APPROXIMATION

A. Electron Impact Excitation

In order to determine the accuracy of the first Born approximation we shall make use not only of experimental data for the total cross section but also of data for the differential cross section. In the first Born approximation the latter may be written as

$$\frac{dQ(K)}{d\Omega} = \frac{2}{K^2} \frac{K_f}{K_i} \frac{1}{\Delta E_{if}} f(K) \tag{6}$$

where f is the generalized oscillator strength and is related to (2) by

$$f(K) = (2/K^2) \Delta E_{if}\, \varepsilon_{if}(K)$$

This theoretical generalized oscillator strength has the important property that it is independent of the energy of the incident particle. By determining the differential cross section experimentally as a function of scattering angle (or K) one may invert Eq. (6) to obtain experimental "generalized oscillator strengths." At low impact energies the experimental generalized oscillator strengths will be energy dependent, but as the energy is increased they will become energy independent and depend only on K, as is the case for the first Born approximation. Thus together with total cross section data, experimental differential cross-section values provide valuable information on the usefulness of the first Born approximation.

1. Hydrogen

Hydrogen has received considerable theoretical attention since it is the only atom for which exact wave functions exist. Suppose that n', l', m' and n, l, m are the principal, azimuthal, and magnetic quantum numbers of the initial and final states of the hydrogen atom, respectively, and let us denote the excitation cross section for the transition $n'l'm' \rightarrow nlm$ by $Q(n'l'm' \rightarrow nlm)$. We are mainly concerned with the average of this cross section over m' and its sum over m, i.e., with the excitation cross section for the transition $n'l' \rightarrow nl$ given by

$$Q(n'l' \rightarrow nl) = (2l' + 1)^{-1} \sum_{m, m'} Q(n'l'm' \rightarrow nlm) \tag{7}$$

A further cross section of relevance is that for the transition $n' \rightarrow n$ between the levels with principal quantum numbers n' and n which is obtained by averaging over l' and summing over l to yield

$$Q(n' \rightarrow n) = (1/n'^2) \sum_{l, l'} (2l' + 1) Q(n'l' \rightarrow nl) \tag{8}$$

First Born approximation cross sections of type (7) have been calculated for the transitions $n's \to ns$, $n's \to np$, $n's \to nd$ for $n' = 1$ to 8 and for a given value of n' for $n = n' + 1$ to 9 by Vainshtein (1961, 1965), Bates and Miskelly (1957), and Somerville (1963). Veldre and Rabik (1965) have extended these calculations to the transitions $1s \to ns$, $1s \to np$, $ns \to (n+1)s$ for $n = 6$ to 11. Milford and his collaborators (Fisher et al., 1960; McCoyd et al., 1960; McCoyd and Milford, 1963; Milford et al., 1960; Scanlon and Milford, 1961) have also carried out extensive investigations. They have considered all transitions between states with principal quantum numbers $n' = 3$ and $n = 4$; all optically allowed transitions between states with $n' = 3$ and $n = 5$, between states with $n' = 4$ and $n = 5$, and between states with $n' = 5$, and $n = 6$; together with the transitions $2s \to 3s, 3p$ and $3d$; $2s \to 4s$; $2p \to 3s, 3p$ and $3d$; $2p \to 4f$; $4s \to 6p$; $4f \to 6g$; $10s \to 11p$, and $10, 9 \to 11, 10$. The most recent calculation is that of Omidvar (1969) who investigated all transitions between states with principal quantum numbers $n' = 3$ and $n = 6$; $n' = 5$ and $n = 6$; $n' = 6$ and $n = 7$; $n' = 7$ and $n = 8$.

For cross sections of type (8) the most extensive calculations are by Omidvar (1965b, 1969). He has considered transitions between the levels $1 \to n$ ($n = 2, \ldots, 15$); $2 \to n$ ($n = 3, \ldots, 12$); $3 \to n$ ($n = 4, \ldots, 11$); $4 \to n$ ($n = 5, \ldots, 10$); $5 \to n$ ($n = 6, \ldots, 9$); $6 \to n$ ($n = 7, 8$); and $7 \to 8$. This work repeats in part the investigations of McCarroll (1957). Boyd (1958) has considered $2s \to n$, ($n = 3, \ldots, 7$) while McCrea and McKirgan (1960) have investigated transitions $2p_m \to n$, ($n = 3, \ldots, 7$), with $m = 0, \pm 1$.

In addition to explicit evaluation of the cross section, considerable attention has also been given to the evaluation of the Bethe coefficients A and B in Eq. (5). Milford and his collaborators have shown that for optically allowed transitions the Bethe approximation provides cross sections which are in close agreement with the first Born approximation down to electron impact energies which are not greatly above the energy where the Born cross section attains its maximum value. McCoyd and Milford (1963) have evaluated Bethe cross sections of the type (7) for all transitions satisfying $n - n' = 1$ or 2 and $l - l' = 1$ with $1 \leq n' \leq 10$. Kingston and Lauer (1966a,b) have extended these calculations to the cases $n - n' = 1$ and 2 for $1 \leq n' \leq 6$ and all permissible values of l and l'. They also evaluated Bethe cross sections $Q(n' \to n)$ for transitions between the levels with principal quantum number n' and n. The most recent evaluation of Bethe cross sections $Q(n' \to n)$ is by Omidvar (1969) who considered the transitions

$1 \to n$ ($n = 2, \ldots, 20$); $\quad 2 \to n$ ($n = 3, \ldots, 20$); $\quad 3 \to n$ ($n = 4, \ldots, 20$);

$4 \to n$ ($n = 5, \ldots, 20$); $\quad 5 \to n$ ($n = 6, \ldots, 16$); $\quad 6 \to n$ ($n = 7, \ldots, 14$);

$7 \to n$ ($n = 8, \ldots, 12$); $\quad 8 \to n$ ($n = 9, \ldots, 11$); $\quad 9 \to 10$;

and by Omidvar and Khateeb (1973) who investigated the transitions $n' \to n$ for $n' = 1$ to 7 and for each specific value of n' all discrete n values.

Few of these Born results may be compared either with experiment or with more sophisticated calculations. Figure 1 compares the Born theory

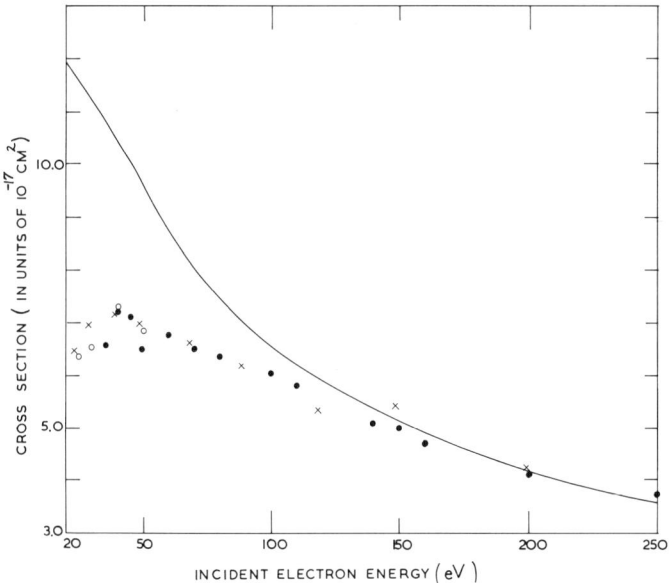

FIG. 1. Total cross section for excitation of atomic hydrogen from the ground state to the 2p state by electron impact. First Born approximation: ——— Vainshtein (1965). Experimental data: ×, Long et al. (1968); ○, Fite et al. (1959); ●, Fite and Brackmann (1958b). The apparent cross section $Q_\perp$ obtained by these experimentalists was converted to the excitation cross section Q using the polarization measurements of Ott et al. (1970). The data of Long et al. were normalized to the Born cross section at 200 eV.

and experiment for the 1s → 2p transition. The first Born approximation cross section is in excellent accord (to better than 10%) with experiment for impact energies greater than 100 eV, but overestimates the cross section with decreasing energy. Tai et al. (1970) compare the first Born approximation with a number of more sophisticated theoretical treatments, and it is again found that the Born approximation is satisfactory for impact energies greater than 100 eV.

For 1s → 2s excitation of hydrogen, the experimental data is uncorrected for cascade effects, and in Fig. 2 we compare experiment with both the 1s → 2s Born cross section and with the composite Born cross section $Q(1s \to 2s) + 0.23Q(1s \to 3p)$ in which the latter term is an allowance for

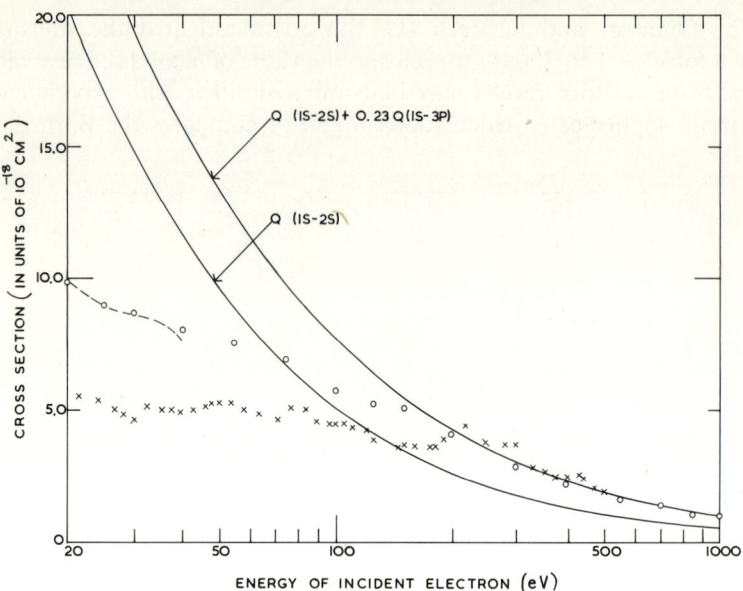

Fig. 2. Total cross section for excitation of atomic hydrogen from the ground state to the 2s state by electron impact. First Born approximation: ——— $Q(1s \to 2s)$, $Q(1s \to 2s) + 0.23 Q(1s \to 3p)$ (without and with allowance for cascade) Vainshtein (1965), Scanlon and Milford (1961). Experiment: ○, Kauppila et al. (1970); – – –, Lichten and Schultz (1959) (normalized to data of Kauppila et al. at 25 eV); ×, Hils et al. (1966) (normalized to Born cross section at 500 eV).

cascade effects as derived by Hummer and Seaton (1961) using Born approximation cross sections for 1s → np transitions together with the appropriate transition probabilities. If we assume that the 1s → np cross sections have the same energy range of validity as that obtained for 1s → 2p above, then the 1s → 2s Born cross section would not be greatly in error for impact energies greater than about 200 eV. Tai et al. (1970), from a comparison of various theoretical treatments, conclude that for this transition the Born approximation is satisfactory for impact energies greater than about 100 eV. It would seem, therefore, that for 1s → 2s excitation, the Born approximation is satisfactory for impact energies greater than about 100–200 eV.

The only other experimental data available for excitation from the ground state of hydrogen is that of Kleinpoppen and Kraiss (1968). They have obtained the composite cross section $Q(1 \to 3)$ for excitation to the $n = 3$ level. Normalizing the experimental data to the Born result at 300 eV, we find that agreement between the Born theory and experiment is excellent down to about 80 eV. Unfortunately, it is not possible to obtain quantitative information from this comparison about the cross sections for the individual

TABLE I

Comparison of First Born Approximation Cross Sections with Experiment for n' to n Transitions in Hydrogen Atoms (Colliding with 5425 eV Electrons)

Transition $n' \to n$	$1 \to 6$	$1 \to 7$	$4 \to 6$	$4 \to 7$	$5 \to 6$	$5 \to 7$	$5 \to 8$	$5 \to 9$	$6 \to 7$	$6 \to 8$	$6 \to 9$
Experiment (Berkner et al., 1965)	<1.3	<0.5	<148	<57	945	210	135[a]		1700	510	190
First Born approximation (Omidvar, 1969)	0.052	0.032	56	18	938	116	37	17	1866	221	69

Cross section (in units of 10^{-18} cm^2)

[a] For $Q(5 \to 8) + Q(5 \to 9)$.

angular momentum excited states. Limited information is available, however, for the 1s → 3d excitation. Woollings and McDowell (1973) have studied this transition using a simplified second Born approximation theory. Their results and the comparison they make with other calculations would indicate that the first Born approximation for 1s → 3d excitation is not in error by more than about 20% for impact energy 100 eV, by about 10% at 200 eV, and 6% at 350 eV.

Thus for excitation from the ground state to the 2s, 2p, and 3d states, the indications are that the first Born approximation is accurate to better than 20% for electron impact energies greater than 100 eV.

The measurement of the cross sections for transitions between excited states is very difficult. However, the work of Berkner et al. (1965) is encouraging. They have obtained cross sections or in some cases upper limits to the cross sections for several transitions between excited states. Their results are compared in Table I with the first Born approximation values of Omidvar (1969). Considering the difficulties associated with the experiment, the agreement is highly satisfactory.

2. Helium

Extensive calculations in the first Born approximation have been performed for electron impact excitation of helium. Recently a large number of Born calculations have been carried out using highly accurate wave functions, and thus we shall not attempt to list all the calculations which have been performed. Rather we shall consider the transitions which have been investigated and for each specific transition record only that calculation which utilizes the most accurate wave functions for both initial and final states. Table II lists the particular transitions considered together with the wave functions utilized; it also states whether the calculations have been performed using the length and/or velocity formulation of the matrix elements. The many parameter-correlated wave functions are of the Hylleraas type and contain at least eighteen terms; when they are utilized for both initial and final states of a given transition, the resulting cross section should be accurate to a few percent. The cross sections for the other transitions will be slightly less accurate. No experimental data is available for transitions from excited states of helium, but there is considerable data available for excitation from the ground state to ^1S, ^1P, and ^1D states.

For excitation of the ground state of helium, experimental data for the total cross section exist for excitation to the n^1S states, $n = 3$ to 6. Figure 3 compares experiment with the first Born approximation for excitation to the 4^1S, 5^1S, 6^1S states. It is convenient to compare the product QE of the cross section Q and the impact energy E. At large values of E, the product QE

TABLE II

First Born Approximation Calculations for e^-, H^+ Excitation of Helium

			Wave functions[b]		
Transition	Projectile	Formulation[a]	Initial state	Final state	Refs.[c]
$1^1S \to N^1S, N = 2-7$[d]	e^-, H^+	L	MP	MP	(1)
$1^1S \to N^1S, N = 8-10$	e^-	L	SO	SO	(2)
$1^1S \to N^1P, N = 2-6$	e^-, H^+	L for $N = 2-4$	MP	MP	(1)
		L, V for $N = 5, 6$	MP	UHF	
$1^1S \to N^1D, N = 3-6$	e^-, H^+	L for $N = 3$	MP	MP	(1)
		L, V for $N = 4, 5, 6$	MP	UHF	
$1^1S \to 4^1F, 5^1F, 5^1G$	e^-	L	CI	CI	(3)
$1^1S \to 4^1F$	H^+	L	SO	SO	(4)
$2^mS \to N^mS, N = 3-7$ $m = 1$ and 3					
$N^mS \to 2^mP, 3^mP,$	e^-, H^+	L	MP	MP	(5)
$4^mP, 3^mD, N = 2-7$					
$2^3S \to 4^3D$	e^-	L	SO	SO	(6)
$2^1S \to 4^1D, 4^1F$	e^-	L	CI	CI	(3)
$2^1P \to 3^1D, 4^1D$	H^+	Impact parameter method	SO	SO	(7)

[a] L, V refer to evaluation of (1) using (2) and (3), respectively (see text).

[b] Abbreviations: MP, many parameter correlated wave functions; UHF, unrestricted Hartree–Fock wave functions; CI, configuration-interaction wave functions; SO, simple analytic separated orbital wave functions.

[c] Key to references: (1) Bell et al. (1968, 1969a); (2) Fox (1965); (3) Vanderpoorten (1970); (4) Gaillard (1966); (5) Kennedy (1968); (6) Moiseiwitsch (1957); (7) McDowell and Pluta (1966).

[d] For H^+, $1^1S \to 7^1S$, excitation, a better calculation is that of Oldham (1969) using formulation L and MP wave functions for both states.

should tend to a constant for optically forbidden transitions but should tend to $A \log E + B$ for optically allowed transitions. It is seen that there is a considerable discrepancy between the experimental data of Van Raan et al. (1971) and that of Moustafa Moussa et al. (1969). The former is larger than the latter at all energies. If we normalize the data of Moustafa Moussa et al. (1969) to theory at 5000 eV, then it is found that for n^1S, $n = 3$ to 6, experiment and the first Born approximation agree to better than about 25% for electron impact energies greater than 500 eV, and that as the energy decreases the experimental data and the Born theory diverge—experiment showing a more rapid fall-off for QE.

Further information on the accuracy of the first Born approximation for excitation of the ground state to the n^1S states may be obtained from experimental generalized oscillator strengths. Figure 4 shows the data of Vriens et al. (1968) for excitation of the 2^1S state for electron impact energies in the

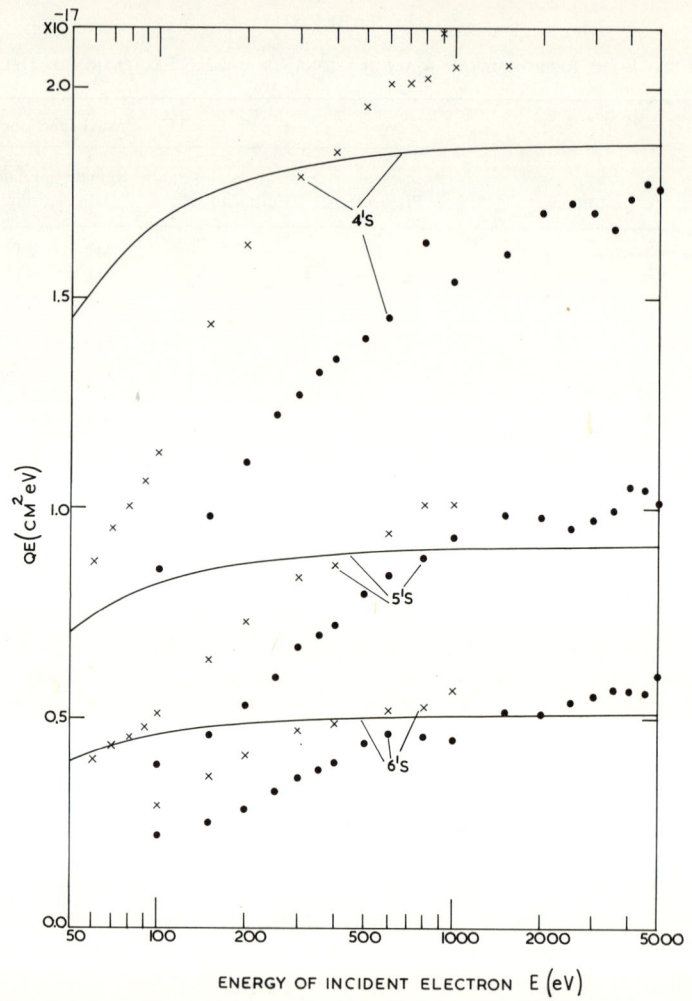

FIG. 3. Total cross sections for excitation of atomic helium from the ground state to the 4^1S, 5^1S, and 6^1S states by electron impact. First Born approximation: ——— Bell et al. (1969a). Experiment: ●, Moustafa Moussa et al. (1969); ×, Van Raan et al. (1971).

range 100–400 eV. As the experimental generalized oscillator strengths are not equal at 300 and 400 eV, it is clear that the Born approximation is not valid at these energies. However, if we draw reasonable curves through the experimental data and plot the generalized oscillator strength against energy for specific values of K (Fig. 5), extrapolation to high energies, although crude, would indicate that the first Born generalized oscillator strengths should be accurate to about 10% at 600 eV and to better than 25% at

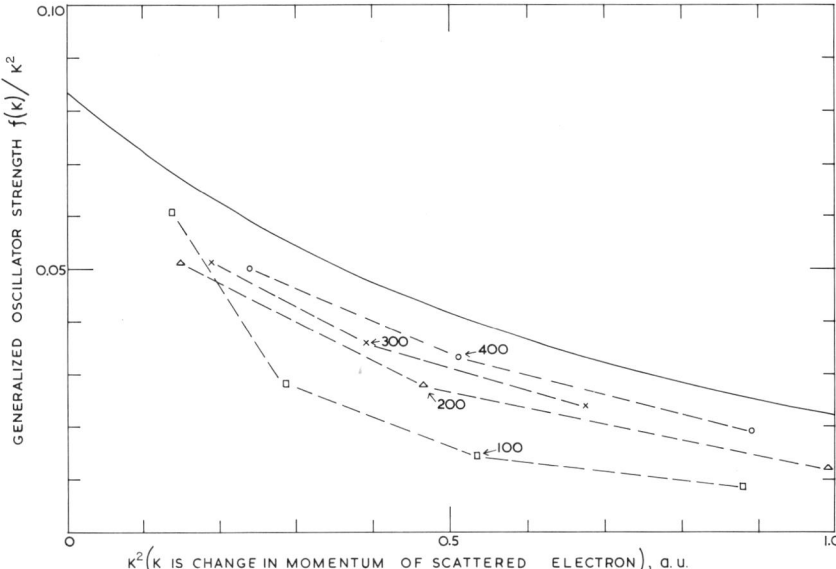

FIG. 4. Generalized oscillator strengths for excitation of atomic helium from the ground state to the 2^1S state by electron impact. First Born approximation: —— Bell et al. (1969a). Experimental data: Vriens et al. (1968) (renormalized using the results of Chamberlain et al., 1970) for incident electron impact energies 100 eV (□), 200 eV (△), 300 eV (×), and 400 eV (○).

400 eV. This experimental evidence, taken together with a comparison of the first Born approximation and more sophisticated theoretical investigations (Berrington et al., 1973; Woollings and McDowell, 1972; Holt et al., 1971) for excitation of the 2^1S state, indicates that for electron impact energies greater than about 500 eV the first Born approximation provides total cross sections for excitation of the ground state of helium to excited ^1S states which are accurate to better than 25%.

For excitation of the ground state of helium to ^1P states, experimental data for the total cross section exists for excitation to the n^1P states, with $n = 2$ to 5. Figure 6 compares experiment with the first Born approximation for excitation to the 2^1P, 3^1P and 4^1P states. There is some scatter in the experimental data, but if we accept the results of Van Eck and de Jongh (1970), de Jongh and Van Eck (1971), or Donaldson et al. (1972), which are in best accord with theory at high impact energies, then small deviations from the first Born approximation appear to arise for impact energies less than about 150 eV. At 150 eV for all these states, the first Born approximation seems to be accurate to about 10%, but with decreasing energy the Born approximation very rapidly becomes much less accurate—the difference

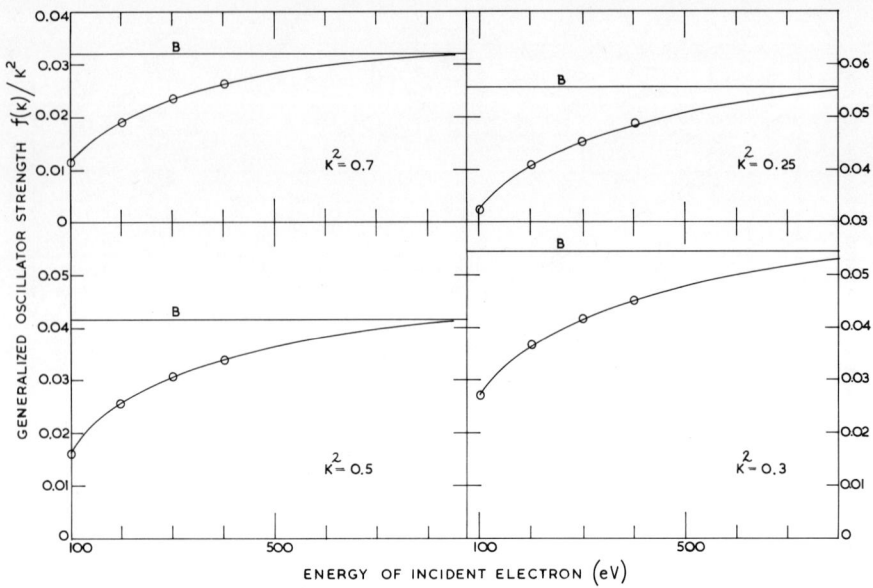

FIG. 5. Generalized oscillator strengths for excitation of the 2^1S state of helium for given values of the square of the change of momentum K of the incident electron plotted against the impact energy of the incident electron. First Born approximation: B, Bell et al. (1969a). Experimental data: —o— estimated values taken from the experimental data given in Fig. 4 for impact energies less than 400 eV with an extrapolation to higher energies (see text).

between the Born theory and experiment being about 25% at 100 eV and about 50% at 60 eV.

Additional information about excitation of the 2^1P and 3^1P states is also available from extensive experimental data on differential cross sections. Some of this data for low impact energies is displayed in Fig. 7 as generalized oscillator strengths. It is quite clear for the 2^1P case that the first Born approximation is highly unsatisfactory for electron impact energy of 55.5 eV, that it is still in error for 81.63 eV, but is highly satisfactory for 150 eV. The 3^1P case is less explicit since Peresse et al. (1970b) do not distinguish in their paper between points corresponding to a given energy in the range covered, 35–150 eV. Although points other than those for 100 and 200 eV lie above the first Born approximation result, the experimental data does have the same shape as the theoretical curve, and it would not be inconsistent for the conclusion reached so far—that the first Born approximation is accurate to 10% at 150 eV for ^1P excitation—to apply also to this comparison of theory and experiment.

More sophisticated calculations have been carried out for $1^1S \rightarrow 2^1P$ excitation (Berrington et al., 1973; Flannery, 1970; Holt et al., 1971). These

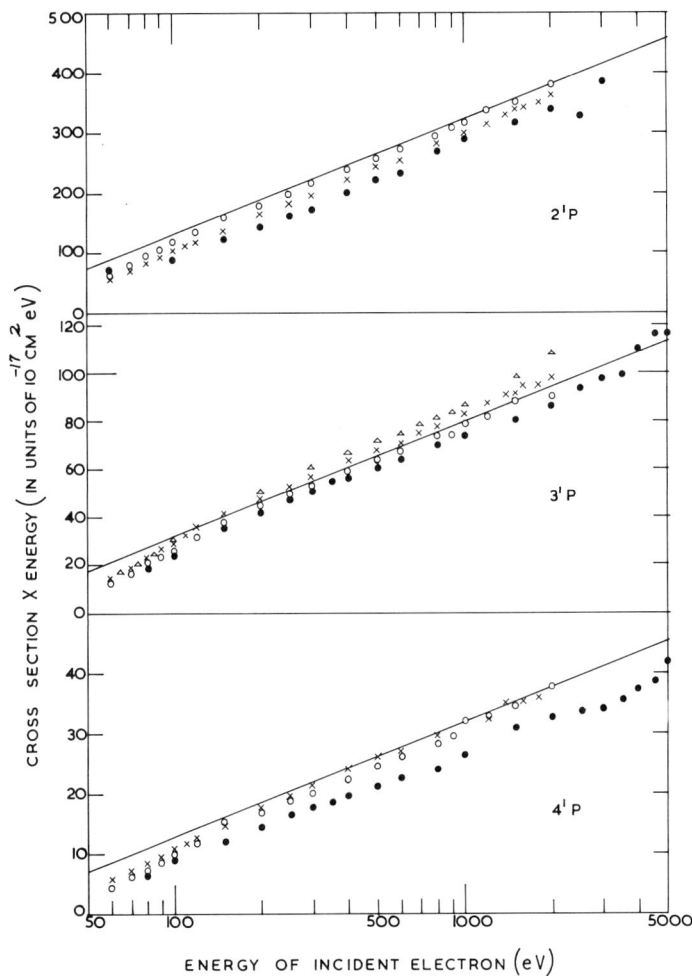

FIG. 6. Total cross sections for excitation of atomic helium from its ground state to the 2^1P, 3^1P, and 4^1P states. First Born approximation: —— Bell *et al.* (1969a). Experimental data: ×, Donaldson *et al.* (1972); △, Van Raan *et al.* (1971); ○, de Jongh and Van Eck (1971), Van Eck and de Jongh (1970); ●, Moustafa Moussa *et al.* (1969).

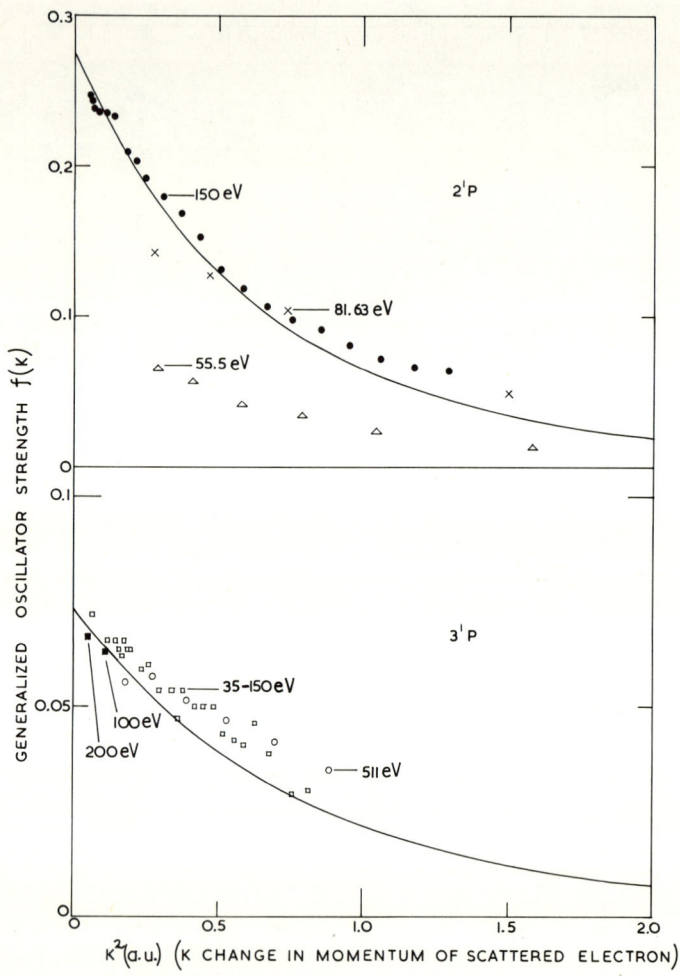

FIG. 7. Generalized oscillator strengths for excitation of atomic helium from its ground state to the 2^1P and 3^1P states by electron impact. First Born approximation:——, Bell et al. (1969a). Experimental data: (the impact energy of the incident electron is indicated for each experiment); ●, Peresse et al. (1970a); × △, Truhlar et al. (1970); ○, Lassettre et al. (1964); ■, Heideman and Vriens (1967); □, Peresse et al. (1970b). The data of Truhlar et al. (1970) have been renormalized according to the work of Rice et al. (1972).

indicate that the first Born approximation is accurate to better than 10% for impact energies greater than 100 eV. Taking all of the theoretical and experimental evidence together, it seems that for electron impact excitation of ^1P states of helium from the ground state, the first Born approximation is accurate to better than about 20% at 100 eV and to better than about 10% for impact energies greater than 150 eV.

For excitation of the ground state of helium to ^1D states, experimental data for the total cross section exists for excitation to the n^1D states, with $n = 3$ to 6. Figure 8 compares experiment with the first Born approximation for excitation to the 4^1D, 5^1D, and 6^1D states. There is some scatter in the experimental data, and for excitation to all of the ^1D states experiment lies above the Born approximation values even at an incident energy of 5 keV. Between about 1 keV and 5 keV, however, theory and experiment have the same shape. If we normalize the experimental data to the first Born approximation at 1 keV, it is found that for excitation to the n^1D states, $n = 3$ to 6 theory and experiment agree to better than 12% for impact energies greater than 300 eV. Below this energy the data of Van Raan et al. (1971) and Moustafa Moussa et al. (1969) tend to disagree with each other and both deviate considerably from the first Born approximation values as the energy decreases. Peresse et al. (1972) have obtained the differential cross section for excitation of the 4^1D state over a range of impact energies 50–200 eV. However, their data were taken at a fixed angle of scattering and so do not provide experimental generalized oscillator strength curves for fixed impact energies. The limited data does suggest, however, that in accordance with the above conclusion, the first Born approximation is considerably in error for impact energies less than 200 eV.

Evidence for the accuracy of the Born approximation for excitation to ^1D states from more sophisticated theoretical treatments is also limited. Woollings and McDowell (1973) have applied a simplified second Born approximation to the study of excitation of the 3^1D and 4^1D states. Taking their "best" results and comparing with the first Born cross section, it is found that for excitation of the 3^1D state, the first Born cross section is accurate to better than 30% and 20% for energies greater than 300 and 400 eV, respectively; while for the 4^1D state it is accurate to better than 15% and 7% for impact energies greater than 200 and 300 eV, respectively. It would appear that the first Born approximation is accurate to lower energies for 4^1D excitation than for 3^1D excitation. Taking all of the theoretical and experimental evidence available at the present time, the most reasonable conclusion that one may reach is that the first Born approximation for excitation of ^1D states is probably accurate to better than about 20% for impact energies greater than 300 eV.

In summary, therefore, we may conclude that from all the evidence available at the present time, the first Born approximation for electron impact

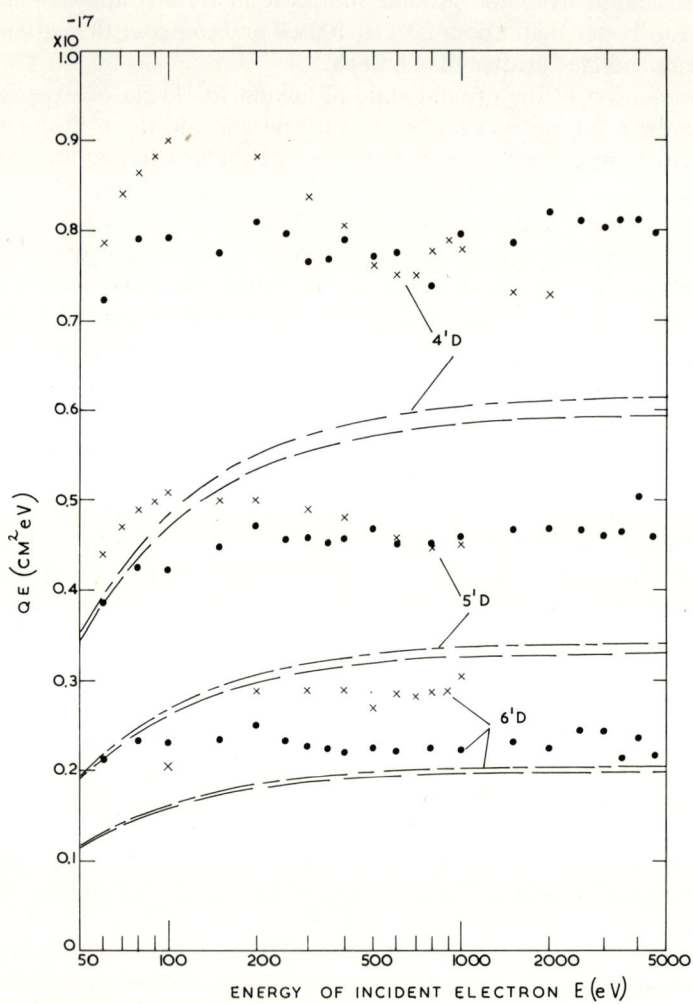

Fig. 8. Total cross sections for excitation of atomic helium from its ground state to the 4^1D, 5^1D, and 6^1D states. First Born approximation: ——— length formulation, — — — velocity formulation (see text and Table II) Bell et al. (1969a). Experimental data: ●, Moustafa Moussa et al. (1969); ×, Van Raan et al. (1971).

excitation of ^{1}S, ^{1}P, and ^{1}D states from the ground state is accurate to better than 20% for impact energies greater than 500 eV, 100 eV, and 300 eV, respectively. We note that of these energies, the highest is for ^{1}S states, and that this is consistent with the work of Lassettre (1970) who, from a study of available experimental data and from theoretical investigations of the differential cross section, concluded that the first Born approximation will be least satisfactory for transitions in which the term symbols are the same in the initial and final states.

3. Alkali Metal Atoms

The ground state of an alkali metal atom has a single s electron outside a core of completely filled electron shells. We shall be concerned with transitions involving this outer s electron.

The first Born approximation has been used to calculate the electron impact excitation cross sections for the following transitions in lithium: 2s → 2p, 2s → np, 2s → ns, 2s → nd, (n = 3–5) (Vainshtein et al., 1964, 1965); 2s → 2p, 2s → 3p (Simsic and Williamson, 1972); and 2s → 2p, 2s → 3p, 3d (Felden and Felden, 1971). In sodium, transitions that have been studied using the Born approximation are: 3s → 3p, 3s → 3d, 3s → 4s, 3s → 4p (Bates et al., 1950); 3s → np, (n = 3–7), 3s → ns, (n = 4–7), 3s → nd, (n = 3–7), 4s → 4p, 5s → 6p, 5s → 7p (Vainshtein et al., 1964, 1965; Vainshtein, 1965); 3s → 3p, 3s → 4p, 4d (Felden and Felden, 1971); 3s → 3p, 3s → 4p (Williamson, 1970; Simsic and Williamson, 1972). Simsic and Williamson (1972) have also applied the first Born approximation to the 4s → 4p transition in potassium, while Vainshtein et al. (1964, 1965) have investigated 4s → 4p, 4s → np, 4s → ns, 4s → n'd; n = 5–7, n' = 3–6. For the rubidium and cesium atoms, the only calculations are those of Vainshtein et al. (1964, 1965). They have studied the rubidium transitions 5s → 5p, 5s → np, 5s → ns (n = 6–8), 5s → nd (n = 4–7), and the cesium transitions 6s → 6p, 6s → np, 6s → ns (n = 7–9), 6s → nd (n = 5–8).

All these calculations have been performed using the length formulation of the Born matrix element and utilize only simple wave functions. Bates et al. (1950) used Hartree–Fock functions. Vainshtein (1965) and Vainshtein et al. (1964, 1965) have employed semiempirical wave functions including exchange, while Williamson (1970) and Simsic and Williamson (1972) have used simple single-particle wave functions based on analytic Slater radial orbitals. Felden and Felden (1971) have employed the coulombic functions of Bates and Damgaard (1949) for the valence electron. Since the wave functions used are very simple, caution must be exercised concerning the accuracy of these calculations. Also, because of the limited experimental data available, we will not consider each atom individually but will concern ourselves first with the resonance transitions of all atoms and then examine the remaining transitions.

Figures 9 and 10 compare the Born cross sections with experimental data for the resonance transitions of lithium through cesium. We note that for lithium the data of Hughes and Hendrickson (1964) is uncorrected for cascade effects. For sodium, the data of Enemark and Gallagher (1972) has been made absolute by utilizing the Bethe formula $QE = A \log E + B$,

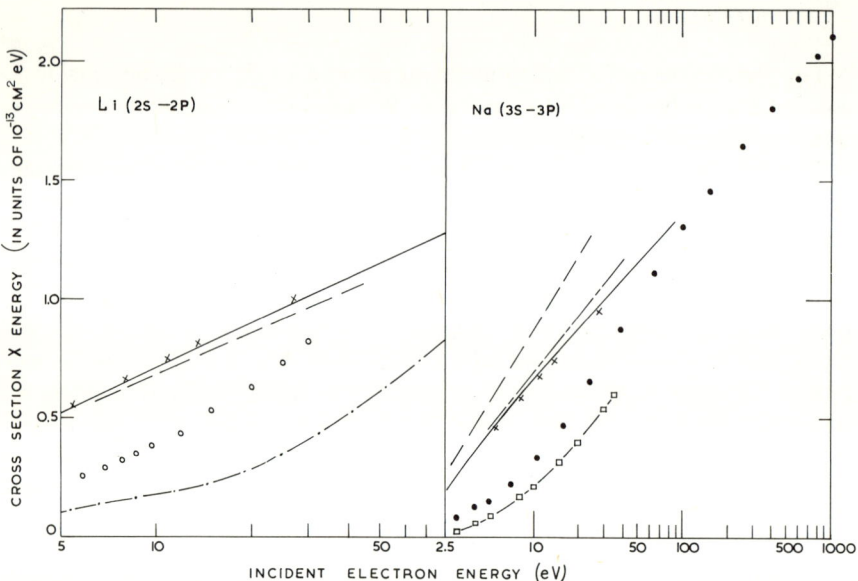

FIG. 9. Total cross sections for excitation of the resonance transitions in lithium and sodium by electron impact. First Born approximation: ———, Vainshtein et al. (1965); — — — Simsic and Williamson (1972), Williamson (1970); — - - — Bates et al. (1950); ×, Felden and Felden (1971). Experimental data: ○, Hughes and Hendrickson (1964); ●, Enemark and Gallagher (1972); □, Zapesochnyi and Shimon (1965); — · —, Aleksakhin et al. (1967).

where A was taken from the optical oscillator strengths given by Weise et al. (1969) and where B was taken to be zero, which is the value suggested by the theoretical data of Vainshtein et al. (1965). We first note that for lithium, sodium, and potassium there is considerable disagreement between the various Born calculations. This difference is due simply to the different wave functions employed in the calculations. If we compare only the Born cross sections of Vainshtein et al. (1965) with experiment (since they are the only authors to have considered all the atoms) then for sodium there is agreement between their results and those of Enemark and Gallagher (1972) to better

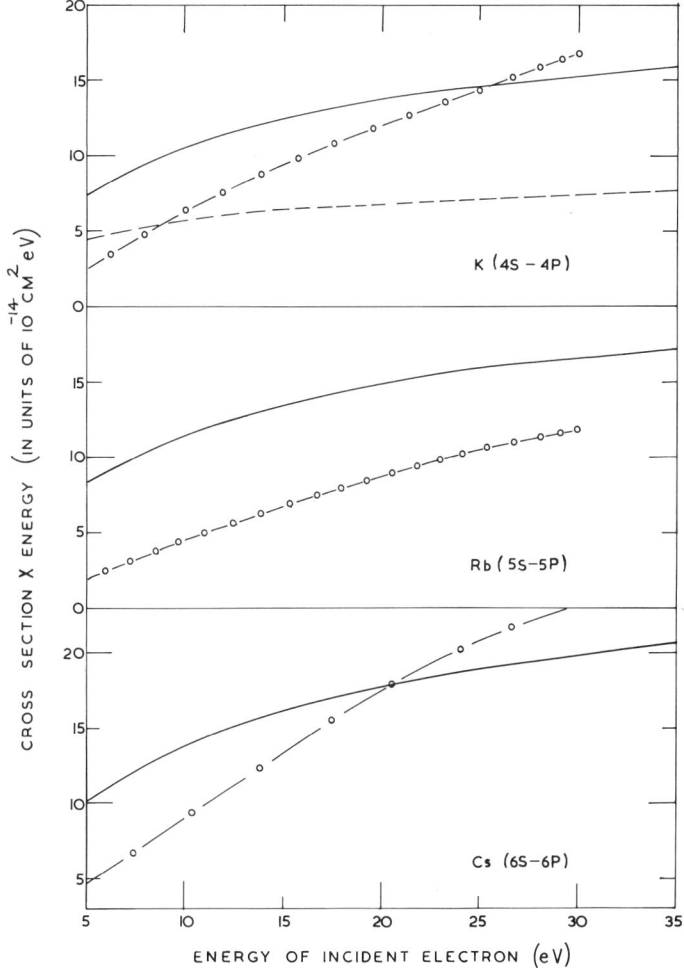

FIG. 10. Total cross sections for excitation of the resonance transitions in potassium, rubidium, and cesium by electron impact. First Born approximation: ———, Vainshtein et al. (1965); — — —, Simsic and Williamson (1972). Experimental data: — ○ —, Zapesochnyi and Shimon (1966c).

than 24% and 10% for impact energies greater than 40 and 60 eV, respectively. For the other atoms, however, theory and experiment have different shapes and the experimental data is both restricted in energy range and in normalization. It is therefore impossible to draw any real conclusion about the accuracy of the first Born approximation for resonance transitions in the alkali metal atoms at the present time. Clearly, further experimental data at higher impact energies would be desirable, but more important first Born approximation calculations are required which utilize accurate wave functions and are performed in the length and velocity formulations.

For nonresonant transitions the Born calculations are extremely sensitive to the wave functions employed. Generally the results of Vainshtein et al. (1964) and Bates et al. (1950) (using similar wave functions) and of Felden and Felden (1971) are in reasonable accord, but the sensitivity of the calculations is revealed by the work of Simsic and Williamson (1972). Figure 11 compares the Born cross sections with experiment for Na(3s → 4p). The curves labeled SW1 and SW2 are the values obtained by Simsic and Williamson (1972) from wave functions using different parameters. The difference between these results and those of Vainshtein et al. (1964) is considerable. Unfortunately no direct comparison between theory and experiment is possible since the only experimental data available is for optical excitation functions [Zapesochnyi and Shimon, 1964, 1965, 1966a,b; Zapesochnyi et al., 1965 (for impact energies less than 35 eV)]. At present, therefore, it is impossible to define the accuracy of the first Born approximation, but further theoretical work (particularly in using more accurate wave functions) is clearly desirable.

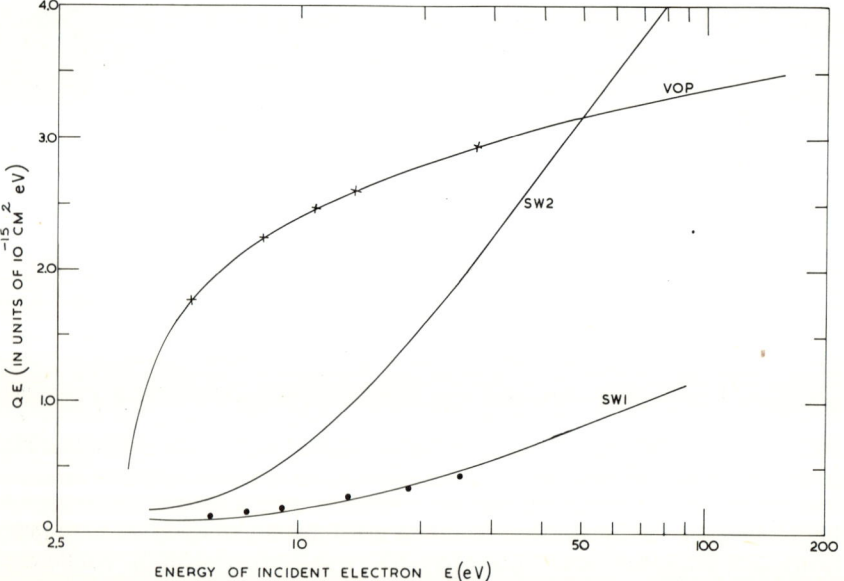

FIG. 11. Total cross section for excitation of the sodium transition $3s \rightarrow 4p$ by electron impact. First Born approximation: VOP, Vainshtein et al. (1964); SW1, SW2, Simsic and Williamson (1972); ×, Felden and Felden (1971). Experimental data: ● Zapesochnyi and Shimon (1965) (optical excitation function).

4. Heavy Rare Gases

Extensive calculations using the first Born approximation have been carried out by Veldre et al. (1965a,b) on the excitation of neon and argon by electron impact.

For neon, total excitation cross sections have been calculated by Veldre et al. (1965b) for transitions between all the terms of the configurations (γ)2p, (γ)3s, (γ)3p, (γ)4s, (γ)3d, (γ)4p, and (γ)5s, where γ signifies the core configuration $1s^2 2s^2 2p^5$. These calculations were performed using the length formulation of the Born matrix element together with wave functions which represented the different electron shells by hydrogenic orbitals. The formulation was also based on the assumption that LS coupling prevails for all configurations of the neon atom, and this is known to be invalid for excited inert gas atoms. Thus the results of Veldre et al. (1965b) may be in considerable error, not only because of inadequate wave functions but also because of the assumed coupling. Realizing this, Veldre et al. (1965a) investigated the effect of different couplings on the neon transitions and also performed Born calculations for transitions in argon between terms arising from the configurations (γ^1)3p, (γ^1)4s, (γ^1)4p, (γ^1)3d, (γ^1)5s, (γ^1)5p, (γ^1)6s, where γ^1 denotes the argon core configuration $1s^2 2s^2 2p^6 3s^2 3p^5$. For electron impact excitation from the ground state to an excited state, LS coupling was used to characterize the ground state, while LS, jl, and jj couplings were chosen for the excited state. In transitions between excited states, the same type of coupling between the optical electron and the core was used for both states, either LS or jl coupling. The wave function of the optical electron was obtained by solving the radial wave equation using Gaspar's potential (1952). An immediate consequence of not employing LS coupling for both initial and final states is that, in the first Born approximation, the cross sections for transitions between terms with different multiplicity, which have zero value when LS coupling is used for both states, may have nonzero values when other coupling approximations are assumed. As examples of the effect of coupling, Veldre et al. (1965a) found for excitation from the ground state of neon to $1s^2 2s^2 2p^5 3p^1 S_0$ and $1s^2 2s^2 2p^5 3p^1 D_2$ that the Born cross sections obtained by using jl coupling for the excited states were factors of three and fifteen, respectively, lower than those obtained from the use of LS coupling for both initial and final states. The reliability of all of these first Born approximation results must therefore remain uncertain due to the unknown effects of coupling assumed, since a pure form of coupling was used and these do not occur in practice, and due to possible inadequacies in the choice of one-electron wave functions describing the states of the optical electron.

A somewhat different and simplified approach to electron impact excitation of the rare gas atoms has been taken by Ganas and Green (1971). They use an independent-particle model and establish averages of the experimental energy levels to arrive at single-particle states. Two parameters are then adjusted so that the independent-particle model potentials accurately characterize the excited state energies. Using the wave functions derived from these potentials, the first Born approximation length matrix element reduces to a one-electron matrix element (corresponding to an electron making the transition $nlm \to n_0 l_0 m_0$) multiplied by a normalization constant which may be determined from the fact that the generalized oscillator strength goes to the optical oscillator strength for zero change in momentum of the incident electron. Cross sections have been determined by Ganas and Green for the transitions $2p \to ns$, $n = 3$–7 in neon; $3p \to ns$, $n = 4$–8 in argon; $4p \to ns$, $n = 5$–9 in krypton; and $5p \to ns$, $n = 6$–10 in xenon. They have, in effect, circumvented the problems of angular momentum coupling and their theoretical data should be compared with experimental data which has been obtained by averaging of fine resolution measurements or by using an incident electron beam of "poor" energy resolution. Such experimental data exists only for the $3p \to 4s$ transition in argon and Fig. 12 compares the

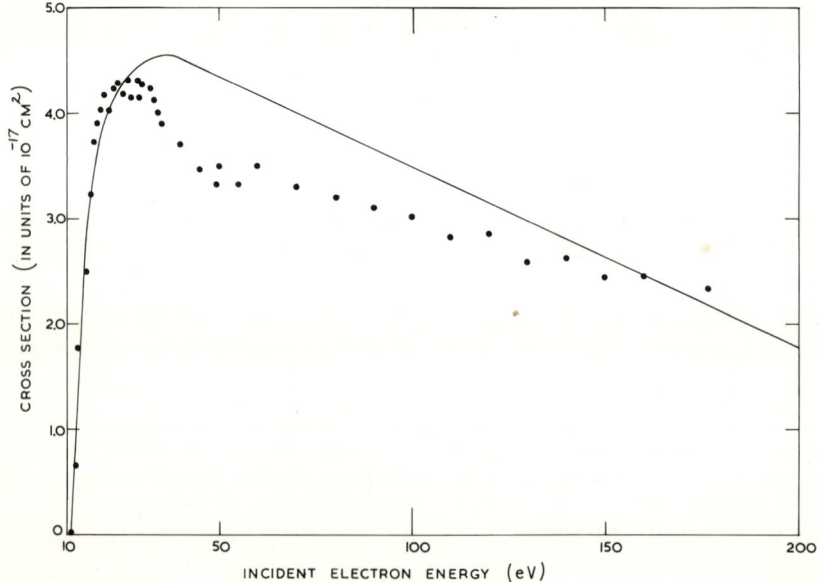

FIG. 12. Total cross section for excitation of the argon $3p \to 4s$ transition by electron impact. First Born approximation: ———, Ganas and Green (1971). Experimental data: ●, Lloyd et al. (1972) (normalized to theory at 160 eV).

Born results of Ganas and Green and Lloyd et al. (1972). The experimental data has been normalized to the Born theory at 160 eV and shows a slower fall-off with increasing energy than theory. Nevertheless, agreement between the Born theory and experiment is highly satisfactory, being better than about 25% over the entire energy range. The only other indication of the accuracy of this use of the first Born approximation comes from the theoretical work of Sawada et al. (1971). They extended the work of Ganas and Green by using the same approach but by considering distortion and exchange effects for Ne(2p $\rightarrow$ 3s) and Ar(3p $\rightarrow$ 4s). Above about 40 eV, distortion and exchange change the first Born approximation results by only a few percent, and this is consistent with the agreement between theory and experiment for Ar(3p $\rightarrow$ 4s) noted above. Further experimental data would be useful in establishing both the accuracy and usefulness of the independent-particle model theory.

5. Mercury

The only application of the first Born approximation to a study of electron impact excitation of mercury seems to be that of McConnell and Moiseiwitsch (1968). They studied the transitions $6^1S_0 \rightarrow 6^1P_1$, $6^1S_0 \rightarrow 6^3P_1$ by treating the mercury atom as though it had a fixed inert xenon-like core surrounded by two $n = 6$ level electrons. Using intermediate coupling they evaluated the first Born approximation cross sections with wave functions derived from orbitals of the Bates and Damgaard (1949) type. No direct experimental comparison may be made with their results. However, for the $6^1S_0 \rightarrow 6^1P_1$ transition experimental generalized oscillator strengths are available for the energy range 90–600 eV (Skerbele and Lassettre, 1973; Hanne and Kessler, 1972). Examination of this data indicates that the Born approximation should give total cross sections which are in error by no more than a few percent for impact energies greater than about 300 eV. Integration of the experimental generalized oscillator strength for an impact energy of 210 eV gives an electron impact excitation cross section value of about $2.0\pi a_0^2$ which may be compared with the Born approximation value of $2.5\pi a_0^2$ derived from an extrapolation of the results of McConnell and Moiseiwitsch. The agreement is satisfactory. Finally we note that, although no theoretical cross sections are available, the experimental generalized oscillator strength data of Skerbele and Lassettre (1973) would indicate that for the transitions $6^1S_0 \rightarrow 7^1P_1$ and $6^1S_0 \rightarrow 7^1S_0$ the first Born approximation should be accurate to a few percent for impact energies greater than 300 eV for excitation of the 7^1P_1 level but will be considerably in error even at 500 eV for 7^1S_0 excitation.

6. Oxygen

Kazaks et al. (1972) have employed the independent-particle model (which was discussed in connection with the rare gases) to calculate first Born approximation cross sections for electron impact excitation of the ground state (^{3}P) of oxygen. Total cross sections are presented for excitation of the 2p electron to 3s, 3p, 3d, 6s, 6p, and 6d. The only experimental data available for comparison is for 2p → 3s (Stone and Zipf, 1971) which is a factor of about two larger than theory. However, the experimental data is not corrected for cascade effects and so, at the present time, it is not possible to determine the accuracy of the Born calculations.

7. Calcium and Magnesium

The only Born calculations for electron impact excitation of these atoms would appear to be those of Simsic and Williamson (1972) for the transitions 4s → 4p in calcium and for 3p → 4s in magnesium. These authors used the length formulation of the Born matrix element and similar wave functions as those discussed for their work on the alkali metal atoms. There does not appear to be any available cross section experimental data and conclusions concerning the accuracy of the Born approximation must await further theoretical and experimental work.

B. Proton Impact Excitation

Since the high-energy Born cross section for electron and proton impact excitation are equal for equal velocities, all of the Bethe cross sections [Eq. (5)] for electron impact excitation may be used for proton impact excitation. Cross sections may also be derived using electron impact cross sections and a relation analogous to that derived by Bates and Griffing (1953) [Eq. (4)]. However, both methods of obtaining proton impact excitation cross sections are high impact energy approximations to the first Born approximation and are of limited value in determining the accuracy of the first Born approximation at lower impact energies. We shall therefore concentrate mainly on explicit evaluation of the Born cross-section formula in this subsection. The only atoms for which Born calculations have been performed are hydrogen, helium, lithium, sodium, and cesium.

1. Hydrogen

As in the case of electron impact, atomic hydrogen has been treated extensively using the first Born approximation. Bates and Griffing (1953) calculated proton impact cross sections for the excitation from the ground state to the 2s, 2p, 3s, 3p, and 3d states. Mittleman (1963) extended these calculations by investigating the transitions 1s → ns, n = 4, 5, and 6; 1s → np, n = 3, 4, 5, and 6; 1s → nd, n = 4, 5, and 6; 1s → nf, n = 5 and 6; and 1s → 6g. More recently, Khandelwal (1969) has considered the transitions 1s → ns, n = 2, 3, ..., 10, 15, 20, using an explicit evaluation of the first Born approximation formula, while Khandelwal and Fitchard (1969) have given analytic expressions for the coefficients (as a function of n) in the high-energy approximation:

$$QE = A \log E + B + (C/E) + (D/E^2) + (F/E^3) \qquad (9)$$

for transitions 1s → ns. For the transitions 1s → np and 1s → nd, Khandelwal and Shelton (1971) have derived analytic expressions for the coefficients A, B, C, and D in (9). The only other work for cross sections of the type (7) is that of Carew and Milford (1963). They have used first Born approximation electron impact cross sections and derived proton impact cross sections using a simplified form of the Bates and Griffing (1953) formula (4) for all transitions between the $n' = 2$ and $n = 3$ levels; between the $n' = 3$ and $n = 4$ levels and for the transitions $4l \to 5l + 1$, $l = 0, 1, 2, 3$; 4s → 6p; 4f → 6g; $5l \to 6l + 1$, $l = 0, 1, ..., 4$; and 10s → 11p.

The first Born approximation formula has also been evaluated explicitly for the transitions to a level; 1s → n, n = 2, 3, ..., 10, 15, 25 by Cheshire and Kyle (1966), 2s → n, 2p → n, n = 3, 4, ..., 10, 12, 15, 25 by Khandelwal and Choi (1968a). At high impact energies these results may be extended by using the analytic expressions (as a function of n) for the coefficients A, B, C, and D in (9) derived by Khandelwal and Choi (1968b) for the 1s → n transitions and for the coefficients A, B, and C derived by Khandelwal and Choi (1969) for the 2s → n and 2p → n transitions. Finally, we record the work of May and Butler (1965) who determined the asymptotic behavior (as a function of impact energy) of the cross section for transitions $(n, l) \to (n + 1, l + 1)$, $(nl) \to (n, l + 1)$ and $(n, l) \to (n + 2, l + 1)$ where $n \gg 1$ and $n \gg (n - l)$.

The only experimental data seems to be that of Stebbings et al. (1965) for 1s → 2p excitation, which is restricted to impact energies less than 30 keV. At such low energies, the Born approximation overestimates the cross section by a factor of about three. Some indication of the accuracy of the first Born approximation may be made by comparison with more sophisticated theoretical treatments of the 1s → 2s and 1s → 2p transitions (Bransden et al., 1972, and references therein). For excitation of the 2p state, the comparison indicates that the first Born approximation is accurate to about 20% for impact energies greater than 90 keV. Remembering that the electron impact excitation cross section had the same accuracy for electron energies greater than 100 eV, it would appear that, for this transition, the first Born approximation is accurate to lower proton impact energies than would have been deduced from electron impact excitation. For excitation of the 2s state, considerable disagreement exists among the various theoretical treatments even at 200 keV. However, it does seem that the first Born approximation is accurate to about 20% for impact energies greater than 200 keV.

Clearly the information available about the accuracy of the first Born approximation for proton excitation of atomic hydrogen is both tentative and limited. Further experimental and theoretical work is desirable.

2. Helium

The excitation of atomic helium by proton impact has received considerable attention from both theorists and experimentalists. We list in Table II those transitions which have been considered in the first Born approximation and for each transition, we reference only that work which utilizes the most accurate wave functions.

For excitation of n^1S states from the ground state, experimental data is available for $n = 3$ to 7. Figures 13 and 14 compare the first Born approximation cross sections with experiment for excitation of the 4^1S, 5^1S and 6^1S, 7^1S states, respectively. For excitation of the 3^1S state experimental data is limited to impact energies less than 200 keV. The disagreement between the various experimental results is apparent, although at the higher impact energies they do agree in shape. To attempt to extract some information, one is forced to normalize the experimental data. Considering only the data of Hasselkamp *et al.* (1971), Thomas and Bent (1967), and Denis *et al.* (1967) (since they are available at high enough impact energies for which normalization to the first Born approximation may be thought acceptable), it is found that if they are renormalized to theory at their highest impact energies, then reasonably consistent experimental curves may be drawn for impact energies greater than about 200 keV. Further, it is then found that the renormalized data and the first Born approximation are in agreement to better than 20% for all states for impact energies greater than 300 keV. For lower impact energies, the first Born approximation underestimates the cross sections. Van den Bos (1969) has performed a close-coupling impact parameter calculation for excitation of the 2^1S state. His results would indicate that for this state an accuracy of 20% for the first Born approximation is achieved only for impact energies greater than 600 keV. Thus for excitation of n^1S states, it is not yet clear as to the actual breakdown of the first Born approximation. However, a reasonable assessment of the present knowledge is that it is accurate to better than 20% for impact energies greater than about 600 keV.

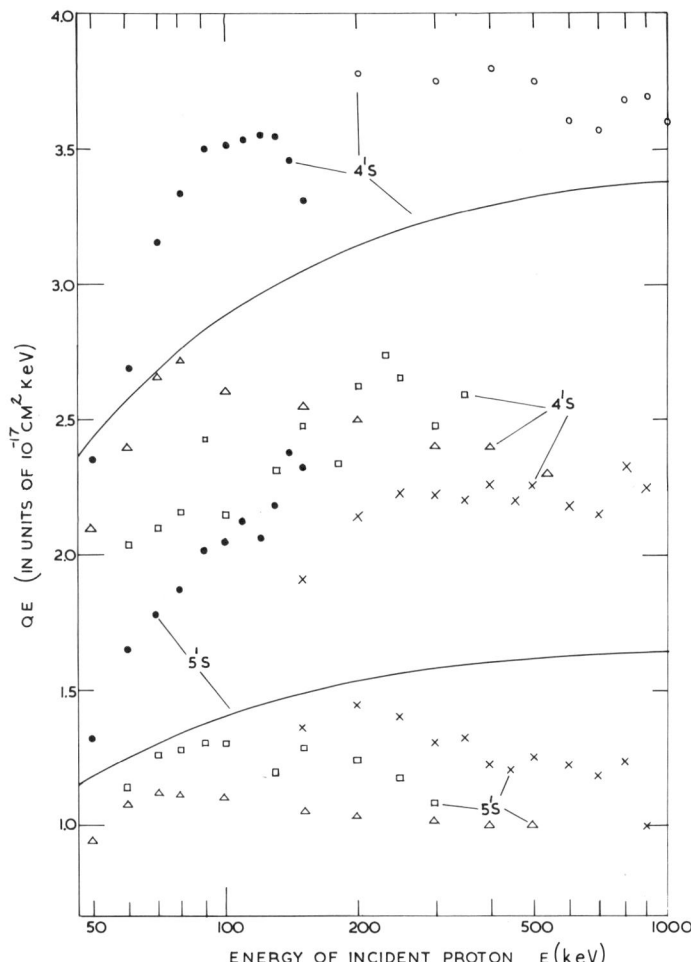

FIG. 13. Total cross sections for excitation of the 4^1S and 5^1S states of atomic helium from the ground state by proton impact. First Born approximation: —— Bell *et al.* (1968). Experimental data: ○, Hasselkamp *et al.* (1971); ●, Van den Bos *et al.* (1968); ×, Thomas and Bent (1967); △, Denis *et al.* (1967); □, Robinson and Gilbody (1967).

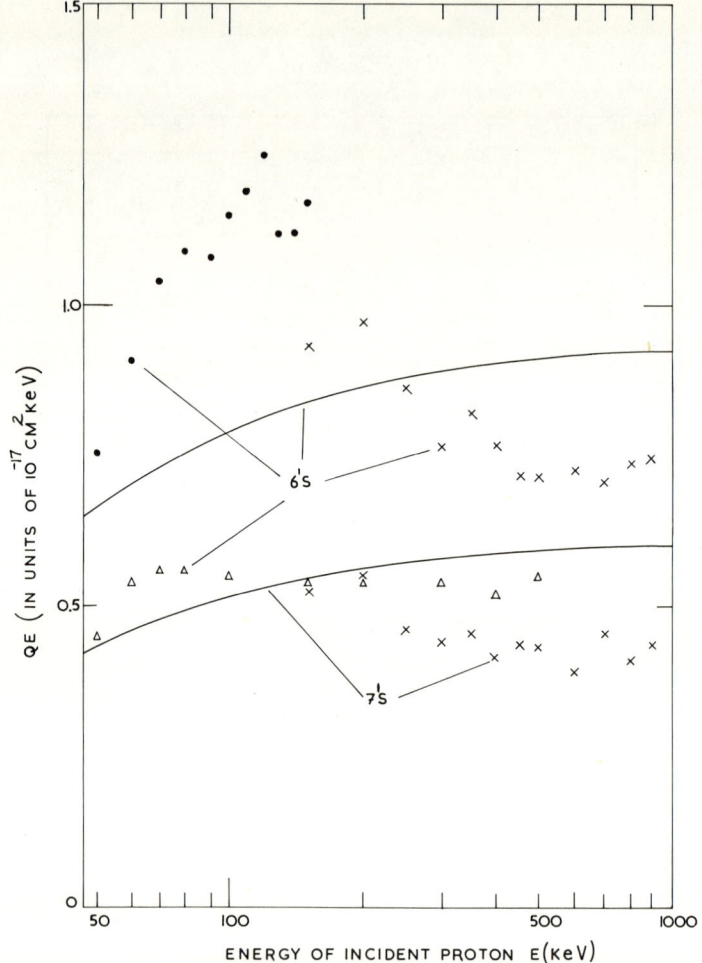

FIG. 14. Total cross sections for excitation of the 6^1S and 7^1S states of atomic helium from the ground state by proton impact. First Born approximation: ——, Bell et al. (1968). Experimental data: ●, Van den Bos et al. (1968); ×, Thomas and Bent (1967); △, Denis et al. (1967).

For excitation of n^1P states from the ground state by proton impact, experimental data is available for $n = 3, 4, 5,$ and 6. Figures 15 and 16 present a comparison of experimental data and first Born approximation calculations. There appears again to be a normalization problem for excitation to the 4^1P and 5^1P states. If we renormalize the data of Thomas and Bent (1967) to agree with the theoretical values above 500 keV and then normalize the data of Van den Bos et al. (1968) to agree with the normalized

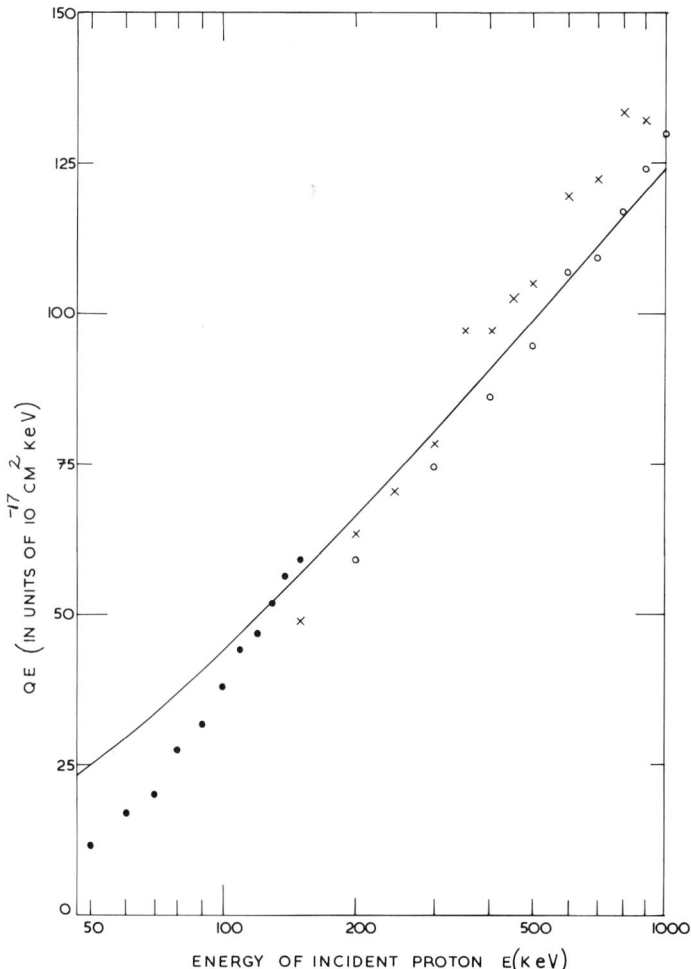

FIG. 15. Total cross section for excitation of the 3^1P state of atomic helium from the ground state by proton impact. First Born approximation:———, Bell et al. (1968). Experimental data: ○, Hasselkamp et al. (1971); ●, Van den Bos et al. (1968); ×, Thomas and Bent (1967).

cross section of Thomas and Bent (1967), it is found that for all of these states, the first Born approximation agrees with the resulting experimental curve to better than 20% for impact energies greater than 200 keV. [Normalization of the data of Hasselkamp et al. (1971) to theory gives results in close accord with the renormalized values of Thomas and Bent over the range 200–1000 keV for the 3^1P state.] At lower impact energies the first Born approximation overestimates the cross sections. The 9-state close-coupling

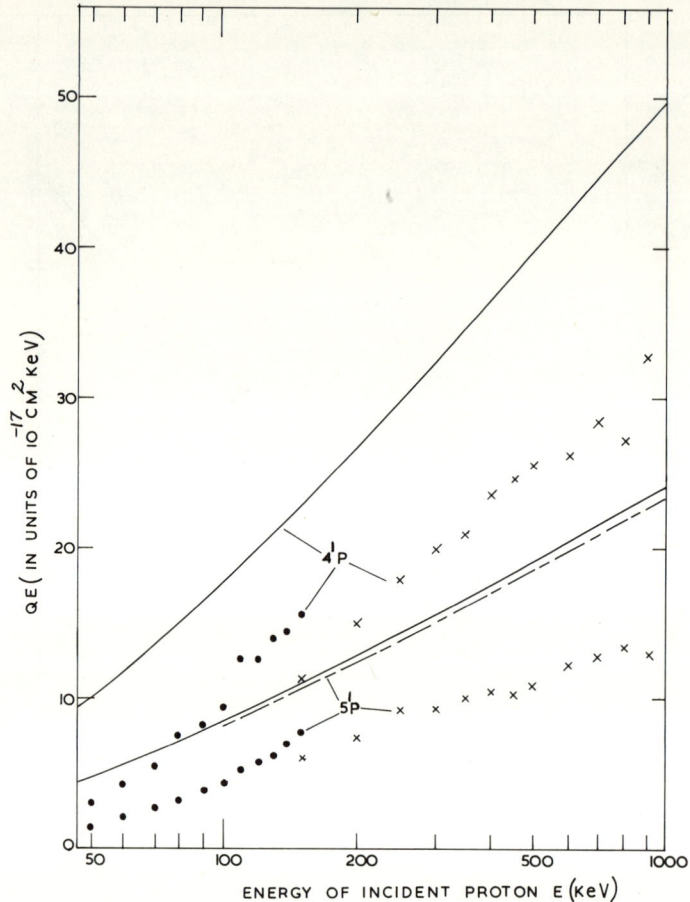

FIG. 16. Total cross sections for excitation of the 4^1P and 5^1P states of atomic helium from the ground state by proton impact. First Born approximation: ———, length formulation; — — —, velocity formulation (see text and Table II), Bell et al. (1968). Experimental data: ●, Van den Bos et al. (1968); ×, Thomas and Bent (1967).

calculations of Van den Bos (1969) would also indicate that the Born cross section had an accuracy of at least 20% for 2^1P and 3^1P states for impact energies greater than 200 keV.

Experimental data is available for excitation of the $n^1D(n = 3$–$6)$ states. Figure 17 compares the first Born approximation cross sections for excitation from the ground state to the 4^1D and 5^1D states with experiment. For excitation of the 3^1D and 6^1D, experimental data is limited to impact energies below 150 keV and 500 keV, respectively. The 4^1D state is particularly important since it is the only ^1D state for which experimental data is avail-

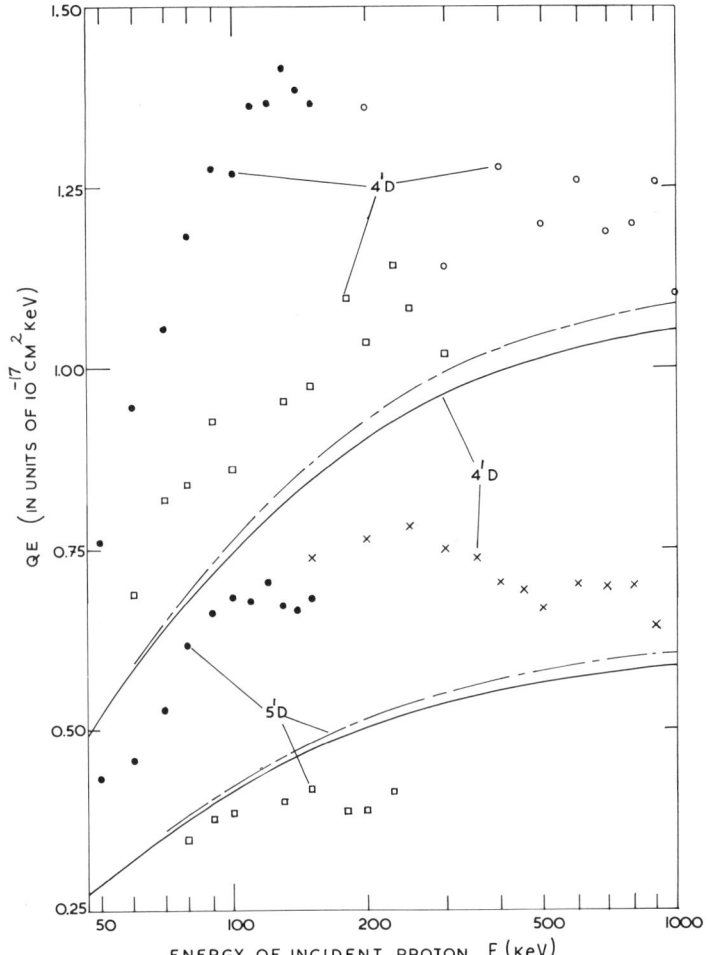

FIG. 17. Total cross sections for excitation of the 4^1D and 5^1D states of atomic helium from the ground state by proton impact. First Born approximation: ———, length formulation; — — —, velocity formulation (see text and Table II), Bell et al. (1968). Experimental data: ○, Hasselkamp et al (1971); ●, Van den Bos et al. (1968); ×, Thomas and Bent (1967); □, Robinson and Gilbody (1967).

able to at least 1000 keV, although the disagreement between the experimental values is considerable. If one normalizes the data of Hasselkamp et al. (1971) and Thomas and Bent (1967) to the first Born approximation (length formulation) of Bell et al. (1968) at 900–1000 keV, then the resulting experimental points are in close accord over the energy range 150–1000 keV. However, drawing a reasonable curve through these points indicates that, while the first Born approximation and renormalized experiment agree to better than 20% for impact energies greater than about 200 keV, the shape of experiment and theory as a function of energy is different. Thus, deviations from the first Born approximation seem to be occurring even at 1000 keV. Unfortunately the close-coupling calculations of Van den Bos (1969) for excitation of the 3^1D state were found to vary considerably as the number of coupled states was altered, and this variation occurred even at 1000 keV. Thus the only conclusion concerning the accuracy of the first Born approximation for excitation of ^1D states, at present, is that if the normalization of theory and experiment at 1000 keV is justified, the first Born approximation is accurate to better than 20% for proton impact energies greater than about 200 keV. Clearly, further experimental data at impact energies greater than 1000 keV would be invaluable.

In summary, the presently available evidence suggests that for proton impact excitation of the ground state of helium, the first Born approximation is accurate to better than 20% for impact energies greater than about 600 keV, 200 keV, and 200 keV for ^1S, ^1P, and ^1D states, respectively. As in the case of electron impact, it seems to break down at higher energies for the ^1S states than for either the ^1P and ^1D states.

3. Alkali Metal Atoms

Only the excitation of the resonance transitions in the alkali metal atoms by proton impact has been investigated using the first Born approximation. Bell and Skinner (1962) used Hartree–Fock wave functions to evaluate the cross section for the 3s → 3p transition in sodium. Mathur et al. (1971) employed self-consistent field wave functions to calculate the first Born approximation cross sections for the transitions 2s → 2p in lithium, 3s → 3p in sodium, and 6s → 6p in cesium. The latter investigation employed the length formulation of wave theory, while the former was performed in the impact parameter treatment. Again the results are found to be sensitive to the wave functions employed. In the energy range (≤ 25 keV) where comparison may be made for sodium, the results of Bell and Skinner are as much as 40% larger than the results of Mathur et al. No comparison may be made with experiment unfortunately. There is clearly a need both for experimental data and for further theoretical work.

4. Summary

In this section we have been concerned with single-electron excitation by electron and proton impact. We have not considered double excitation mainly because of the lack of experimental data but also because the few calculations which have been performed have utilized simple wave functions, and it is known that it is essential to include correlation terms in sophisticated wave functions in order to obtain accurate first Born approximation cross sections (for a discussion of double excitation transitions, see Gillespie, 1970).

For single excitation processes there is very little information concerning the accuracy of the first Born approximation. A lack of reliable experimental data is partially responsible but problems also still exist in the evaluation of the first Born approximation cross section. Apart from hydrogen and helium, there is a need for calculations using accurate wave functions and/or for calculations to be performed using the length and velocity formulations of the Born matrix element. For the inert gases there is also the problem of the type of coupling to be used. However, for the helium atom in particular, there seem to be fairly well-defined energy limits above which the Born approximation is sufficiently accurate, and further there is considerable evidence that it is poorest for transitions between states which have the same term symbols.

III. Ionization of Atoms by Electron and Proton Impact

In this section we first give some of the formulas which arise in the first Born approximation for electron and proton ionization of atoms. The Born results for electron ionization are then discussed atom by atom and finally the results for proton ionization are given. As we are primarily concerned with exploring the range of validity of the Born approximation, we only discuss those Born calculations for which there are experimental results.

In order to avoid confusion about the interpretation of experimental ionization cross section results, it must be remembered that when an electron or proton collides with a many electron atom A we will have a one electron ejection process (10) with cross section Q_1

$$p + A \longrightarrow p + A^+ + e \qquad (10)$$

and also an n electron ejection process (11) with cross section Q_n

$$p + A \longrightarrow p + A^{n+} + ne \qquad (11)$$

It is important to distinguish between two measured ionization cross sections; the gross ionization cross section, determined by the total charge of the ions or the total number of ejected electrons

$$Q_g = \sum nQ_n \qquad (12)$$

and the counting ionization cross section, determined by the number of ions produced independent of their charge

$$Q_c = \sum Q_n \qquad (13)$$

The Born approximation calculations normally only give Q_1, the cross section for the one-electron ejection process (10). The measured cross sections Q_g and Q_c should only be compared with the Born cross section if $Q_c \approx Q_g$, in which case $Q_g \approx Q_c \approx Q_1$. If $Q_g \not\approx Q_c$ special experimental techniques must be used to obtain values for the single ionization cross section Q_1.

In the Born approximation the differential cross section for the ejection of a single atomic electron with energy in the range from ε to $\varepsilon + d\varepsilon$ is

$$\sigma(\varepsilon) = M_R^2 \frac{k\kappa}{q} \iint |f(\mathbf{k}; \boldsymbol{\kappa})|^2 \, d\hat{\mathbf{k}} \, d\hat{\boldsymbol{\kappa}} \qquad (14)$$

The kinetic energy of the ejected electron $\varepsilon = \tfrac{1}{2}k^2$ and the conservation of energy gives

$$q^2 - \kappa^2 = 2M_R(I + \varepsilon) \qquad (15)$$

Here I is the ionization potential of the atom, $\mathbf{k}$ is the momentum of the ejected electron, $\boldsymbol{\kappa}$ is the final momentum of the incident particle and $\mathbf{q}$ is the initial momentum of the incident particle. The magnitude of $\mathbf{q}$ is $|\mathbf{q}| = M_R(2E/M_P)^{1/2}$, with E and M_P being the energy and mass of the incident particle and M_R, the reduced mass of the colliding particles. $f(\mathbf{k}; \boldsymbol{\kappa})$ is defined, in the length formulation, by

$$f(\mathbf{k}; \boldsymbol{\kappa}) = -\frac{2}{K^2} \int \phi_\mathbf{k}^* \left[\sum_{t=1}^{N_A} \exp(i\mathbf{K} \cdot \mathbf{r}_t) \right] \phi_i \, d\mathbf{r} \qquad (16)$$

where the change in momentum of the incident particle $\mathbf{K} = \mathbf{q} - \boldsymbol{\kappa}$ and where ϕ_i the ground state wave function is normalized to unity and $\phi_\mathbf{k}$ the wave function of the ejected electron and the residual ion is normalized to $\delta(\mathbf{k} - \mathbf{k}')$. The velocity formulation of $f(\mathbf{k}; \boldsymbol{\kappa})$ is the same as that for excitation (3).

The total cross section for ionization is

$$Q = \int_0^{\varepsilon_{\max}} \sigma(\varepsilon) \, d\varepsilon \qquad (17)$$

For proton ionization (15) gives

$$\varepsilon_{\max} = (q^2/2M_R) - I \qquad (18)$$

Early calculations on electron ionization also use (18). However almost all recent calculations on electron ionization follow the suggestion of Rudge and Seaton (1965) and take

$$\varepsilon_{max} = \tfrac{1}{2}[(q^2/2M_R) - I] \tag{19}$$

This limit ensures that the faster of the two electrons moving away from the ion is always described by the plane wave. When this modified limit (19) is used in (17) the results are termed "modified Born" or BII. When the upper limit is taken as (18), the results are called "unmodified Born" or BI. Unless otherwise stated we will use (19) as the upper limit for electron ionization.

The scattering of the incident particle through a given angle is most conveniently discussed in terms of the generalized continuous oscillator strength

$$\frac{df(K)}{d\varepsilon} = \frac{2k(I+\varepsilon)}{K^2} \int \left| \int \phi_\mathbf{k}^* \left[\sum_{t=1}^{N_A} \exp(i\mathbf{K} \cdot \mathbf{r}_t) \right] \phi_i \, d\mathbf{r} \right|^2 d\hat{\mathbf{k}} \tag{20}$$

which is similar to the generalized oscillator strength for excitation (6).

The angular distribution of the ejected atomic electron is usually given in terms of the double differential cross section I_D, differential in electron momentum k, and electron solid angle. In the Born approximation it is given by

$$I_D = (\kappa/q)(kM_R)^2 \int |f(\mathbf{k} : \boldsymbol{\kappa})|^2 \, d\hat{\boldsymbol{\kappa}} \tag{21}$$

For every atomic system, except hydrogen, it is not possible to obtain exact wave functions to calculate the scattering amplitude (16). The error produced in the Born matrix element by the use of approximate wave functions may be assessed by repeating the calculation using a variety of wave functions for both the initial and final states. The variation of the results as the accuracy of the wave function is increased gives a good indication of the accuracy of the calculations.

An estimate of the error produced by the use of approximate wave functions may also be obtained by carrying out the calculations using the velocity formulation of the Born matrix element.

A. Electron Ionization

1. Hydrogen

Atomic hydrogen is the only atomic system for which the initial and final wave functions are known exactly, and the Born approximation to the total electron ionization cross section can be evaluated very accurately. Extensive calculations have been carried out by Omidvar (1965a) for electron ionization of the ground state of hydrogen and also for ionization of excited states of hydrogen. In these calculations Omidvar took the upper limit of (17) to be given by (18). In Fig. 18 we compare his unmodified Born (BI) cross sections for ionization of the ground state of hydrogen with the modified Born (BII) results which use the upper limit (19) (Rudge and Schwartz, 1966). The BI and BII results do not differ significantly for incident electron energies which are more than ten times the ionization potential, at the maximum of the Born cross section the BII results are about 20% below the BI results and at

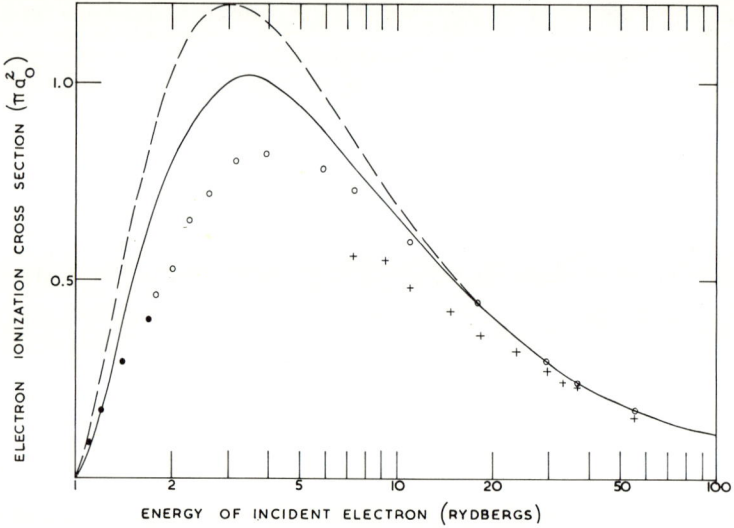

FIG. 18. Cross section for electron ionization of atomic hydrogen. Calculated cross section: ———, the modified Born approximation, BII (Rudge and Schwartz 1966); — — — the unmodified Born approximation, BI (Omidvar, 1965a). Measured cross sections: ○, Fite and Brackmann (1958a); +, Rothe et al. (1962); ●, McGowan and Clarke (1968).

the ionization threshold the two cross sections differ by about a factor of two.

Also plotted in Fig. 18 is a selection of the experimental data for this cross section. As hydrogen is in its molecular state in normal conditions, it is difficult to carry out measurements on hydrogen atoms. Therefore it is not surprising that there are significant differences between some of the results. Because of this lack of agreement in the measurements it is not possible to make a definitive statement about the accuracy of the Born cross section for the electron ionization of hydrogen.

However, to obtain some indication of the accuracy of the Born approximation we plot in Fig. 19 the ratio of the Born results to the measured cross sections of Fite and Brackmann (1958a) and McGowan and Clarke (1968). It is seen that the Born approximation is valid at very large impact energies where the velocity of the incident electron is much larger than the velocity associated with the atomic electron. The Born approximation is less than 10% in error if the incident velocity is more than three times the average velocity of the atomic electron. As the energy of the incident electron decreases, the difference between the Born and the measured cross sections

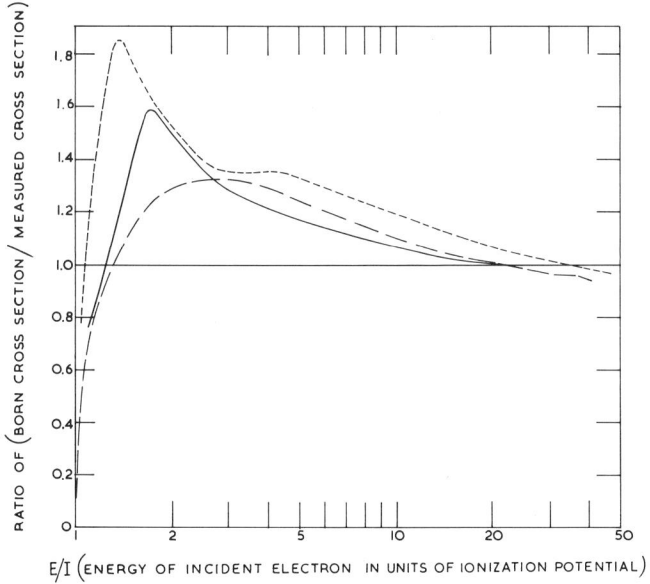

FIG. 19. Ratio of (Born ionization cross section/measured ionization cross section). For atomic hydrogen ———; helium — —; neon - - -.

increases, the ratio of the two cross sections reaching a maximum of about 1.6 where the incident energy is about 1.7 times the ionization potential. At very low impact energies the Born approximation falls below the measured cross section. Since the Born cross section decreases as $2.5x^{3/2}$ (cross section in πa_0^2, $x = E - I$ in rydbergs) and the measured cross section decreases approximately as $0.9x$, the Born approximation gives results which differ from the measured cross sections by less than a factor of two if the energy of the incident electron is more than 1.04 times the ionization potential.

2. Helium

The study of the electron ionization of atomic helium is particularly important, for helium is comparatively easy to study experimentally, and since helium is only a two electron system it is possible to obtain approximate wave functions which represent the system quite adequately. A large number of authors have calculated the Born cross section for the ionization of helium using a variety of wave functions. Some of the more recent calculations are compared in Fig. 20. In the calculations of Peach (1965) and Sloan

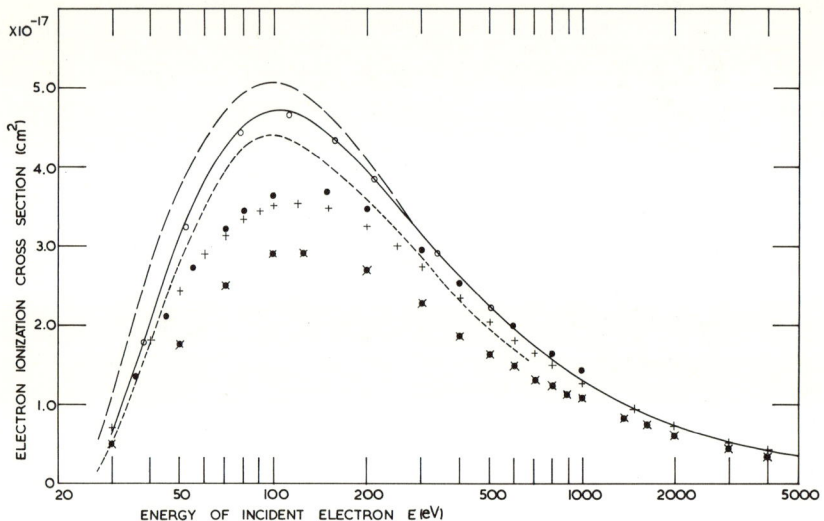

FIG. 20. Cross sections for electron ionization of atomic helium. Born cross sections; ———, Bell and Kingston (1969b); – – –, Peach (1965); — —, McGuire (1971); ○, Sloan (1965). Experimental cross sections: +, Smith (1930); ✖, Schram (1966), Adamczyk et al. (1966); ●, Rapp and Golden (1965).

(1965) the ground state was a Hartree–Fock function. However, for the free state Peach employed Coulomb wave functions, whereas Sloan used the much more accurate polarized orbitals wave functions. The difference between the two calculations shows the effect of improving the free-state wave function. Economides and McDowell (1969) repeated this calculation with free-state functions similar to those used by Sloan but with an open shell wave function for the ground state. Their results are in close agreement with Sloan's above 200 eV. The most accurate results have been obtained by Bell and Kingston (1969b) who used Sloan's polarized orbitals free-state functions and a six parameter ground state wave function. Sloan's cross section agrees with Bell and Kingston's results to better than 1%. Also plotted on the graph are the results of McGuire (1971). His wave functions are based on the central potential of Herman and Skillman (1963). At high energies his results are in very good agreement with those of Bell and Kingston.

To test the accuracy of their results, Bell and Kingston also carried out calculations using the velocity formulation of the Born matrix element.

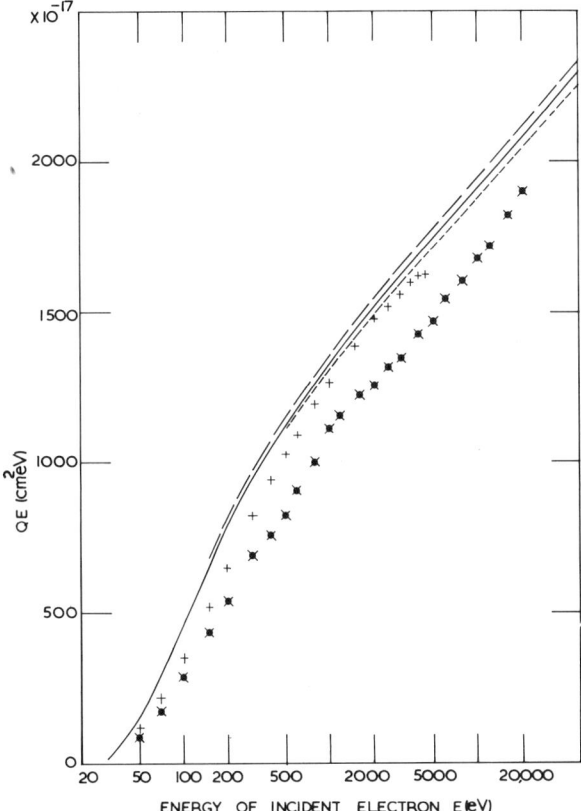

FIG. 21. Product QE of the electron ionization cross section Q of helium and electron energy E. Theoretical cross sections; —— Born length formulation of Bell and Kingston (1969b); — —, Born velocity formulation of Bell and Kingston (1969b); - - -, Bethe results of Kim and Inokuti (1971). Experimental cross sections: +, Smith (1930); ✶, Schram (1966), Adamczyk et al. (1966).

Figure 21 compares the length and velocity results at high energies. They differ by less than 2% at high energies but at very low energies they differ by almost 6%. The difference between the length and velocity results is a good indication of the error introduced by the use of approximate wave functions.

An independent check on the accuracy of any Born ionization cross-section calculation may be obtained by considering the high-energy behavior of the cross section which is

$$Q = A(\log E/E) + B/E. \tag{22}$$

If Q is in πa_0^2 and E is in atomic units then

$$A = 2.303 \int_0^\infty (I + \varepsilon)^{-1} \frac{df}{d\varepsilon} d\varepsilon \tag{23}$$

where $df/d\varepsilon$ is the continuous oscillator strength for the equivalent optical ionization transition. The high energy fall off $5.59[E^{-1} \log E]10^{-15}$ cm^2 of the ionization cross section of Bell and Kingston is in good agreement with the accurate value of $5.34[E^{-1} \log E]10^{-15}$ cm^2 obtained from a study of accurate theoretical oscillator strengths (Bell and Kingston, 1967). This value is in very good agreement with the Bethe results of $5.39[E^{-1} \log E]10^{-15}$ cm^2 given by Kim and Inokuti (1971). When comparing these asymptotic forms it must be remembered that the Born calculations only refer to single ionization, but the other results also include a contribution of about 1% from simultaneous excitation ionization (Bell et al., 1973a)

$$e + He \longrightarrow He^+(nl) + 2e \tag{24}$$

and also a very small contribution from double ionization. These would suggest that the cross sections of Bell and Kingston are about 5% too large at high energies. The Bethe results of Kim and Inokuti are displayed in Fig. 21. It is seen that they are in better agreement with Bell and Kingston at low energies than at high energies.

In Figs. 20 and 21 we have also compared some experimental measurements of the electron ionization cross section with the Born theory. It is seen that there are quite large differences between the various measurements; for example, a selection of the results (in units of 10^{-17} cm^2) at $E = 1000$ eV, are 1.40 (Rapp and Golden, 1965), 1.27 (Smith, 1930), 1.10 (Schram, 1966), and 0.89 (Gaudin and Hagemann, 1967). For helium the double ionization cross section is small and it cannot account for the difference between the measurements. As the general shape of the various experimental cross sections are all similar we must conclude that the absolute calibration of the cross sections is still in doubt.

To obtain some idea of the accuracy of the Born approximation we plot in Fig. 19 the ratio of the Born cross section (Bell and Kingston, 1969b) to the measured cross section of Rapp and Golden (1965). These measurements were chosen because they agree with the calculations at high energies. As the energy of the incident electron decreases, the difference between the theoretical and measured cross sections increases and the ratio reaches a maximum of about 1.3 when the energy is three times the ionization potential. At lower impact energies the ratio drops rapidly and the Born cross section decreases as $0.97x^{3/2}$ (cross section in πa_0^2, $x = E - I$ in units of ionization potential) while the experimental cross section decreases as $0.34x$. The Born approximation gives results which differ from the measured total electron ionization cross section by less than a factor of two if the energy of the incident electron is more than 1.04 times the ionization potential.

For helium reliable measurements (Opal et al., 1971) and Born calculations (Bell et al., 1970b) are available for the differential cross section $\sigma(\varepsilon)$ (14) for electron ionization. These are compared in Fig. 22. At large incident energies ($E = 2000$ eV) theory and experiment are in excellent agreement particularly at large energies of ejection, at small energies of ejection the

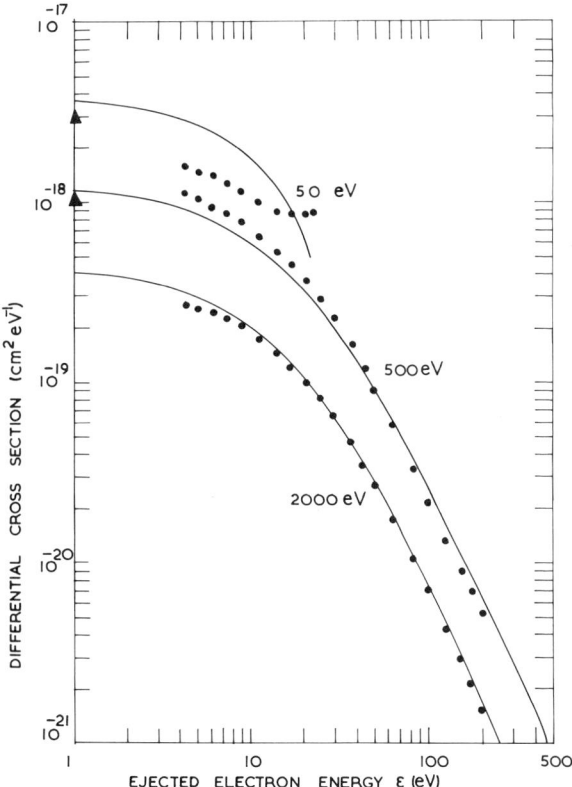

FIG. 22. Differential ionization cross section for e + He collisions, for incident electron energies of 50, 500, and 2000 eV. Solid curves: Born approximation (Bell et al., 1970b); data points: ●, Opal et al. (1971); ▲, Grissom et al. (1972).

measured differential cross section lies about 10% below the Born calculation. The agreement between theory and experiment is less good at $E = 500$ eV, where the Born differential cross section lies above the measured cross section at large ejected energies, but the Born lies below the measured differential cross section at small ejected energies. By projecting their measurements to $\varepsilon = 0$, Opal et al. (1971) obtained a total cross section of 2.4×10^{-17} cm^2, which is larger than most of the recent measurements of the total cross section. The value of $\sigma(\varepsilon)$ obtained by Grissom et al. (1972) at $\varepsilon = 0$ lies below the Born calculation and is consistent with the recent measurements of the total cross section which also lie below the Born results. There is serious disagreement at $E = 50$ eV between the measurements of Opal et al. (1971) and the Born calculations. The total cross section predicted from their measured cross section is 1.3×10^{-17} cm^2 which is smaller than the cross section measured by other workers. The value of $\sigma(\varepsilon)$ obtained by Grissom et al. (1972) at $\varepsilon = 0$ is about 20% below the Born differential cross section, and this is consistent with the total ionization cross section measurements of Smith (1930) and Rapp and Golden (1965) (see Fig. 20).

For ionization, the scattering of the incident electron through a given angle is most conveniently discussed in terms of the generalized continuous oscillator strength $df(K)/d\varepsilon$ (20). This quantity is dependent only on the change of momentum K and the energy of the ejected electron ε; it is independent of the energy of the incident particle. The incident energy at which the experimental results disagree with the generalized continuous oscillator strength provides a good indication of the accuracy of the Born approximation for total cross sections. Continuous generalized oscillator strengths for helium have been measured by Silverman and Lassettre (1964) and Lassettre et al. (1964) for incident electron energies of approximately 500 eV. The measured differential cross section (Fig. 22) and total cross sections (Fig. 20) would suggest that the Born approximation is valid at this energy. Figure 23 shows that for small energies of ejection the measured generalized continuous oscillator strengths agree with the Born calculations of Bell and Kingston (1970). However, for an energy loss of about 60 eV there is serious disagreement between the Born theory and experiment. The increase in the measured $df(K)/d\varepsilon$ is due to the first $^1P^0$ resonance of helium which is at $\Delta E = 60.27$ eV. Ionization occurs through the process

$$e + He \longrightarrow He(2s, 2p) + e$$
$$He(2s, 2p) \longrightarrow He^+(1s) + e \qquad (25)$$

Born calculations which take account of this process have not yet been carried out.

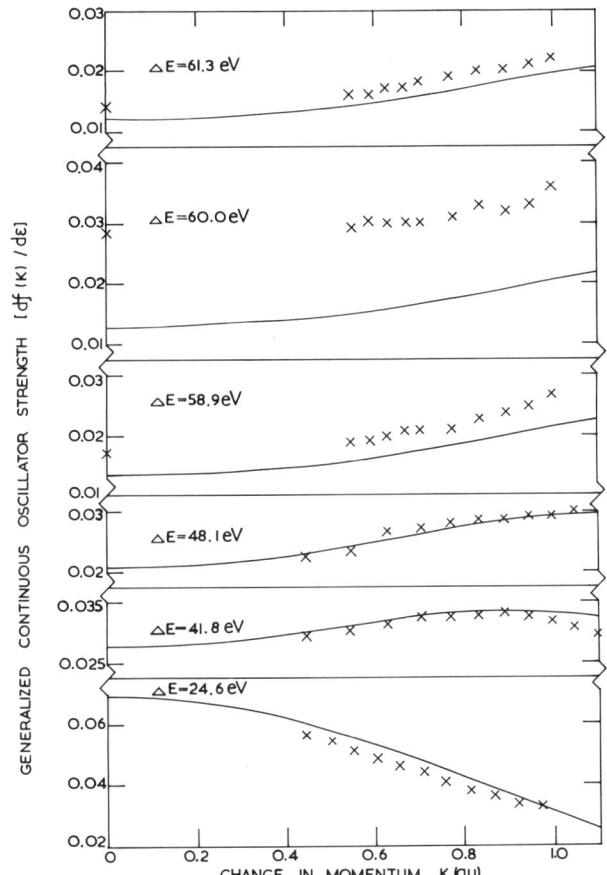

FIG. 23. Helium generalized oscillator strengths, per electron volt in energy, for various values of the energy loss ΔE. ———, Born calculations (Bell and Kingston, 1970); ×, experimental values for incident electron energy of approximately 500 eV (Silverman and Lassettre, 1964; Lassettre et al., 1964).

3. Lithium

Several workers have obtained the Born cross section for the electron ionization of lithium. Calculations by Peach (1965) and McDowell (1969) give the cross section for ionization of the 2s electron only. The results given by McDowell are the most accurate yet obtained as he uses Hartree–Fock functions for both his ground and free states. Peach also used a Hartree–Fock ground state function but with a Coulomb free-state function. Figure 24 compares these two results, and it is seen that the use of the more

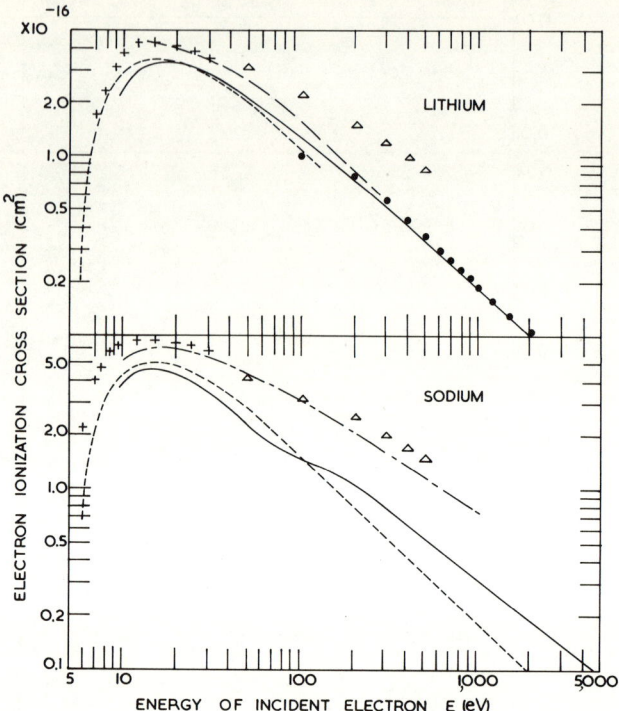

FIG. 24. Cross sections for electron ionization of lithium and sodium. Born cross sections: ———, McGuire (1971); - - -, Peach (1965, 1970) for outer electron only; —— ——, McDowell (1969) for outer electron only; —— -, Omidvar et al. (1972). Measured cross sections: +, Zapesochnyi and Aleksakhin (1969); △, McFarland and Kinney (1965); ●, Jalin et al. (1973).

accurate Hartree–Fock free-state function increases the cross section by about 40% at $E = 100$ eV.

Calculations by McGuire (1971) and Omidvar et al. (1972) give the cross section for ionization of the (2s) shell and the (1s) shell. The (1s) shell only increases the total cross section by a small amount at high energies. At low energies ($E < 100$ eV) the inner shell ionization is unimportant and all four theoretical calculations only give the cross section for ionization of the 2s electron. In this region Fig. 24 illustrates that the calculations of McGuire and Peach are in good accord, but they both lie considerably below the more accurate results of McDowell. The calculations of Omidvar et al. lie above the results of McDowell.

At high energies ($E > 100$ eV) McDowell's cross section for ionization of the (2s) shell appears to tend to McGuire's cross section for ionization of both the (1s) and (2s) shells. If McDowell had taken account of the (1s) shell ionization his results would probably be between 10 and 20% higher than those of McGuire at large incident energies.

Some of the experimental measurements of the cross section for electron ionization of lithium are displayed in Fig. 24. Recent measurements by Jalin *et al.* (1973) suggest that the double ionization cross section is small, and so it is not necessary to distinguish between the gross ionization cross section Q_g (12) and the counting ionization cross section Q_c (13). At large incident energies there is very good agreement between the measurements of Jalin *et al.* and the calculations of McGuire, but at $E = 100$ eV the data of Jalin *et al.* lies about 20% below McGuire's cross section and about 40% below McDowell's cross section. The measurements of McFarland and Kinney (1965) are about 2.5 times McGuire's results at $E = 400$ eV. Near the maximum in the cross section the experimental cross section of Zapesochnyi and Aleksakhin (1969) is in good agreement with the calculation of McDowell.

The reasonably good agreement at high energies between all the Born cross sections for the electron ionization of lithium would suggest that the Born results of McGuire cannot be greatly in error at high energies.

4. Nitrogen

The ground state of nitrogen has a $(1s)^2 (2s)^2 (2p)^3$ ^4S configuration. The inner (1s) shell contributes very little to the ionization cross section. Since the ionization potential of the (2p) shell is only slightly lower than that of the (2s) shell, and since there are two 2s and only three 2p electrons, the Born electron ionization cross sections for the two shells are almost equal (Peach, 1971).

Figure 25 compares the Born ionization cross sections of Peach (1968, 1970, 1971) and McGuire (1971), and it is seen that the two sets of results differ considerably. Both the ground and free electron wave functions used by McGuire are based on the central potential of Herman and Skillman (1963). These wave functions are less accurate than Hartree–Fock wave functions. Peach uses a Hartree–Fock wave function to describe the active bound electron and a Coulomb wave, with $Z = 1$, to describe the ejected electron. We have the situation that Peach's bound state function is better than McGuire's bound state function, but Peach's free state function is less reliable than McGuire's free state function. Hence from a consideration of the wave functions used by Peach and McGuire we cannot say that one set of calculations is more reliable than the other.

The ground state wave functions used by Omidvar *et al.* (1972) in their calculation of the Born electron ionization cross sections are less accurate than Peach's ground state wave functions. As Omidvar *et al.* used the same free state function as Peach did we would expect their results to be less accurate than those of Peach.

There is a large difference between the two measured ionization cross sections for nitrogen, which are displayed in Fig. 25. The discrepancy is not

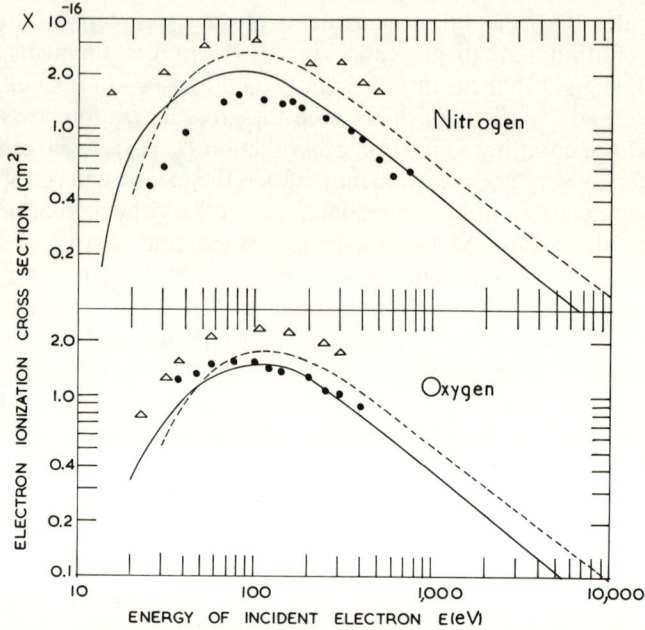

FIG. 25. Cross sections for electron ionization of nitrogen and oxygen. Calculated Born cross sections: ———, McGuire (1971); - - -, Peach (1968, 1970, 1971). Measured cross sections for nitrogen: ●, Smith *et al.* (1962); △, Peterson (1964). Measured cross sections for oxygen: ●, Fite and Brackmann (1959); △, Boksenberg (1961).

due to the difference between the gross Q_g and the counting Q_c cross section, for Peterson (1964) measures the single ionization cross section and Smith *et al.* (1962) state that multiple ionization is small. The difficulty is not resolved by the present theoretical calculations for they differ almost as much as the measurements. It is worth pointing out that Peterson's cross section for the electron ionization of argon is about a factor of two greater than other measurements at $E = 500$ eV.

5. Oxygen

As the ground state of oxygen has two electrons in the (2s) shell and four electrons in the (2p) shell it is not surprising that in the Born approximation the contribution to the ionization cross section from the (2p) shell is only a little larger than the contribution from the (2s) shell. The Born ionization cross sections of Peach (1968, 1970, 1971) and McGuire (1971) differ considerably (see Fig. 25).

The measured cross sections of Fite and Brackmann (1959) and Rothe *et al.* (1962) are in good agreement with each other, but they do not agree with the measurements of Boksenberg (1961). As Fite and Brackmann only measure single ionization and Rothe *et al.* include multiple ionization, it may be assumed that multiple ionization is small. The difference between the measurements of Fite and Brackmann and Boksenberg is illustrated in Fig. 25. It is seen that the theoretical calculations are too inaccurate to be used as a guide to the accuracy of the measured cross section.

6. *Neon*

The importance of inner-shell ionization in the electron ionization of neon is displayed in Fig. 26, where we compare Peach's (1968, 1970, 1971) total Born ionization cross section with the contribution which arises from the (2p) shell. The results of Peach differ significantly from those of McGuire (1968) particularly at low energies. This difference must be attributed chiefly

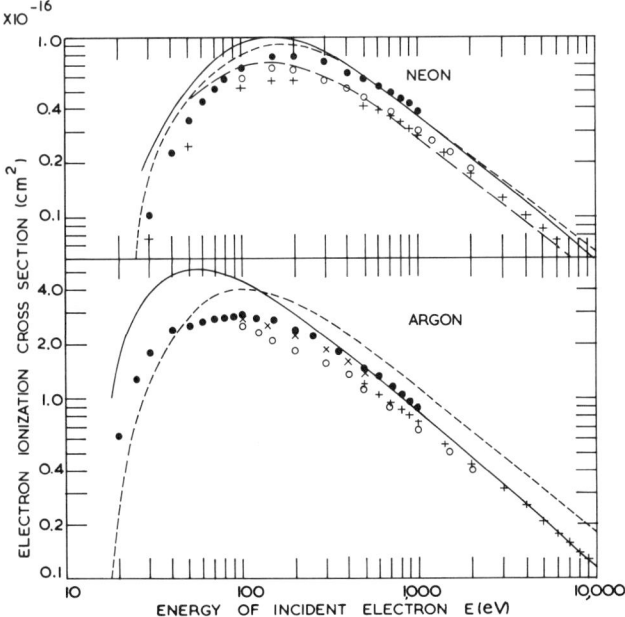

FIG. 26. Cross sections for electron ionization of neon and argon. Born cross sections: ——, McGuire (1971); – – –, Peach (1968, 1970, 1971); ——, Peach (1968, 1970, 1971) for (2p) shell only for neon. Measured cross sections for single electron ejection (10): ○, Gaudin and Hagemann (1967); +, Schram (1966), Adamczyk *et al.* (1966) below 500 eV for neon. Measured gross ionization cross sections (12) for sum of single and multiple electron ejection: ●, Rapp and Golden (1965); ×, Schram (1966).

to large differences in the magnitude of the calculated Born matrix elements, although the difference in the ionization potentials used in the two calculations must also contribute to the divergence at low energies. Between 300 and 3000 eV the two calculated cross sections are in good agreement, but at large energies Peach's cross section falls off as $2.1 \times 10^{-14}[E^{-1} \log E]$ cm^2 compared with $1.9 \times 10^{-14}[E^{-1} \log E]$ for McGuire. Since the wave functions used by McGuire give a photoionization cross section which agrees with the measured photoionization cross section (McGuire, 1968), it is not surprising that when the experimental photoionization cross section is substituted into (23) (Kingston, 1965) we obtain a high energy fall-off $1.83 \times 10^{-14}[E^{-1} \log E]$ cm^2 which agrees well with McGuire. The good agreement of McGuire's high-energy cross section with this independent check would suggest that his calculations are more reliable than those of Peach at high impact energies. At low impact energies, however, the ionization potential used by Peach is much closer to the true value than that used by McGuire, and hence McGuire's results will be greatly in error at low energies.

Various experimental electron ionization cross sections are also compared in Fig. 26. The measured cross sections for single electron ejection (10) (Schram, 1966; Adamczyk et al., 1966; Gaudin and Hagemann, 1967) are in good agreement except at the maximum in the cross section. Agreement among the measured gross ionization cross sections is not so good. At 1000 eV the gross ionization results of Rapp and Golden are about 35% greater than the single ionization results of Schram. Multiple ionization only contributes about 8% to Schram's gross ionization cross section, hence there is a real discrepancy of about 25% between the gross ionization cross sections of Schram and Rapp and Golden. Recent measurements by Fletcher and Cowling (1973) are in good agreement with Rapp and Golden at high energies.

In order to gain more information about the accuracy of the Born approximation we plot in Fig. 19 the ratio of the Born results of McGuire to the measured cross section of Rapp and Golden. The use of these particular sets of cross sections for this comparison is somewhat arbitrary, and the comparison must be taken only as a very approximate indication of the accuracy of the Born approximation. At high energies the ratio is almost one, and as the energy decreases, the ratio increases, reaching a maximum value of 1.9 at an energy of 1.4 times the ionization potential. At lower energies the ratio drops rapidly. We note that at low energies the use of the ionization potential as the unit of energy reduces the large difference between the Born and measured cross section which arose from the difference between the theoretical ionization potential used by McGuire and the experimental ionization potential. Just above threshold the measured cross section is given by $0.28x$

(cross section in πa_0^2, $x = E - I$ in units of the ionization potential) and the Born cross section by $1.0 x^{3/2}$. This threshold behavior shows that for these two cross sections if the energy of the incident electron is more than 1.02 times the ionization potential then the Born cross section will be accurate to within a factor of two.

The Born differential cross section (14) for electron ionization of neon has also been calculated by Omidvar et al. (1972). For an incident electron energy of 500 eV their Born calculations are about half of the measured differential cross section of Opal et al. (1971) at low energies of ejection, but the calculations are twice as big as the measurements at high energies of ejection. Although these two differential cross sections differ greatly it is found that the total cross sections derived from them only differ by about 20%.

7. Sodium

The Born cross section for electron ionization of the 3s electron of sodium has been calculated by Peach (1970) using a Hartree–Fock ground state and a Coulomb free state. This calculation has also been performed by Bates et al. (1965) with approximate Hartree–Fock functions for the ground state and the $l = 0$ and 1 partial waves of the free wave and Coulomb functions for the $l > 2$ partial waves. The two calculations are in good agreement particularly at high energies; at the maximum in the cross section the results of Bates et al. are about 10% larger than those of Peach.

In his Born calculation of the cross section for electron ionization of sodium, McGuire (1971) takes account of inner shell ionization but at low energies the (3s) shell is dominant and his calculation may be compared directly with Peach and Bates et al. Figure 24 shows that McGuire's results differ by about 10% from Peach's in the region of the maximum. This difference increases with the energy until the inner-shell cross section becomes important at about $E = 70$ eV.

Omidvar et al. (1972) also take account of innershell ionization in their Born calculation of the electron ionization of sodium. In the region of the maximum their results are larger than the other calculations, and at $E = 1000$ eV their cross section is about two times bigger than McGuire's. However, if Peach's cross section for the ionization of the 2p (Peach, 1970) is modified as suggested by Peach (1971), it is found that the sum of Peach's cross section for the (3s) shell plus the (2p) shell is in close agreement with McGuire's cross section at $E = 1000$ eV.

The measured cross sections for electron ionization of sodium are displayed in Fig. 24. At the maximum in the cross section the measurements are greater than all the calculated cross sections. The very large disagreement between the measurements of McFarland and Kinney (1965) and the calculations of McGuire for sodium is very similar to the disagreement between these cross sections for lithium. However, in the case of lithium the calculations of Omidvar et al. (1972) were in good agreement with McGuire's calculations, but for sodium the calculations of Omidvar et al. are in good agreement with the measurements of McFarland and Kinney. For sodium the accuracy of the Born cross sections is as yet too uncertain to use the Born cross section to normalize the experimental data.

8. Magnesium

The situation for the electron ionization of magnesium is similar to that for sodium. The Born ionization cross section at high energies of Omidvar et al. (1972) agrees with the measurements of Okuno et al. (1970), but they are about a factor of two greater than the Born cross section calculated by Peach (1968, 1970, 1971).

9. Argon

In the Born calculations of the electron ionization cross section for argon by McGuire (1971) and Peach (1968, 1970, 1971), only the outer two (3s) and (3p) shells are considered. As the contribution from the other inner shells is quite small, their neglect should not produce a significant error. Figure 26 compares the two calculations, and it is seen that they differ greatly. The difference is greatest at small impact energies; this is due in part to the different ionization potentials used in the calculations, but the major difference must be in the magnitude of the Born matrix elements. At very large impact energies the two calculated cross sections also disagree, McGuire's calculations fall off approximately as $3.0 \times 10^{-14}[E^{-1} \log E]$ cm^2 compared with $5.7 \times 10^{-14}[E^{-1} \log E]$ cm^2 for Peach's cross section. Using experimental values of the continuous oscillator strength of argon in (23), Kingston (1965) obtained an asymptotic behavior of $4.5 \times 10^{-14}[E^{-1} \log E]$ cm^2. This result is more accurate than those of Peach and McGuire, and it suggests that at high energies both cross section calculations are incorrect.

Recent measurements of the ionization of argon are in quite good agreement (Fig. 26). For single-electron ejection the measured cross sections of Schram (1966) and Gaudin and Hagemann (1967) are in good agreement. The results from both these experiments also show that for argon multiple ionization is important at large incident energies, the multiple ionization cross section contributing about 12% to the gross ionization cross section Q_g

at 1000 eV. Multiple ionization accounts for most of the difference seen in Fig. 26 between the single ionization cross sections Q_1 (Schram, 1966; Gaudin and Hagemann, 1967) and the gross ionization cross sections Q_g (Rapp and Golden, 1965; Schram, 1966).

No valid conclusion can yet be drawn about the Born approximation from a comparison of the Born and measured cross sections for the electron ionization of argon. The measured cross sections are in good agreement with each other but the difference between the Born approximation calculations is too great to allow any conclusion to be drawn about the range of validity of the Born approximation.

10. Other Atoms

In this section when it was possible we compared the Born electron ionization cross section with experiment. Some atoms were not considered because no experimental data are available and other atoms were not considered because we did not think that the measurements or the calculations were sufficiently accurate to make meaningful comparisons. For completeness we give some of the Born calculations for the electron ionization cross section which have been carried out recently. Peach (1965, 1968, 1970, 1971) obtained Born ionization cross sections for all the elements in the periodic table up to argon; McGuire (1971) considers all the elements in the first row of the periodic table; Omidvar *et al.* (1972) also gives results for carbon, potassium, zinc, and krypton.

B. Proton Ionization

Experimental cross sections for electron or proton ionization of an atom are often obtained by measuring only the current produced by slow moving ions. For electronic collisions, slow ions are only produced by ionization but for proton collisions, these slow moving ions are also produced by charge transfer

$$H^+ + A \longrightarrow H + A^+$$

This process is much less important than direct ionization for incident proton energies above about 300 keV. At lower energies the charge transfer cross section increases rapidly as the energy decreases, and it is important to take account of charge transfer.

The Born calculation of the ionization cross sections by electron and by proton impacts are very similar, and at high impact energies the cross sections for the two particles are the same at equal velocities. However, at low impact energies the approximation made in the Born method for electrons and for protons is fundamentally different. Neither an electron nor a proton will produce ionization if the energy of the particle is less than the ionization potential of the atom. For electron impacts this implies that ionization will not be produced if the velocity of the incident particle is below the velocity that may be associated with the bound electron. However, because of the large mass of the proton, it may ionize the atom when its velocity is very much lower than the velocity that may be associated with the bound electron.

1. Hydrogen

The cross section for proton ionization of atomic hydrogen has been measured by Fite et al. (1960) and Gilbody and Ireland (1964). Their cross sections are compared in Fig. 27 with the Born calculation of Bates and

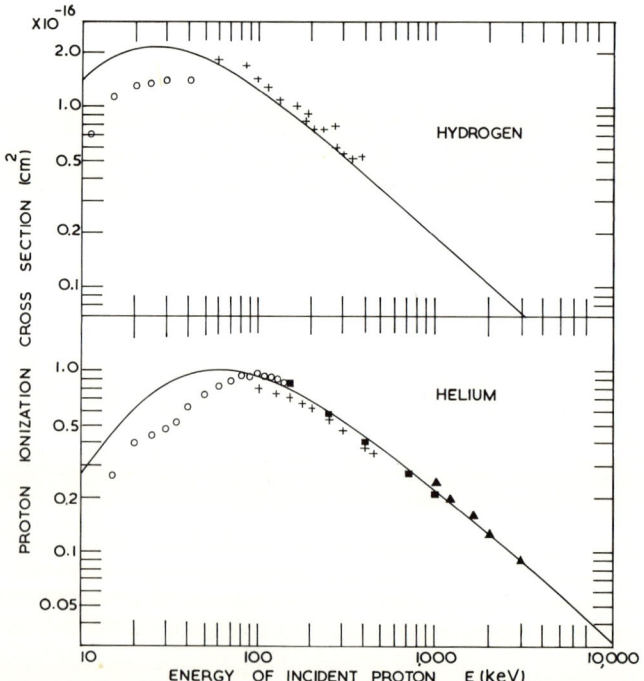

FIG. 27. Cross sections for proton ionization of atomic hydrogen and helium. For Hydrogen: Calculated Born cross section: ——, Bates and Griffing (1953). Measured cross section: ○, Fite et al. (1960); +, Gilbody and Ireland (1964). For Helium: Calculated Born cross section: ——, Bell and Kingston (1969a). Measured cross section: ○, de Heer et al. (1966); ■, McDaniel (1964); ▲, Pivovar and Levchenko (1967); +, Gilbody and Lee (1963).

Griffing (1953). There is good agreement between the theory and experiment of Gilbody and Ireland for proton energies above 60 keV. This good agreement between the Born theory and experiment must be treated with some caution for the measurements of Gilbody and Ireland do not appear to tend to those of Fite *et al.* at low energies.

In a recent theoretical paper Bates and Tweed (1974) used the impact parameter treatment of the Born approximation to calculate the probability $P(D, C)$ of excitation or ionization of hydrogen by proton impact. At a proton energy of 100 keV they found that $P(D, C)$ was 0.56 for head on collisions and was 0.36 and 0.18 at impact parameters $\rho = 1$ and 2 a.u., respectively. Since $P(D, C)$ must be small if the Born approximation is valid, they suggest that the agreement between the Born calculation and the measurements of Gilbody and Ireland (1964) must be in part fortuitous.

2. *Helium*

Since helium is a two-electron system, we must use approximate wave functions to describe the initial and the final states of the system; it is therefore not possible to evaluate the Born matrix element exactly. In Fig. 28

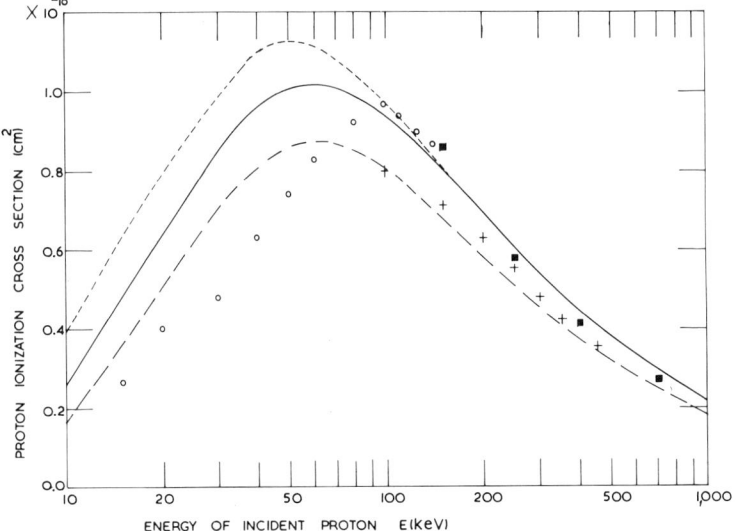

FIG. 28. Cross section for proton ionization of helium. Born cross section: ———, Bell and Kingston (1969a); — —, Peach (1965); - - -, McGuire (1971). Experimental cross sections: ○, de Heer *et al.* (1966); ■, McDaniel (1964); +, Gilbody and Lee (1963).

we compare the proton ionization cross sections calculated using three different sets of wave functions (Peach, 1965; Bell and Kingston, 1969a; McGuire, 1971). It is seen that at high energies the calculations of McGuire and of Bell and Kingston are in good agreement, but they are larger than the results of Peach. Below 150 keV the results of McGuire are larger than those of Bell and Kingston and at low energies there is great disagreement among the three calculations; for example, at 10 keV McGuire's results are more than twice Peach's calculations. In order to judge the magnitude of the error produced by approximate wave functions, Bell and Kingston calculated the cross section using both the length and velocity forms of the Born matrix element. The difference between the results from the two formulations increases as the energy decreases, and at 10 keV the results differed by 14%; this is consistent with the large differences between the calculations displayed in Fig. 28 and this is a good indication of the accuracy of their results. These alternative formulations used by Bell and Kingston gave results which differed by less than 3% for proton energies above 80 keV. An independent check on the high-energy fall-off of the cross section (Bell and Kingston, 1969a), using optical oscillator strengths, would suggest that at high energies the ionization cross section given by Bell and Kingston is about 5% too large.

Some of the measured cross sections for the proton ionization of helium are displayed in Figs. 27 and 28. The largest difference between these measurements is about 20%. The Born cross section agrees with the experimental data to better than 20% for proton energies above 80 keV. At lower energies the Born calculations considerably overestimate the cross section.

A great amount of useful information about the differential cross section for proton ionization of helium has been obtained by Rudd and his co-workers (Rudd and Jorgensen, 1963; Rudd et al., 1966). Their measurements are compared in Figs. 29a and 29b with the Born differential cross section of Bell et al. (1970b). At proton energies of 200 and 300 keV there is reasonably good agreement between theory and experiment over the full experimental range of ejection energies. In both these cases theory and experiment are in good agreement at low energies of ejection, the measurements then become larger than theory as the ejection energy increases and finally at large energies of ejection the measurements fall below theory. A similar pattern is also found for the measurements of Rudd et al. at an incident energy of 100 keV, but in this case for very large energies of ejection the measurements are almost a factor of two below the Born calculations. The measurements of Park and Schowengerdt (1969) at 100 keV are also plotted in Fig. 29a; their results are about half of the measurements of Rudd et al. for ejection energies up to 40 eV.

Comparison of the Born total ionization cross section with the measured cross section suggests that the Born cross section is too large at a proton

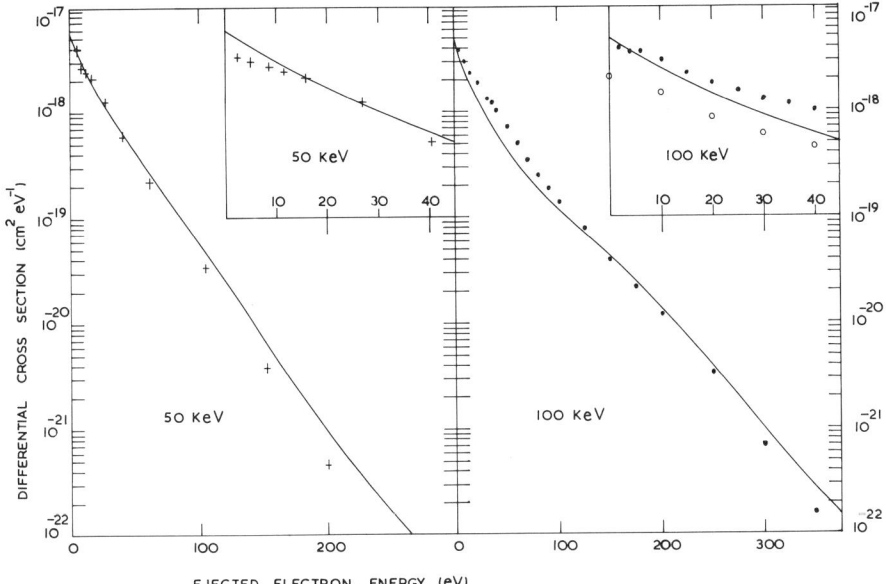

FIG. 29a. Differential ionization cross section for $H^+ + He$ collisions. Solid curves Born approximation (Bell et al., 1970b); Experimental data points: +, Rudd and Jorgensen (1963); ●, Rudd et al. (1966); ○, Park and Schowengerdt (1969). The energy of the incident proton is indicated in each section of the figure. (Note: the inserts show the regions near the origin with energy scale magnified.)

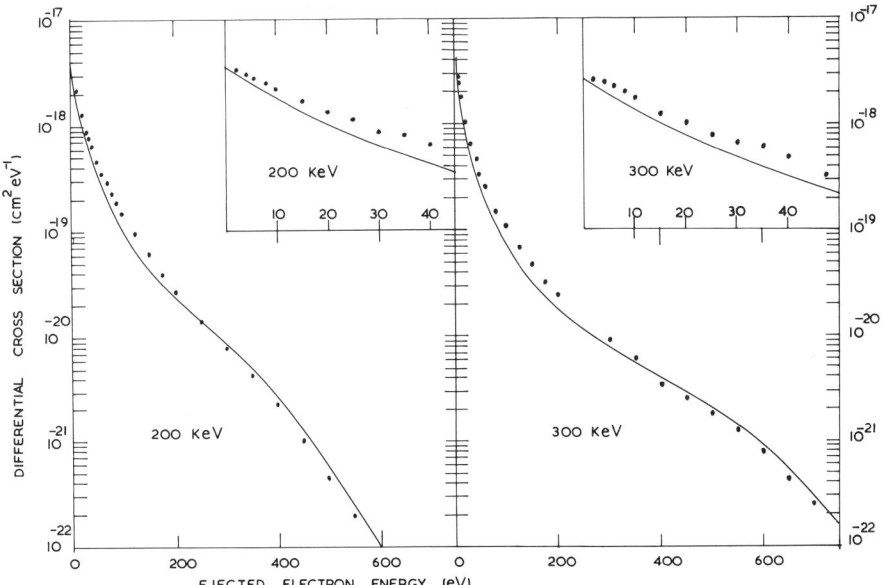

FIG. 29b. Differential ionization cross section for $H^+ + He$ collisions. Notation same as Fig. 29a.

energy of 50 keV. The comparison of the differential cross section at 50 keV (Fig. 29a) shows that at this incident energy the measured differential cross section lies below the Born differential cross section at almost all energies of ejection.

The differential cross section measurements of Rudd and his co-workers were obtained by integrating the measured double-differential cross sections I_D (21) over all angles of ejection. Born double-differential cross sections have been calculated by Oldham (1967). These are compared in Fig. 30 with

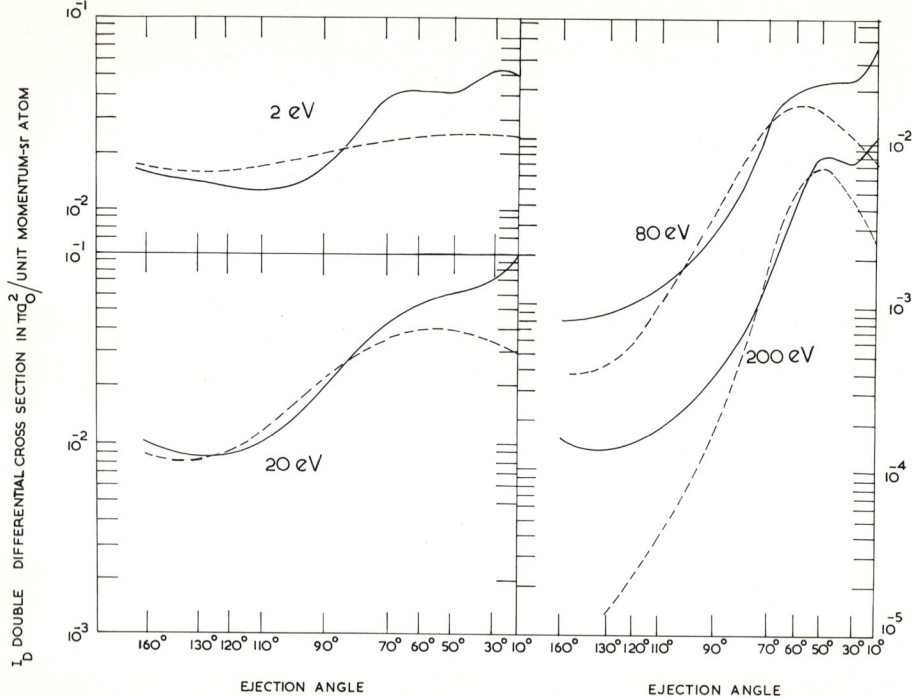

FIG. 30. Double differential cross section I_D for ionization of helium by a proton of energy 300 keV. Born calculations of Oldham (1967) ———; measurements of Rudd et al. (1966) ———. The energy of the ejected electron is given beside each curve.

the measurements of Rudd et al. (1966) for an incident proton energy of 300 keV. At this energy the total Born ionization cross section is in quite good agreement with the measured ionization cross section. However, Fig. 30 shows that there is serious disagreement between the Born and the measured double-differential cross section, particularly at small and large angles of ejection and at large energies of ejection.

3. Neon

For the proton ionization of neon, Fig. 31 shows that the Born cross sections of Peach (1968, 1970, 1971) and McGuire (1971) are in excellent agreement except at low energies. The cross sections differ slightly at very high energies, but, as in the case of electron ionization, independent calculations using experimental photoionization cross sections would suggest that at very high energies the calculations of McGuire are slightly more accurate than those of Peach.

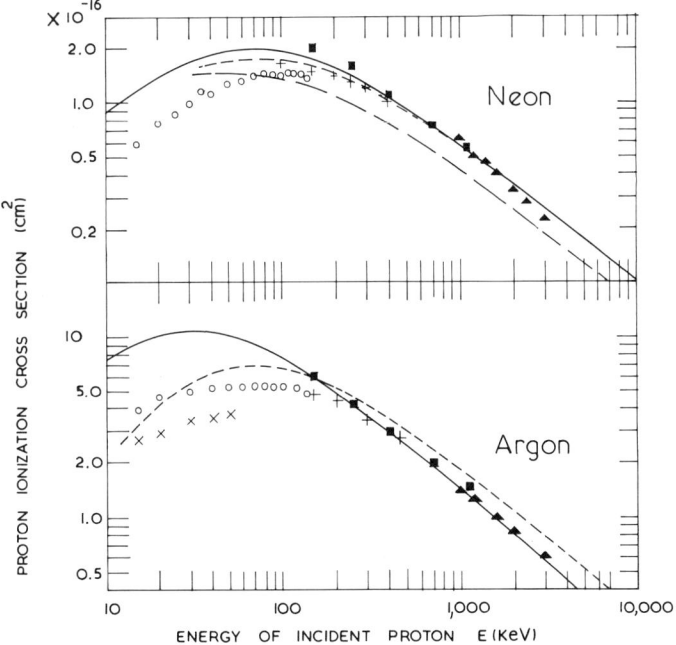

FIG. 31. Cross sections for proton ionization of neon and argon. Born cross section: ———, McGuire (1971); – – –, Peach (1968, 1970, 1971); — —, Peach (1968, 1970, 1971) for (2p) shell only for neon. Measured cross section: ○, de Heer *et al.* (1966); ■, McDaniel (1964); ▲, Pivovar and Levchenko (1967); +, Gilbody and Lee (1963); ×, Afrosimov *et al.* (1967).

In measuring the ionization cross section, de Heer *et al.* (1966) and Gilbody and Lee (1963) took proper account of charge transfer. McDaniel (1964) did not consider the charge transfer process and this may account for the difference between his measurements and the cross sections of de Heer *et al.* and Gilbody and Lee. The charge transfer cross section decreases rapidly as the energy increases and at $E > 300$ keV charge transfer may be ignored.

Before comparing theory and experiment, it should be remembered that the experimentalists have measured gross ionization cross section Q_g (12), but the theoreticians have calculated the ionization cross section for the ejection of a single electron. Electron ionization measurements would suggest that multiple ionization only contributes about 10% to the gross ionization cross section. The experimental cross section for the ejection of one electron may be about 10% below the measured gross ionization cross section displayed in Fig. 31.

Figure 31 shows that there is very good agreement between theory and experiment for $E > 300$ keV, but at lower energies the measured cross section is much lower than the Born cross section.

4. Argon

In their Born calculations of the proton ionization of argon both Peach (1968, 1970, 1971) and McGuire (1971) only considered the (3s) and (3p) shells. The contribution from the other inner shells will be small compared with the large differences which exist between the two calculations at both large and small energies (see Fig. 31). It is not possible to say that one of these calculations is more accurate than the other.

As in the case of neon, charge transfer may account for the difference between the measurements of McDaniel (1964) and those of de Heer *et al.* (1966) and Gilbody and Lee (1963). The difference between the low-energy measurements of de Heer *et al.* and Afrosimov *et al.* (1967) arises because de Heer *et al.* measured the gross ionization cross section Q_g, while Afrosimov *et al.* measured the cross section Q_1 for the ejection of one electron. Since the Born calculations only give the cross sections for single ionization they should only be compared with the measurements of Afrosimov *et al.* The importance of multiple ionization at low energies suggests that multiple ionization may also be important at higher energies, and it is possible that the good agreement between the measured gross ionization cross sections and McGuire's Born cross section for $E > 300$ keV is fortuitous.

5. Other Atoms

Finally we note that Peach (1965, 1968, 1970, 1971) has used the Born approximation to calculate the proton ionization cross section for all of the elements in the first two rows of the periodic table, and McGuire (1971) gives Born ionization cross sections for all the elements in the periodic table up to sodium. In this section we have only discussed the Born calculations for which there is experimental data.

IV. Atom–Atom and Ion–Atom Collisions

In this section we consider the application of the first Born approximation to atom (or ion)–atom collisions, in which at least one of the colliding systems is excited, ionized, or de-excited. Let an atom (or ion) of nuclear charge Z_A, mass M_A, having N_A electrons collide with an atom of nuclear charge Z_B, mass M_B, having N_B electrons. Suppose A makes the inelastic transition from initial state i to final state f, while B goes from state p to state q. The derivation of the cross section then follows closely that for nuclei–atom collisions (cf. Mott and Massey, 1965): the wave function of the complete system is expanded in the orthonormal set obtained from products of the separated (unperturbed) atomic wave functions; all but direct coupling is neglected; electron exchange is ignored; and the incident relative motion of the nuclei is described by a plane wave. It may readily be shown that the first Born approximation cross section for this process is

$$Q({}^{i\,\to\,f}_{p\,\to\,q}) = \frac{8\pi}{V^2}\int_{K_1}^{K_2}\frac{dK}{K^3}\,|Z_B\,\delta_{pq} - I_B^{pq}(K)|^2\,|I_A^{if}(K)|^2 \tag{26}$$

with

$$I_C^{rs}(K) = \int d\tau_C\,\phi_r^*(C)\sum_{t=1}^{N_c}e^{i\mathbf{K}\cdot\mathbf{r}_{ct}}\phi_s(C) \tag{27}$$

where $\phi_s(C)$ denotes the wave function of atom C in state s, V is the incident relative velocity of the colliding systems; and $K_1 = |K_i - K_f|$, $K_2 = K_i + K_f$, where $K_i = M_R V (M_R = M_A M_B/M_A + M_B)$, and by energy conservation $K_f^2 = K_i^2 + 2M_R \Delta E(\text{if; pq})$, where $\Delta E(\text{if; pq})$ is the difference in the electronic energies of the initial and final states of the system.

Thus the evaluation of atom–atom cross sections in the Born approximation requires the same matrix elements as are necessary for electron–atom collisions in the first Born approximation, and again errors not already inherent in the Born approximation may be introduced through the use of poor wave functions.

A further difficulty is encountered, however, for atom–atom collisions. In general it is not possible to measure the cross section $Q({}^{i\,\to\,f}_{p\,\to\,q})$ for a process in which both atoms undergo specific transitions. Measurements may only be obtained for the sum of the processes in which one atom (say, A) undergoes the transition i → f, while the other atom may be left in any final state, the cross section, $Q(i \to f)$, being given by $\sum_q Q({}^{i\,\to\,f}_{p\,\to\,q})$. The evaluation of this summation requires considerable labor and a further approximation, the closure approximation, is widely used to reduce the labor involved. This approximation makes use of the fact that the upper limit K_2 in the integration (26) may be taken to be independent of q as it is infinite for all practical purposes. Hence, if one replaces $\Delta E(\text{if}; pq)$ by some average excitation energy, $\Delta E(\text{if}; p)$, independent of q, then the lower limit K_1 of (26) is also independent of q and so in evaluating $\sum_q Q(^{i\ \to f}_{p\ \to q})(q \neq p)$ the summation may be carried out before integration. In the integrand of (26) the summation arises as $\sum_q |I_B^{pq}(K)|^2$ $(q \neq p)$ and since the wave functions ϕ_q form a complete set we may use the closure property to carry out the summation

$$\sum_{\substack{q \\ q \neq p}} |I_B^{pq}(K)|^2 = N_B + \sum_{\substack{n \\ n \neq m}}^{N_B} \sum_m^{N_B} \int d\tau_B\, \phi_p^*(B) e^{i\mathbf{K} \cdot (\mathbf{r}_{Bm} - \mathbf{r}_{Bn})} \phi_p(B)$$

The right-hand side of this equation may be evaluated using only a knowledge of the initial state wave function ϕ_p of atom B. Thus by employing closure with an average excitation energy, we can obtain an approximate atom–atom collision cross section for $\sum_q Q(^{i\ \to f}_{p\ \to q})$ without the labor of calculating $Q(^{i\ \to f}_{p\ \to q})$ for each q.

There seems to be no *a priori* reasoning for the choice of the average excitation energy $\Delta E(\text{if}; p)$, and many values have been chosen including, for example, the ionization potential of atom B or the logarithmic mean energy used in Bethe's theory for the stopping power of heavy particles in matter. At low impact energies, $Q(^{i\ \to f}_{p\ \to p})$ dominates the summation and closure produces no significant error, while at high energies, where $K_1 \to 0$, the use of closure is certainly valid. However, in the intermediate range, results obtained from closure may differ considerably from those obtained by direct evaluation of the first Born approximation cross sections. Bell *et al.* (1973b) have shown that this difference may be as large as a factor of five for certain processes. Thus caution must be exercised in drawing conclusions about the first Born approximation when comparing experiment with results obtained using closure.

A. Excitation

We shall review in this section theoretical work for processes for which the first Born approximation has been employed to investigate the excitation of a *discrete* state of one of the colliding partners.

The only atom for which exact matrix elements (27) may be obtained is atomic hydrogen, and considerable effort has been put into the calculation of cross sections for the processes

$$H(i) + H(p) \longrightarrow H(f) + H(q) \tag{28}$$

since the first calculations by Bates and Griffing (1953, 1954, 1955) for the following cases (using the notation (i → f; p → q)): for single excitation collisions (1s → 1s; 1s → 2s, 2p, 3s, 3p, 3d); for double excitation collisions (1s → 3s, 3p, 3d; 1s → 3s, 3p, 3d); (1s → 2s, 2p; 1s → 2s, 2p, 3s, 3p, 3d). The papers of Pomilla and Milford (1966) and Pomilla (1967) contain all other data available for processes (28). Pomilla and Milford present results in the center-of-mass energy range 2 keV–100 MeV for the transitions: (1s → 2s, 2p; $nl → n + 1l'$) for all possible Δl for $n = 1–7$ and for $l = n–1$, $l' = n$ for $n = 8, 9, 10$; (2s → 3p; ns → $n + 1$, p) for $n = 1–7$; (2s → 3p; $n, n - 1 → n + 1, n$) for all $n \leq 10$; (4s → 5p; 4s → 5p); (4s → 5p; 7s → 8p); (1s → 2s, 2p; 1s → nl) for all possible Δl for $n = 3–10$; (1s → 2s, 2p; $2l → nl'$) for all possible Δl for $n = 4–6$; (1s → 2s, 2p; $nl → n'l'$) for all possible Δl when $n = 4–6$ and $n' = 6–9$. Pomilla has extended this work to evaluate cross sections for (1s → 1s; $nl → n'l'$) for $n \leq 7$ and all possible Δl and presents tables of total cross sections for the transitions $n → n'$ obtained by appropriate summing and averaging over l, l' of these individual contributions. He tabulates results in the energy range 1 eV–10 MeV for $n → n + 1$, $n = 1–7$; $1 → n'$, $n' = 3–10$; $2 → n'$, $n' = 4–6$; $4 → n'$, $n' = 6–9$; $5 → n'$, $n' = 7–9$; $6 → n'$, $n' = 8–9$. This recent work of Pomilla has confirmed the early observation of Bates and Griffing (1954) that for single-transition collisions between hydrogen atoms the Born cross sections are larger than those for corresponding double-transition collisions at incident energies E_H less than 10 keV. He also finds that the single-transition cross sections are proportional to $1/E_H$ for $E_H \gtrsim 10$ keV and are proportional to n^{-3} for energies $\gtrsim 10$ keV. [This latter result was shown analytically by May (1965) by considering $n \gg 1$ and high energies.] Pomilla and Milford find that for processes in which both H atoms undergo $\Delta n = 1$ transitions double inelastic collisions between excited state atoms have cross sections much larger than corresponding ground-state–excited-state collisions, the more highly excited the incident atoms are the larger the cross sections.

Unfortunately, no experimental data exist for atomic hydrogen to enable conclusions to be drawn as to the energy region of validity of the Born results. Some indication of the validity of the Born approximation may be drawn from further theoretical studies for a limited number of the processes. For the single transition processes (1s → 1s; 1s → 2s, 2p, 3s, 3p, 3d) Flannery (1969a) has performed close coupling impact parameter two-state calculations and finds that the first Born approximation and close coupling results agree to better than 10% for excitation of the s, p, and d states for incident energies E_H greater than about 20, 10, and 6 keV, respectively. Such cross sections behave as $1/E_H$ for $E_H \gtrsim 10$ keV, and thus the indications are that for single-transition cross sections the first Born approximation results are valid for $E_H \gtrsim 20$ keV. For double-transition processes the data is even more restricted. Levy (1969a) has performed a two-state close coupling impact parameter treatment for (1s → 2s, 2p; 1s → 2s, 2p). For the processes (1s → 2p; 1s → 2p), (1s → 2s, 1s → 2p) in which at least one atom makes an optically allowed transition, first Born and close coupling results agree to better than 10% for $E_H \gtrsim 75$ keV. For the doubly "forbidden" process (1s → 2s; 1s → 2s) 10% agreement is only reached above 200 keV. Further work is clearly desirable and the only conclusion possible at present is that it appears that the first Born approximation is valid to much lower energies for single-transition processes than for double-transition reactions in hydrogen–hydrogen collisions.

The colliding partners atomic hydrogen–helium provide more information about the usefulness of the first Born approximation, since, recently, both experimental data and highly accurate Born matrix elements for helium have become available. Direct evaluation of (26) has been carried out for the processes:

$$H(1s) + He(1^1S) \longrightarrow H(2s, 2p) + [He] \qquad (29)$$

Bell *et al.* (1973c),

$$H(1s) + He(1^1S) \longrightarrow [H] + He(2,3,4,5^1S; 2,3,4^1P; 3,4,5^1D) \qquad (30)$$

Levy (1969c), where [ ] denotes that all final states of the atom (discrete and continuum) have been considered.

For the processes (29) Bell *et al.* have used both length and velocity formulation Born matrix elements for helium and estimate that the error arising from the use of approximate wave functions is less than 1% for the total cross sections $Q(1s \to nl)$ obtained by summing the individual contributions (1s → nl; $1^1S \to N^1L$) over all possible final helium states (including the continuum). Experimental data is available over the hydrogen impact energy range 1–120 keV for excitation of both the 2s and 2p states of hydrogen. For 2s excitation, comparison of the first Born approximation results with both experiment (Fig. 32) and more sophisticated calculations (Bell *et al.*, 1973c) indicates that it is a satisfactory approximation for impact energies greater than about 10 keV. Further, it is noted that in the approximate energy range 10–30 keV, the single-transition process dominates and thus agreement with experiment indicates that for $E_H \gtrsim 10$ keV the Born approximation for the single-transition process is satisfactory—a conclusion consistent with that arrived at above for H–H collisions.

Figure 33 presents the first Born approximation cross section $Q(1s \to 2p)$ together with experimental data. Unfortunately there is considerable disagreement among the experimental results, and it is difficult to define a good lower energy limit for the region of validity of the Born approximation. However, agreement between theory and the most recent experimental data

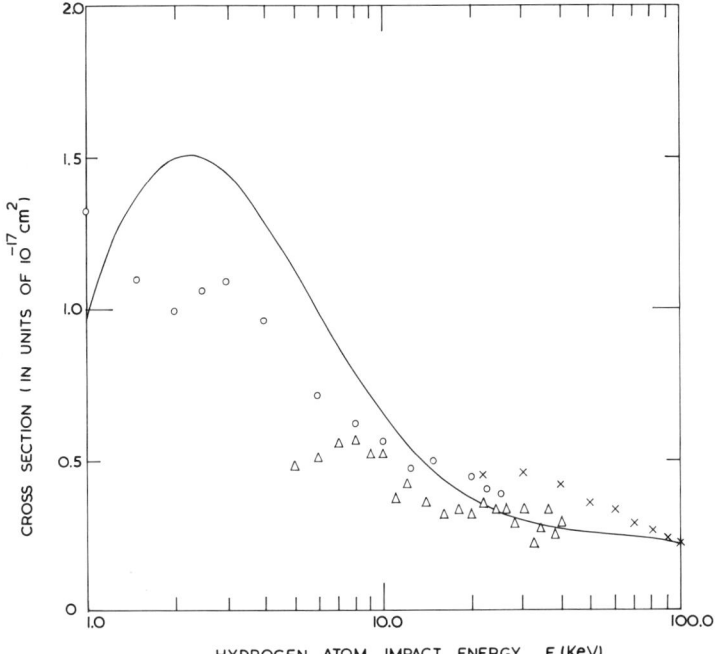

FIG. 32. Total cross section for excitation of the 2s state of hydrogen by collision with helium. Solid curve: first Born approximation (Bell *et al.*, 1973c). Experimental data: △, Orbeli *et al.* (1970); ×, Hughes and Choe (1972b); ○, Birley and McNeal (1972).

is excellent for impact energies greater than 40 keV. This region is of considerable interest since the total cross section shows a second maximum. This second maximum arises because single-transition collisions are more effective at low impact energies while double-transition collisions become significant with increasing energy and are more effective at high energies. The agreement between theory and experiment, shown in Fig. 33, is satisfactory confirmation of the interplay between single and double transitions.

Levy (1969c) has also used accurate helium matrix elements for direct evaluation of (26) for the processes (30). Experimental data are available in the incident hydrogen atom energy range 10–35 keV for excitation of the 4^1S, 5^1S, 3^1P, 4^1D, and 5^1D states of helium (Van Eck *et al.*, 1964) and over a wider range 20–100 keV for the 4^1S and 4^1D states (Blair and Gilbody, 1973). Unfortunately there is considerable uncertainty in the absolute values of these cross sections, and it is difficult to draw conclusions from a comparison of theory and experiment. Further experimental data are desirable, particularly at higher impact energies.

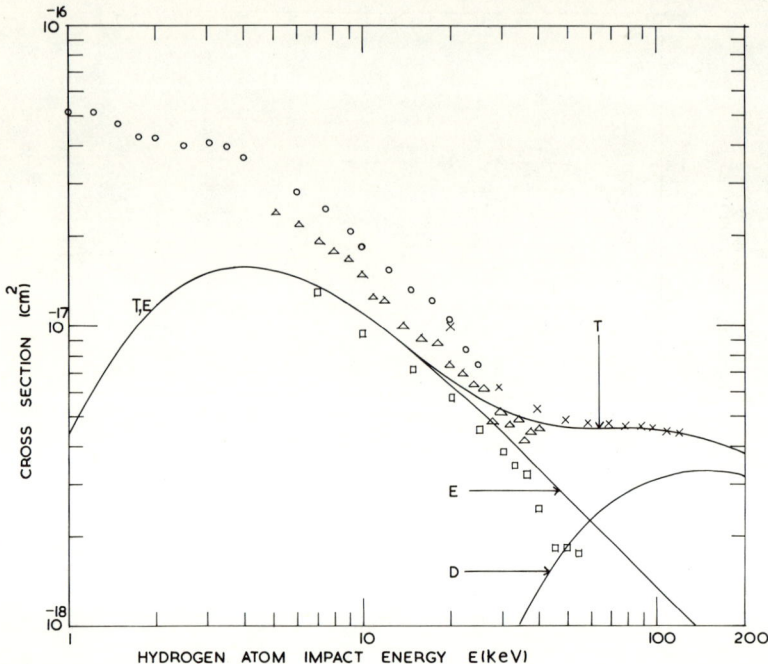

FIG. 33. Total cross section for excitation of the 2p state of hydrogen by collision with helium. Solid curves: first Born approximation (Bell et al., 1973c); Curve T denotes the total cross section; curve E denotes the single transition contribution; curve D denotes the double transition contributions. Experimental data: □, Dose et al. (1968); △, Orbeli et al. (1970); ×, Hughes and Choe (1972a); ○, Birely and McNeal (1972).

All other applications of the first Born approximation have been concerned either with specific transitions of both atoms or have involved the use of closure.

The collisions involving specific transitions of both atoms cannot be compared with experiment. Reactions that have been studied are

$$H(1s) + He^+(1s) \longrightarrow H(2s,2p) + He^+(1s)$$
Flannery (1969b)
$$He(2^3S) + Ne(2p^6\ ^1S) \longrightarrow He(2^3P, 3^3P) + Ne(2p^6\ ^1S)$$
Adler and Moiseiwitsch (1957)
$$He(2^3S) + H(1s) \longrightarrow He(2^3P, 3^3P) + H(1s, 2s, 2p)$$
Adler and Moiseiwitsch (1957)
$$He^+(1s) + He(1^1S) \longrightarrow He^+(1s) + He(2^1P)$$
Moiseiwitsch and Stewart (1954)

Collisions that have been investigated using closure are: $1^1S \to N^1L$, $N = 2\text{–}5, L = 0, 1, 2$ excitation of helium by collision with target He, Ne, Ar, and Kr (Levy, 1970); $2^3S \to 2^3P$, $2^3S \to 3^3P$ excitation of helium by Ne, H (Adler and Moiseiwitsch, 1957); 1s $\to$ 2s, 1s $\to$ 2p excitation of H by Ne, Ar, Kr (Levy, 1969b); 1s $\to nl, n = 2, 3, l = 0 \to n - 1$ excitation of H by Li, Na, K, Rb, Cs (Mathur *et al.*, 1972); 1s $\to n, n = 2, 3, 8, 15$ excitation of H by Li (Cheshire and Kyle, 1965). In these collisions the target atom was taken to be initially in its ground state, and comparison between theory and experiment may as yet only be made for excitation of the 2s and 2p states of H by Ne, Ar, and Kr (Birley and McNeal, 1972; Orbeli *et al.*, 1970; Hughes and Choe, 1972a,b) and for excitation of the 4^1S and 4^1D states of He by He (Blair *et al.*, 1973). For excitation of H, comparison with the available experimental data (5–120 keV) shows that for excitation of both 2s and 2p and for all targets, the first Born approximation greatly overestimates (by as much as an order in magnitude) the cross section even at impact energy 120 keV. Figure 34 illustrates this by comparing theory and experiment for excitation of hydrogen to the 2s state by collision with neon, argon, and krypton. By contrast it is found that the theoretical cross section for He(4^1S) excitation by He is in rough accord with experiment (20–100 keV) while the first Born approximation for He(4^1D) excitation by He underestimates the cross section by as much as a factor of ten. It seems unlikely that at these fairly low impact energies the error is caused by the use of closure. The difference between theory and experiment is caused by the failure of the first Born approximation to describe properly the single-transition process in which the target remains in its initial state. Further evidence of such a breakdown and further discussion of the point is presented below.

In summary, it does not appear possible to define generally a lower energy limit to the range of validity of the first Born approximation for excitation of an atom to a discrete state by collision with another atom. For the collision between ground state H atoms and single-transition processes, this limit may be as low as 20 keV; for the excitation of H by He to the 2s and 2p states, the limit seems to be as low as 15–40 keV; but for other colliding atoms the limit is not yet definable and may well be as large as several hundred thousand electronvolts, or even in the mega electron volt range.

B. Electron Loss and Ionization

We now turn our attention to applications of the first Born approximation to the study of ionization of one of the colliding partners. It is customary to refer to the ionization of the projectile atom in an experiment as an electron-loss process, though the first Born approximation does not, of course, distinguish the particles as projectile and target. For these ionization processes (26) defines a differential cross section for ejection of an electron with a certain energy, and the total cross section is obtained by integrating

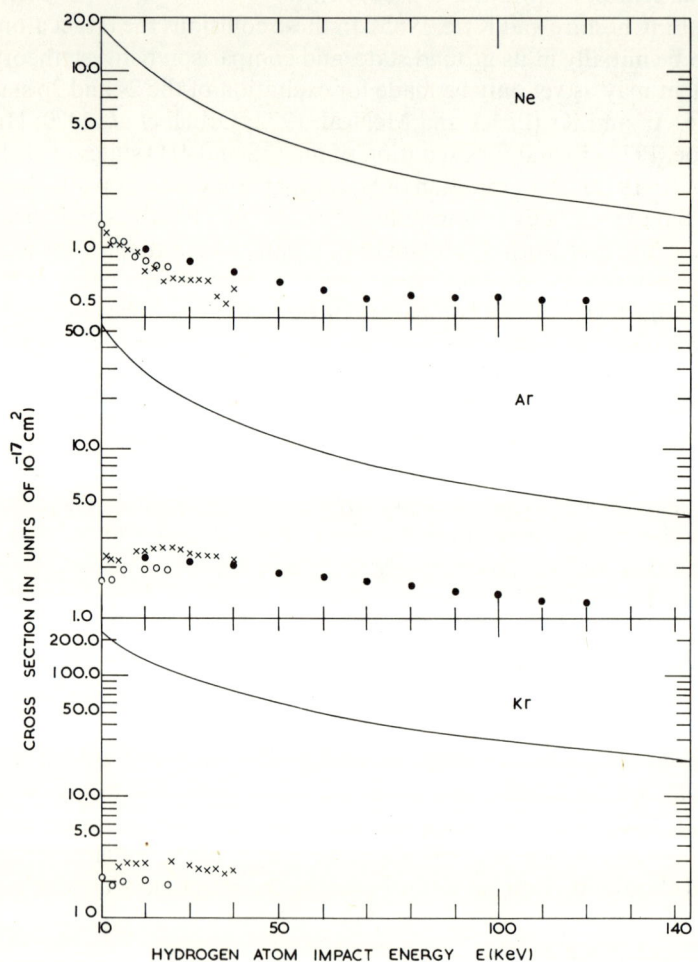

FIG. 34. Total cross sections for excitation of the 2s state of hydrogen by collision with neon, argon, and krypton. Solid curve: first Born approximation using closure (Levy, 1969b). Experimental data: ×, Orbeli *et al.* (1970); ●, Hughes and Choe (1972b); ○, Birely and McNeal (1972).

over all energies of the ejected electron (assuming the appropriate normalization of the continuum wave function).

The first calculations performed were by Bates and Griffing (1953, 1954, 1955) for the reactions:

$$H(1s) + H(1s) \longrightarrow H^+ + e^- + H(1s,2s,2p,3s,3p,3d \text{ or } c) \qquad (31)$$

where c denotes the continuum; and by Boyd et al. (1957) for the processes

$$He^+(1s) + H(1s) \longrightarrow He^{++} + e^- + H(1s,2s,2p,3s,3p,3d \text{ or } c) \qquad (32a)$$

$$He^+(1s) + H(1s) \longrightarrow He^+(1s) + H^+ + e^-. \qquad (32b)$$

Since then all calculations have been concerned with obtaining a *total* cross section for processes

$$A(i) + B \text{ (ground state)} \longrightarrow A^+ + e^- + [B] \qquad (33)$$

where *total* cross section in what follows is to be interpreted as the sum of the cross sections from all individual processes (33) over all final states of atom B. The reactions (33) will be denoted in what follows by (A(i); B).

We shall first consider those investigations which have carried out the summation explicitly (without the use of closure). Omidvar and Kyle (1970) have repeated and extended the calculations of Bates and Griffing for processes (31) and have evaluated total cross sections for (H(nl); H) for $n = 1 \to 4, 0 \leq l \leq n - 1$. Bell et al. (1969b, 1970a) and Bell and Kingston (1971) have considered the processes: (H(1s); He), (He(1^1S); H), (He(1^1S): He), (He$^+$(1s); He), (He(1^1S); He$^+$), (H(2s); H), and (H(2s); He). Experimental data exist for most of the above reactions, the exceptions being (H(nl); H) for $n > 1$ for which no data exist and for (H(1s); He$^+$), (He$^+$(1s); H), and (H(2s); H) for which data exist for molecular hydrogen rather than for atomic hydrogen. The relevant experimental papers are referenced in the above papers with the exceptions of the work of Pedersen and Hvelplund (1971) for (He(1^1S); He) and of Puckett et al. (1969, and references therein) for (He(1^1S); He), (H(1s); He), (He(1^1S); H), and (He(1^1S); He$^+$). For (H(2s); H) the experimental data is so limited in energy range (10–30 keV) that comparison of theory and experiment provides no indication of the region of validity of the first Born approximation. However, for the other colliding partners, the following table gives tentative values for the lower limit of the incident relative velocity above which experiment and theory are in reasonable accord.

For (i) and (iii) in the table the limits are derived by assuming that molecular hydrogen behaves as two hydrogen atoms. For He$^+$–He collisions it is also noted that comparison of theory and experiment for the total electron production cross section (Bell et al., 1970a) gives a lower limit of about 2 a.u. While all of these limits must be regarded with caution, it does

	A	B	Relative velocity (a.u.)
(i)	H(1s)	H	~ 1.8
(ii)	He$^+$(1s)	H	~ 1.5
(iii)	H(1s)	He$^+$	~ 1.5
(iv)	H(1s)	He	~ 2.3
(v)	He(1^1S)	H	~ 2.5
(vi)	He(1^1S)	He	~ 3.0
(vii)	He$^+$(1s)	He	~ 3.5
(viii)	He(1^1S)	He$^+$	~ 1.5
(ix)	H(2s)	He	~ 2.0

seem that for ground-state atom–atom collisions between light elements the first Born approximation is successful in describing one-electron ionization events for incident relative velocities greater than about 2 a.u.

All other calculations have employed closure to evaluate the total cross sections. Dmitriev et al. (1966) have investigated the ionization of ground-state hydrogenic ions of nuclear charge $Z = 1$ to 10^4 by collision with atomic hydrogen and helium. Lodge (1969) has considered (H(n); H, Li, Na) for $n = 1, 2, 3, \ldots, \infty$, but Omidvar and Kyle (1970) have repeated his (H(n); H) calculations for $n = 1, 2, 3$, and 4 and feel that his results contain numerical errors. Levy (1969b) has evaluated the total cross section for ionization of H(1s) by He, Ne, Ar, Kr, C, N, and O while Tripathi et al. (1972) have performed calculations for ionization of He(1^1S) by He, Ne, Ar, and Kr.

McGuire (1971) has considered various ionization processes occurring between ground state He colliding with He, Ne, and Ar. Having previously considered processes involving collisions between light atoms, we turn our attention to the collision of H and He with the heavier atoms. As was found in Section (IV,A) for excitation, for all such collisions the theoretical cross section again considerably overestimates and differs by even an order of magnitude from experiment. It is not until one gets into the mega electron volt range that theory begins to show any agreement with experiment. This conclusion is illustrated in Fig. 35, where theory and experiment are compared for electron loss from hydrogen in collision with neon, argon, and krypton. Insight into the failure of theory may be obtained, for example, from the calculations of McGuire (1971). In Fig. 36 we plot his cross sections for the single-transition processes

$$\text{He}(1^1\text{S}) + \text{Ar}(^1\text{S}), \text{Ne}(^1\text{S}) \longrightarrow \text{He}^+ + e^- + \text{Ar}(^1\text{S}), \text{Ne}(^1\text{S})$$

(the cross sections being divided by the square of the nuclear charges Z of the heavy particle). The experimental data of Puckett et al. (1969) show the difference between theory and experiment even at 1 MeV. The interesting

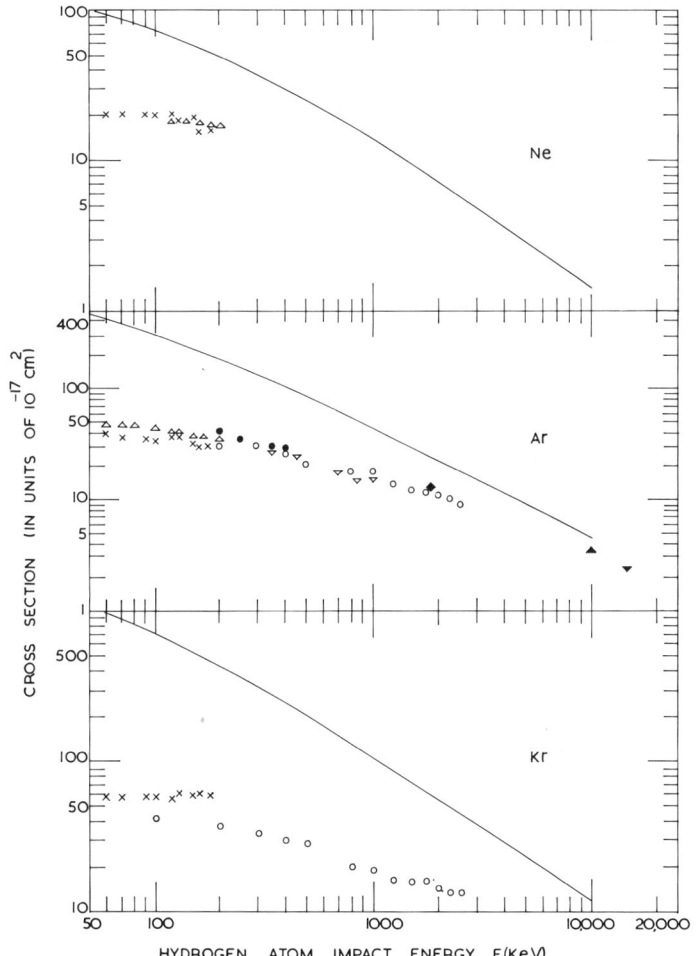

FIG. 35. Total cross sections for electron loss from hydrogen by collision with neon, argon, and krypton. Solid curve: first Born approximation using closure (Levy, 1969b). Experimental data: ×, Solov'ev et al. (1962); △, Stier and Barnett (1956); ○, Toburen et al. (1968); ●, Puckett et al. (1969); ▽, Barnett and Reynolds (1958); ▲, Berkner et al. (1964); ▼, Smythe and Toevs (1965); ◆, Welsh et al. (1967).

feature of these calculations is that the cross sections are nearly scaling as Z^2. Referring to (26) this means in effect that in the elastic matrix element for the heavy particle $I(K)$ is so small that it is effectively zero. This implies that the first Born approximation does not take into account the structure of the heavy particle during the collision. Thus, as in the case of excitation, it would seem that the first Born approximation is not a useful approximation for

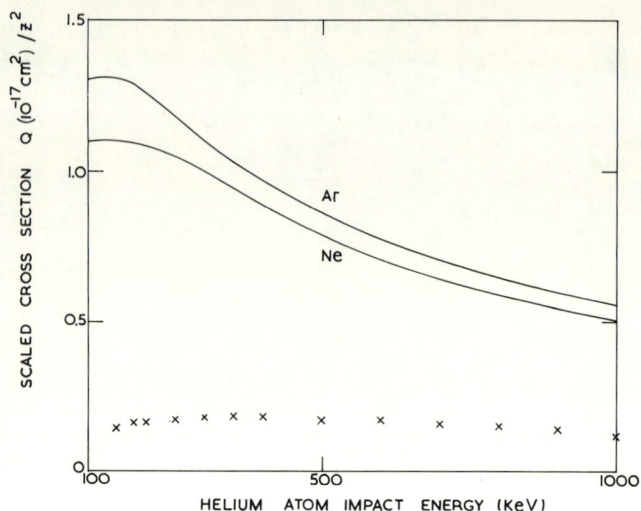

FIG. 36. Scaled cross sections for electron loss from helium atoms colliding with argon and neon (Z is the charge of heavy particle). Solid curves: first Born approximation for the single transition process in which the heavy particle remains unexcited, McGuire (1971). Experimental data: ×, Puckett et al. (1969), total cross section for all processes in which the argon atom is left in any final state.

collisions involving a heavy particle, except in the high mega electron volt region.

In summary therefore, for ionization events occurring in the collision of light particles, the first Born approximation seems adequate for incident relative velocities greater than about 2 a.u., but for collisions involving a heavy particle it is not adequate until the mega electron volt region, due to an inadequate description of the heavy particle structure during the collision.

C. Metastable Atom Destruction

To complete this review of atom–atom collisions, we finally record the Born calculations of Bell et al. (1973b) on the destruction of metastable hydrogen H(2s) by collision with He(1^1S) in which a total cross section was obtained by summing explicitly all contributions from all possible hydrogen and helium transitions, and the closure calculations of Levy (1971) on the destruction of H(2s) by collision with ground state He, Ne, Ar, and Kr. All theoretical cross sections are found to be larger than experiment. However, the limited energy range of experimental data and the present experimental uncertainties make comparison between theory and experiment unprofitable at this time.

ACKNOWLEDGMENT

This research reported was accomplished with the support of the U.S. Office of Naval Research, under Contract N00014-69-C-0035.

REFERENCES

Adamczyk, B., Boerboom, A. J. H., Schram, B. L., and Kistemaker, J. (1966). *J. Chem. Phys.* **44**, 4640.
Adler, J., and Moiseiwitsch, B. L. (1957). *Proc. Phys. Soc., London, Sect. A* **70**, 117.
Afrosimov, V. V., Marnaev, Y. A., Panov, M. N., and Uroshevich, V. (1967). *Sov. Phys.—Tech. Phys.* **12**, 512.
Aleksakhin, I. S., Zapesochnyi, I. P., and Shpenik, O. B. (1967). *Int. Conf. Phys. Electron At. Collision Phenomena, 5th, 1967 Abstr.* p. 499.
Arthurs, A. M. (1961). *Proc. Cambridge Phil. Soc.* **57**, 904.
Barnett, C. F., and Reynolds, H. K. (1958). *Phys. Rev.* **109**, 355.
Bates, D. R. (1962). "Atomic and Molecular Processes." Academic Press, New York.
Bates, D. R., and Damgaard, A. (1949). *Phil. Trans. Roy. Soc. London, Ser. A* **242**, 101.
Bates, D. R., and Griffing, G. (1953). *Proc. Phys. Soc., London, Sect. A* **66**, 961.
Bates, D. R., and Griffing, G. (1954). *Proc. Phys. Soc., London, Sect. A* **67**, 663.
Bates, D. R., and Griffing, G. (1955). *Proc. Phys. Soc., London, Sect. A* **68**, 90.
Bates, D. R., and Miskelly, D. (1957). *Proc. Phys. Soc., London, Sect. A* **70**, 539.
Bates, D. R., and Tweed, R. J. (1974). *Proc. Phys. Soc., London (At. Mol. Phys.)* [2] **7**, 117.
Bates, D. R., Fundaminsky, A., Leech, J. W., and Massey, H. S. W. (1950). *Phil. Trans. Roy. Soc. London, Ser. A* **243**, 93.
Bates, D. R., Boyd, A. H., and Prasad, S. S. (1965). *Proc. Phys. Soc., London* **85**, 1121.
Bell, K. L., and Kingston, A. E. (1967). *Proc. Phys. Soc., London* **90**, 901.
Bell, K. L., and Kingston, A. E. (1969a). *Proc. Phys. Soc., London (At. Mol. Phys.)* [2] **2**, 653.
Bell, K. L., and Kingston, A. E. (1969b). *Proc. Phys. Soc., London (At. Mol. Phys.)* [2] **2**, 1125.
Bell, K. L., and Kingston, A. E. (1970). *Proc. Phys. Soc., London (At. Mol. Phys.)* [2] **3**, 1300.
Bell, K. L., and Kingston, A. E. (1971). *Proc. Phys. Soc., London (At. Mol. Phys.)* [2] **4**, 162.
Bell, K. L., Kennedy, D. J., and Kingston, A. E. (1968). *Proc. Phys. Soc., London (At. Mol. Phys.)* [2] **1**, 1037.
Bell, K. L., Kennedy, D. J., and Kingston, A. E. (1969a). *Proc. Phys. Soc., London (At. Mol. Phys.)* [2] **2**, 26.
Bell, K. L., Dose, V., and Kingston, A. E. (1969b). *Proc. Phys. Soc., London (At. Mol. Phys.)* [2] **2**, 831.
Bell, K. L., Dose, V., and Kingston, A. E. (1970a). *Proc. Phys. Soc., London (At. Mol. Phys.)* [2] **3**, 129.
Bell, K. L., Freeston, M. W., and Kingston, A. E. (1970b). *Proc. Phys. Soc., London (At. Mol. Phys.)* [2] **3**, 959.
Bell, K. L., Kingston, A. E., and Taylor, I. R. (1973a). *Proc. Phys. Soc., London (At. Mol. Phys.)* **6**, 1228.
Bell, K. L., Kingston, A. E., and McIlveen, W. I. (1973b). *Proc. Phys. Soc., London (At. Mol. Phys.)* [2] **6**, 1237.
Bell, K. L., Kingston, A. E., and McIlveen, W. I. (1973c). *Proc. Phys. Soc., London (At. Mol. Phys.)* [2] **6**, 1246.
Bell, R. J., and Skinner, B. G. (1962). *Proc. Phys. Soc., London* **80**, 404.
Berkner, K. H., Kaplan, S. N., and Pyle, R. V. (1964). *Phys. Rev.* **134**, A1461.

Berkner, K. H., Kaplan, S. N., Paulikas, G. A., and Pyle, R. V. (1965). *Phys. Rev.* **138**, A410.
Berrington, K. A., Bransden, B. H., and Coleman, J. P. (1973). *Proc. Phys. Soc., London (At. Mol. Phys.)* [2] **6**, 436.
Bethe, H. (1930). *Ann. Phys. (Leipzig)* [5] **5**, 325.
Birely, J. H., and McNeal, R. J. (1972). *Phys. Rev. A* **5**, 257.
Blair, W. G. F., and Gilbody, H. B. (1973). *Proc. Phys. Soc., London (At. Mol. Phys.)* [2] **6**, 483.
Blair, W. G. F., McCullough, R. W., Simpson, F. R., and Gilbody, H. B. (1973). *Proc. Phys. Soc., London (At. Mol. Phys.)* [2] **6**, 1265.
Boksenberg, A. (1961). Ph.D. Thesis, University of London.
Born, M. (1926). *Z. Phys.* **38**, 803.
Boyd, T. J. M. (1958). *Proc. Phys. Soc., London* **72**, 523.
Boyd, T. J. M., Moiseiwitsch, B. L., and Stewart, A. L. (1957). *Proc. Phys. Soc., London, Sect. A* **70**, 110.
Bransden, B. H., Coleman, J. P., and Sullivan, J. (1972). *Proc. Phys. Soc., London (At. Mol. Phys.)* [2] **5**, 546.
Carew, J., and Milford, S. N. (1963). *Astrophys. J.* **138**, 772.
Chamberlain, G. E., Mielczarek, S. R., and Kuyatt, C. E. (1970). *Phys. Rev. A* **2**, 1905.
Cheshire, I. M., and Kyle, H. L. (1965). *Phys. Lett.* **17**, 115.
Cheshire, I. M., and Kyle, H. L. (1966). *Proc. Phys. Soc., London* **88**, 785.
Crothers, D. S. F., and Holt, A. R. (1966). *Proc. Phys. Soc., London* **88**, 75.
de Heer, F. J., Schutten, J., and Moustafa Moussa, H. R. (1966). *Physica (Utrecht)* **32**, 1766.
de Jongh, J. P., and Van Eck, J. (1971). *Abstr., Int. Conf. Phys. Electron. At. Collision Phenomena, 7th, 1971* p. 701.
Denis, A., Dufay, M., and Gaillard, M. (1967). *C. R. Acad. Sci.* **264**, 440.
Dmitriev, I. S., Zhileikin, Ya. N., and Nikolaev, V. S. (1966). *Sov. Phys.—JETP* **22**, 352.
Donaldson, F. G., Hender, M. A., and McConkey, J. W. (1972). *Proc. Phys. Soc., London (At. Mol. Phys.)* [2] **5**, 1192.
Dose, V., Gunz, R., and Meyer, V. (1968). *Helv. Phys. Acta* **41**, 264.
Economides, D. G., and McDowell, M. R. C. (1969). *Proc. Phys. Soc., London (At. Mol. Phys.)* [2] **2**, 1323.
Enemark, E. A., and Gallagher, A. (1972). *Phys. Rev. A* **6**, 192.
Felden, M. M., and Felden, M. A. (1971). *Phys. Lett. A* **37**, 88.
Fisher, L., Milford, S. N., and Pomilla, F. R. (1960). *Phys. Rev.* **119**, 153.
Fite, W. L., and Brackmann, R. T. (1958a). *Phys. Rev.* **112**, 1141.
Fite, W. L., and Brackmann, R. T. (1958b). *Phys. Rev.* **112**, 1151.
Fite, W. L., and Brackmann, R. T. (1959). *Phys. Rev.* **113**, 815.
Fite, W. L., Stebbings, R. F., and Brackmann, R. T. (1959). *Phys. Rev.* **116**, 356.
Fite, W. L., Stebbings, R. F., Hummer, D. G., and Brackmann, R. T. (1960). *Phys. Rev.* **119**, 663.
Flannery, M. R. (1969a). *Phys. Rev.* **183**, 231.
Flannery, M. R. (1969b). *Proc. Phys. Soc., London (At. Mol. Phys.)* [2] **2**, 1044.
Flannery, M. R. (1970). *Proc. Phys. Soc., London (At. Mol. Phys.)* [2] **3**, 306.
Fletcher, J., and Cowling, I. R. (1973). *Proc. Phys. Soc., London (At. Mol. Phys.)* [2] **6**, L258.
Fox, M. A. (1965). *Proc. Phys. Soc., London* **86**, 789.
Frame, J. W. (1931). *Proc. Cambridge Phil. Soc.* **27**, 511.
Gaillard, M. (1966). *C. R. Acad. Sci., Ser. B* **263**, 549.
Ganas, P. S., and Green, A. E. S. (1971). *Phys. Rev. A* **4**, 182.
Gaspar, R. (1952). *Acta Phys.* **2**, 151.
Gaudin, A., and Hagemann, R. (1967). *J. Chim. Phys.* **64**, 1209.
Gilbody, H. B., and Ireland, J. V. (1964). *Proc. Roy. Soc., Ser. A* **277**, 137.
Gilbody, H. B., and Lee, A. R. (1963). *Proc. Roy. Soc., Ser. A* **274**, 365.
Gillespie, E. S. (1970). M.Sc. Thesis, Queen's University of Belfast, Belfast, Northern Ireland.

Grissom, J. T., Compton, R. N., and Garrett, W. R. (1972). *Phys. Rev. A* **6**, 977.
Hanne, F., and Kessler, J. (1972). *Phys. Rev. A* **5**, 2457.
Hasselkamp, D., Hippler, R., Scharmann, A., and Schartner, K. H. (1971). *Z. Phys.* **248**, 254.
Heideman, H. G. M., and Vriens, L. (1967). *J. Chem. Phys.* **46**, 2911.
Herman, F., and Skillman, S. (1963). "Atomic Structure Calculations." Prentice-Hall, Englewood Cliffs, New Jersey.
Hils, D., Kleinpoppen, H., and Koschmieder, H. (1966). *Proc. Phys. Soc., London* **89**, 35.
Holt, A. R., Hunt, J., and Moiseiwitsch, B. L. (1971). *Proc. Phys. Soc., London (At. Mol. Phys.)* [2] **4**, 1318.
Hughes, R. H., and Choe, S. S. (1972a). *Phys. Rev. A* **5**, 656.
Hughes, R. H., and Choe, S. S. (1972b). *Phys. Rev. A* **5**, 1758.
Hughes, R. H., and Hendrickson, G. G. (1964). *J. Opt. Soc. Amer.* **54**, 1494.
Hummer, D. G., and Seaton, M. J. (1961). *Phys. Rev. Lett.* **6**, 471.
Inokuti, M. (1971). *Rev. Mod. Phys.* **43**, 297.
Jalin, R., Hagemann, R., and Botter, R. (1973). *J. Chem. Phys.* **59**, 952.
Kauppila, W. E., Ott, W. R., and Fite, W. L. (1970). *Phys. Rev. A* **1**, 1099.
Kazaks, P. A., Ganas, P. S., and Green, A. E. S. (1972). *Phys. Rev. A* **6**, 2169.
Kennedy, D. J. (1968). Ph.D. Thesis, Queen's University of Belfast, Belfast, Northern Ireland.
Khandelwal, G. S. (1969). *Proc. Phys. Soc., London (At. Mol. Phys.)* [2] **2**, 151.
Khandelwal, G. S., and Choi, B. H. (1968a). *Proc. Phys. Soc., London (At. Mol. Phys.)* [2] **1**, 1218.
Khandelwal, G. S., and Choi, B. H. (1968b). *Proc. Phys. Soc., London (At. Mol. Phys.)* [2] **1**, 1220.
Khandelwal, G. S., and Choi, B. H. (1969). *Proc. Phys. Soc., London (At. Mol. Phys.)* [2] **2**, 308.
Khandelwal, G. S., and Fitchard, E. E. (1969). *Proc. Phys. Soc., London (At. Mol. Phys.)* [2] **2**, 1118.
Khandelwal, G. S., and Shelton, J. E. (1971). *Proc. Phys. Soc., London (At. Mol. Phys.)* [2] **4**, L109.
Kim, Y-K., and Inokuti, M. (1971). *Phys. Rev. A* **3**, 665.
Kingston, A. E. (1965). *Proc. Phys. Soc., London* **85**, 467.
Kingston, A. E., and Lauer, J. E. (1966a). *Proc. Phys. Soc., London* **87**, 399.
Kingston, A. E., and Lauer, J. E. (1966b). *Proc. Phys. Soc., London* **88**, 597.
Kleinpoppen, H., and Kraiss, E. (1968). *Phys. Rev. Lett.* **20**, 361.
Lassettre, E. N. (1970). *J. Chem. Phys.* **53**, 3801.
Lassettre, E. N., Krasnow, M. E., and Silverman, S. M. (1964). *J. Chem. Phys.* **40**, 1242.
Levy, H. (1969a). *Phys. Rev.* **184**, 97.
Levy, H. (1969b). *Phys. Rev.* **185**, 7.
Levy, H. (1969c). *Phys. Rev.* **187**, 130.
Levy, H. (1970). *Proc. Phys. Soc., London (At. Mol. Phys.)* [2] **3**, 1501.
Levy, H. (1971). *Phys. Rev. A* **3**, 1987.
Lichten, W., and Schultz, S. (1959). *Phys. Rev.* **116**, 1132.
Lloyd, C. R., Teubner, P. J. O., Weigold, E., and Hood, S. T. (1972). *Proc. Phys. Soc., London (At. Mol. Phys.)* [2] **5**, L44.
Lodge, J. G. (1969). *Proc. Phys. Soc., London (At. Mol. Phys.)* [2] **2**, 322.
Long, R. L., Cox, D. M., and Smith, S. J. (1968). *J. Res. Nat. Bur. Stand., Sect. A* **72**, 521.
McCarroll, R. (1957). *Proc. Phys. Soc., London, Sect. A* **70**, 460.
McCarroll, R., and Salin, A. (1966). *C. R. Acad. Sci. Ser. B* **263**, 329.
McConnell, J. C., and Moiseiwitsch, B. L. (1968). *Proc. Phys. Soc., London (At. Mol. Phys.)* [2] **1**, 406.
McCoyd, G. C., and Milford, S. N. (1963). *Phys. Rev.* **130**, 206.
McCoyd, G. C., Milford, S. N., and Wahl, J. J. (1960). *Phys. Rev.* **119**, 149.
McCrea, D., and McKirgan, T. V. M. (1960). *Proc. Phys. Soc., London* **75**, 235.

McDaniel, E. W. (1964). "Collision Phenomena in Ionized Gases," Chapter 6. Wiley, New York.
McDowell, M. R. C. (1969). "Case Studies in Atomic Collision Physics," Chapter 2. North-Holland Publ., Amsterdam.
McDowell, M. R. C., and Pluta, K. M. (1966). *Proc. Phys. Soc., London* **89**, 793.
McFarland, R. A., and Kinney, J. D. (1965). *Phys. Rev. A* **137**, 1058.
McGowan, J. W., and Clarke, E. M. (1968). *Phys. Rev.* **167**, 43.
McGuire, E. J. (1968). *Phys. Rev.* **175**, 20.
McGuire, E. J. (1971). *Phys. Rev. A* **3**, 267.
Mathur, K. C., Tripathi, A. N., and Joshi, S. K. (1971). *Phys. Lett. A* **35**, 139.
Mathur, K. C., Tripathi, A. N., and Joshi, S. K. (1972). *Phys. Rev. A* **6**, 1248.
May, R. M. (1965). *Phys. Lett.* **14**, 198.
May, R. M., and Butler, S. T. (1965). *Phys. Rev.* **138**, A1586.
Milford, S. N., Morrissey, J. J., and Scanlon, J. H. (1960). *Phys. Rev.* **120**, 1715.
Mittleman, M. H. (1963). *Phys. Rev.* **129**, 190.
Moiseiwitsch, B. L. (1957). *Mon. Not. Roy. Astron. Soc.* **117**, 189.
Moiseiwitsch, B. L. (1966). *Proc. Phys. Soc., London* **87**, 885.
Moiseiwitsch, B. L., and Smith, S. J. (1968). *Rev. Mod. Phys.* **40**, 238.
Moiseiwitsch, B. L., and Stewart, A. L. (1954). *Proc. Phys. Soc., London, Sect. A* **67**, 1069.
Mott, N. F., and Massey, H. S. W. (1965). "The Theory of Atomic Collisions," 3rd ed. Oxford Univ. Press, London and New York.
Moustafa Moussa, H. R., de Heer, F. J., and Schutten, J. (1969). *Physica (Utrecht)* **40**, 517.
Okuno, Y., Okuno, K., Kaneko, Y., and Kanomata, I. (1970). *J. Phys. Soc. Jap.* **29**, 164.
Oldham, W. J. B. (1967). *Phys. Rev.* **161**, 1.
Oldham, W. J. B. (1969). *Phys. Rev.* **181**, 463.
Omidvar, K. (1965a). *Phys. Rev.* **140**, A26.
Omidvar, K. (1965b). *Phys. Rev.* **140**, A38.
Omidvar, K. (1969). *Phys. Rev.* **188**, 140.
Omidvar, K., and Khateeb, A. H. (1973). *Proc. Phys. Soc., London (At. Mol. Phys.)* [2] **6**, 341.
Omidvar, K., and Kyle, H. L. (1970). *Phys. Rev. A* **2**, 408.
Omidvar, K., Kyle, H. L., and Sullivan, E. C. (1972). *Phys. Rev. A* **5**, 1174.
Opal, C. B., Beaty, E. C., and Peterson, W. J. (1971). Report No. 108. J.I.L.A., University of Colorado, Boulder.
Orbeli, A. L., Andreev, E. P., Ankudinov, V. A., and Dukelskii, V. M. (1970). *Sov. Phys.—JETP* **30**, 63.
Ott, W. R., Kauppila, W. E., and Fite, W. L. (1970). *Phys. Rev. A* **1**, 1089.
Park, J. T., and Schowengerdt, F. D. (1969). *Phys. Rev.* **185**, 152.
Peach, G. (1965). *Proc. Phys. Soc. London* **85**, 709.
Peach, G. (1968). *Proc. Phys. Soc., London (At. Mol. Phys.)* [2] **1**, 1088.
Peach, G. (1970). *Proc. Phys. Soc., London (At. Mol. Phys.)* [2] **3**, 328.
Peach, G. (1971). *Proc. Phys. Soc., London (At. Mol. Phys.)* [2] **4**, 1670.
Pedersen, E. H., and Hvelplund, P. (1971). *Int. Conf. Phys. Electron. At. Collision Phenomena, 7th, 1971 Abstr.* p. 522.
Peresse, J., Le Nadan, A., and Pochat, A. (1970a). *C. R. Acad. Sci., Ser. B* **271**, 31.
Peresse, J., Le Nadan, A., and Gelebart, F. (1970b). *C. R. Acad. Sci., Ser. B* **271**, 100.
Peresse, J., Pochat, A., and Gelebart, F. (1972). *Phys. Lett. A* **41**, 135.
Peterson, J. R. (1964). *In* "Atomic Collision Processes" (M. R. C. McDowell, ed.), p. 465. North-Holland Publ., Amsterdam.
Pivovar, L. I., and Levchenko, Y. Z. (1967). *Sov. Phys.—JETP* **25**, 27.
Pomilla, F. R. (1967). *Astrophys. J.* **148**, 559.
Pomilla, F. R., and Milford, S. N. (1966). *Astrophys. J.* **144**, 1174.

Puckett, L. J., Taylor, G. O., and Martin, D. W. (1969). *Phys. Rev.* **178**, 271.
Rapp, D., and Golden, P. (1965). *J. Chem. Phys.* **43**, 1464.
Rice, J. K., Truhlar, D. G., Cartwright, D. C., and Trajmar, S. (1972). *Phys. Rev. A* **5**, 762.
Robinson, J. M., and Gilbody, H. B. (1967). *Proc. Phys. Soc., London* **92**, 589.
Rothe, E. W., Marino, L. L., Neynaber, R. H., and Trujillo, S. M. (1962). *Phys. Rev.* **125**, 582.
Rudd, M. E., and Jorgensen, T., Jr. (1963). *Phys. Rev.* **131**, 666.
Rudd, M. E., Sautter, C. A., and Bailey, C. L. (1966). *Phys. Rev.* **151**, 20.
Rudge, M. R. H., and Schwartz, S. B. (1966). *Proc. Phys. Soc., London* **88**, 563.
Rudge, M. R. H., and Seaton, M. J. (1965). *Proc. Roy. Soc., Ser. A* **283**, 262.
Sawada, T., Purcell, J. E., and Green, A. E. S. (1971). *Phys. Rev. A* **4**, 193.
Scanlon, J. H., and Milford, S. N. (1961). *Astrophys. J.* **134**, 724.
Schram, B. L. (1966). Ph.D. Thesis, University of Amsterdam, Holland.
Silverman, S., and Lassettre, E. N. (1964). *J. Chem. Phys.* **40**, 1265.
Simsic, P., and Williamson, W. (1972). *J. Chem. Phys.* **57**, 4617.
Skerbele, A., and Lassettre, E. N. (1973). *J. Chem. Phys.* **58**, 2887.
Sloan, I. H. (1965). *Proc. Phys. Soc., London* **85**, 435.
Smith, A. C. H., Caplinger, E., Neynaber, R. H., and Trujillo, S. M. (1962). *Phys. Rev.* **127**, 1647.
Smith, P. T. (1930). *Phys. Rev.* **36**, 1293.
Smythe, R., and Toevs, J. W. (1965). *Phys. Rev.* **139**, A15.
Solov'ev, E. S., Il'in, R. N., Oparin, V. A., and Fedorenko, N. V. (1962). *Sov. Phys.—JETP* **15**, 459.
Somerville, W. B. (1963). *Proc. Phys. Soc., London* **82**, 446.
Starostin, A. N. (1967). *Sov. Phys.—JETP* **25**, 80.
Stebbings, R. F., Young, R. A., Oxley, C. L., and Ehrhardt, H. (1965). *Phys. Rev.* **138**, A1312.
Stier, P. M., and Barnett, C. F. (1956). *Phys. Rev.* **103**, 896.
Stone, E. J., and Zipf, E. C. (1971). *Phys. Rev. A* **4**, 610.
Tai, H., Bassel, R. H., Gerjuoy, E., and Franco, V. (1970). *Phys. Rev. A* **1**, 1819.
Thomas, E. W., and Bent, G. (1967). *Phys. Rev.* **164**, 143.
Toburen, L. H., Nakai, M. Y., and Langley, R. A. (1968). *Phys. Rev.* **171**, 114.
Tripathi, A. N., Mathur, K. C., and Joshi, S. K. (1972). *Phys. Rev. A* **7**, 109.
Truhlar, D. G., Rice, J. K., Kuppermann, A., Trajmar, S., and Cartwright, D. C. (1970). *Phys. Rev. A* **1**, 778.
Vainshtein, L. A. (1961). *Opt. Spectrosc. (USSR)* **11**, 163.
Vainshtein, L. A. (1965). *Opt. Spectrosc. (USSR)* **18**, 538.
Vainshtein, L. A., Opykhtin, V., and Presnyakov, L. (1964). P. N. Lebedev Institute of Physics Report A-53 (see Moiseiwitsch and Smith, 1968, for revised data).
Vainshtein, L. A., Opykhtin, V., and Presnyakov, L. (1965). *Sov. Phys.—JETP* **20**, 1542 (see Moiseiwitsch and Smith, 1968, for revised data).
Van den Bos, J. (1969). *Phys. Rev.* **181**, 191.
Van den Bos, J., Winter, G. J., and de Heer, F. J. (1968). *Physica (Utrecht)* **40**, 357.
Vanderpoorten, R. (1970). *Physica (Utrecht)* **48**, 254.
Van Eck, J., and de Jongh, J. P. (1970). *Physica (Utrecht)* **47**, 141.
Van Eck, J., de Heer, F. J., and Kistemaker, J. (1964). *Physica (Utrecht)* **30**, 1171.
Van Raan, A. F. J., de Jongh, J. P., Van Eck, J., and Heideman, H. G. M. (1971). *Physica (Utrecht)* **53**, 45.
Veldre, V. Ya., and Rabik, L. L. (1965). *Opt. Spectrosc. (USSR)* **19**, 265.
Veldre, V. Ya., Lyash, A. V., and Rabik, L. L. (1965a). *In* "Atomic Collisions III" (V. Ya. Veldre, ed.), p. 85. Akad. Nauk. Latv. SSR Inst. Fiz., Riga (English Translation: J.I.L.A. Inform. Cent. Rep. No. 3, p. 73. University of Colorado, Boulder).
Veldre, V. Ya., Lyash, A. V., and Rabik, L. L. (1965b). *Opt. Spectrosc. (USSR)* **19**, 182.
Vinogradov, A. V. (1967). *Opt. Spectrosc. (USSR)* **22**, 361.

Vriens, L., Simpson, J. A., and Mielczarek, S. R. (1968). *Phys. Rev.* **165**, 7.
Weise, W. L., Smith, M. W., and Miles, B. M. (1969). "Atomic Transition Probabilities—Sodium through Calcium," U.S. Nat. Bur. Stand.—Nat. Stand. Ref. Ser. 22, Vol. II, p. 2. US Govt. Printing Office, Washington, D.C.
Welsh, L. M., Berkner, K. H., Kaplan, S. N., and Pyle, R. V. (1967). *Phys. Rev.* **158**, 85.
Williamson, W. (1970). *J. Chem. Phys.* **53**, 1944.
Woollings, M. J., and McDowell, M. R. C. (1972). *Proc. Phys. Soc., London* (*At. Mol. Phys.*) [2] **5**, 1320.
Woollings, M. J., and McDowell, M. R. C. (1973). *Proc. Phys. Soc., London* (*At. Mol. Phys.*) [2] **6**, 450.
Zapesochnyi, I. P., and Aleksakhin, I. S. (1969). *Sov. Phys.—JETP* (English Transl.) **28**, 41.
Zapesochnyi, I. P., and Shimon, L. L. (1964). *Opt. Spectrosc.* (*USSR*) **16**, 504.
Zapesochnyi, I. P., and Shimon, L. L. (1965). *Opt. Spectrosc.* (*USSR*) **19**, 268.
Zapesochnyi, I. P., and Shimon, L. L. (1966a). *Opt. Spectrosc.* (*USSR*) **20**, 421.
Zapesochnyi, I. P., and Shimon, L. L. (1966b). *Opt. Spectrosc.* (*USSR*) **20**, 525.
Zapesochnyi, I. P., and Shimon, L. L. (1966c). *Opt. Spectrosc.* (*USSR*) **21**, 155.
Zapesochnyi, I. P., Shimon, L. L., and Soshnikov, A. K. (1965). *Opt. Spectrosc.* (*USSR*) **19**, 480.

PHOTOELECTRON SPECTROSCOPY

W. C. PRICE

King's College
London, England

I. Introduction 131
   A. Introductory Survey 131
   B. Technical Development 133
II. X-Ray Photoelectron Spectroscopy 134
III. Ultraviolet Photoelectron Spectroscopy 136
   A. General Features 136
   B. Photoelectron Spectra of Hydrids Isoelectronic with the Inert Gases ... 139
   C. Photoelectron Spectra of the Halogen Derivatives of Methane 143
   D. Photoelectron Spectra of "Multiple" Bonded Diatomic Molecules 144
   E. Spectra and Structure of Some "Multiple" Bonded Polyatomic Molecules 149
   F. Spectra and Structure of Triatomic Molecules 151
   G. Methods of Orbital Assignment of Bands in Photoelectron Spectra 154
IV. Physical Aspects of Photoelectron Spectroscopy 155
   A. The Intensities of Photoelectron Spectra 155
   B. Selection Rules 157
   C. Dependence of Cross Section on Photoelectron Energy 157
   D. Autoionization 159
   E. Angular Distribution of Photoelectrons 164
   F. Electron–Molecule Interactions 167
   G. Transient Species 168
V. Conclusion 169
   References 170

I. Introduction

A. INTRODUCTORY SURVEY

Within the last decade it has become clear that one of the most successful ways of investigating the nature of the electron configurations of atomic systems is to use monoenergetic photons to eject the electrons from their orbitals and to measure the kinetic energy and other properties of the photoelectrons produced by such monochromatic irradiation. If the energy which an electron acquires in this process is $\frac{1}{2}mv^2$, then the binding or ionization energy of the particular orbital from which it was ejected is given by

$$E = h\nu - \tfrac{1}{2}mv^2 \qquad (1)$$

While in certain simple cases such as isolated atoms and a few molecules, orbital ionization energies can be obtained by conventional (photon) spectroscopy by extrapolating excited states in Rydberg series to the ionization limit, this has not proved to be a technique which can be generally applied. This is because Rydberg states are frequently diffuse or distorted by admixture with valence states of the same symmetry, and, especially in the case of excitation from inner orbitals, the absorption bands corresponding to excited states of polyatomic systems become broad and indeterminate. When, however, the electron is excited into the ionization continuum as in photoelectron spectroscopy, many of these disturbing effects can no longer operate and a photoelectron spectrum can be obtained which reveals the orbital structure of the system in a very direct and readily understandable manner. For this reason photoelectron spectroscopy has been taken up with great enthusiasm by chemists and physicists alike. The literature on the subject has expanded enormously and a *Journal of Electron Spectroscopy* has been established which is largely devoted to this field. The proceedings of three conferences and five books dealing with various aspects of photoelectron spectroscopy have been published (see Reference list).

In order to use Eq. (1) to obtain acceptable values of orbital ionization energies it is necessary to measure the kinetic energy of the photoelectron with a precision comparable with optical spectroscopy. Only in recent years has this become possible as a result of the development of sensitive electron detectors and improved electron energy analyzers (both electrostatic and electromagnetic) which are capable of giving high accuracy and good resolution. The reward for these developments has been that, by using the methods of photoelectron spectroscopy, it has become possible to pick out an electron from each orbital of a molecule or a solid—whether it is an inner (core) or outer (valence shell) electron—and to determine accurately the energy with which it is bound into the system. Many other features which reveal the function of the electron in the electronic structure of the system also appear in the photoelectron spectrum. It was previously not possible to obtain this information for inner electrons by optical spectroscopy except in a few simple cases because of the interaction of their excited states with the ionization continua of the outer electrons. By using a high energy photon the energy of which did not coincide with that of a super excited bound state, it was found that the inner electron could be ejected past the outer electrons and that the interaction was largely eliminated. For example, it had been possible by conventional spectroscopy to study only the outer two nonbonding electrons in the water molecule, and it was not until the advent of photoelectron spectroscopy that any detailed information became available on the other eight electrons in this relatively simple and vitally important molecule.

B. Technical Development

Photoelectron spectroscopy is conveniently divided into the two major fields in which it has been developed over the last decade. The first uses X-ray irradiation and is largely concerned with core electrons. It was pioneered by Siegbahn and his co-workers at Uppsala, who used their experience on β-ray spectrometry to make the necessary technical development. Extensive details of the work of this school are given in two major publications (Siegbahn et al., 1967, 1969). The second field employs ultraviolet radiation, and the earliest experiments were carried out by Vilesov et al. (1961, 1962) who used monochromatized radiation from a hydrogen discharge as a photon source and a retarding potential analyzer to measure the photoelectron energies. As they employed a lithium fluoride window to isolate the monochromator, it was not possible to use photons of energies greater than 11 eV and consequently only very limited exploration of orbital structure was possible. In 1962, Turner and Al-Joboury (1962, 1963) used the strong helium resonance line as a source of high energy photons (21.2 eV) together with differential pumping to avoid the limitations imposed by the lithium fluoride window. They were thus able to photoionize electrons from most of the valence orbitals of common molecules and the work of Turner and his associates (1970) quickly established the field of molecular photoelectron spectroscopy on a firm basis. The range was later extended by Price and collaborators who employed the shorter wavelength line of ionized helium (30.4 nm, 40.83 eV) as a photon source which gave deeper penetration into the valence shell (Price, 1968; Potts et al., 1970). The X-ray and ultraviolet irradiation techniques are conveniently abbreviated as XPS and UPS, respectively. While the former gives information on all electrons, the latter gives much more highly resolved spectra for the more loosely bound outer electrons since the helium linewidth (about 0.001 eV) is less than that of the X-ray lines (about 0.5 eV) used by Siegbahn. Further, because the photoelectron velocity is much lower in UPS than XPS, the precision of its determination is correspondingly greater, and more highly resolved spectra often showing detailed vibrational structure are obtained when ultraviolet photons are used.

A schematic diagram of the equipment used in photoelectron spectroscopy is shown in Fig. 1. The energy spectrum of the photoelectrons emitted by the sample is obtained by plotting the count rate of electrons arriving at the detector against a continuously increasing energy setting of the electron spectrometer. Thus photoelectron energy increases from left to right along the abscissa and binding energy increases from right to left. This presentation is used in all published XPS spectra and in some UPS spectra. There is an increasing tendency in the latter case, however, to plot binding energy

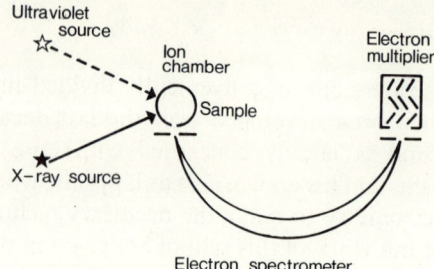

Fig. 1. Schematic diagram of photoelectron spectrometer with ultraviolet or X-ray source.

(ionization potential) increasing from left to right, since the energy scale is always calibrated against gases of spectroscopically known ionization potential. This occurs because the photoelectron energy is affected by contact potentials and other factors and its absolute value cannot be determined accurately from the voltages applied to the analyzer.

II. X-Ray Photoelectron Spectroscopy

While this review is intended to deal mainly with ultraviolet photoelectron spectroscopy, the brief account of X-ray photoelectron spectroscopy which follows is necessary to emphasize the relationship between the two and to discuss common features in subsequent sections. Most X-ray photoelectron spectroscopy has used the soft $K\alpha$ X-ray lines of Mg or Al (1254 and 1487 eV, respectively) because these are narrower than, for example, the lines of Cu $K\alpha$ and they give photoelectrons whose energies are lower and can therefore be measured more accurately. These lines permit examination of the electronic structure down to 1000 eV which covers the 1s shells of the light atoms. The XPS lines from the K shells of the atoms in the second period of the periodic table are shown in Fig. 2. Their binding energies increase from lithium (55 eV) to neon (867 eV), the differences between successive elements gradually increasing from 56 eV (Li–Be) to 181 eV (F–Ne). The photoelectron lines are thus well separated from one another and can be used as a method of identifying the elements present in the sample. For the heavier atoms the binding energies of electrons in L, M, and higher inner shells can be used for this purpose and thus all the elements of the periodic table can be identified by X-ray photoelectron spectroscopy. It was for this reason that Siegbahn called his method ESCA, the letters standing for "electron spectroscopy for chemical analysis." However, this is a rather restrictive title in view of the wider use now being made of the spectra.

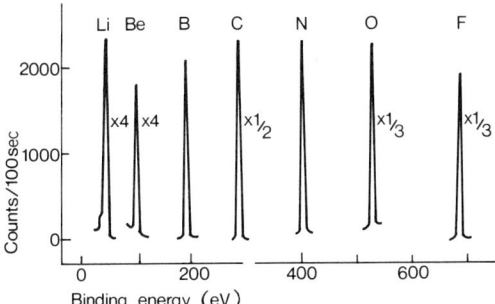

FIG. 2. Photoelectron lines from the K shells of elements of the second period when irradiated with Al Kα plotted against binding energy (i.e., 1487 eV minus the photoelectron energy.)

Although the binding energy of an electron in an inner shell is mainly controlled by the charge Ze within the shell, it is also affected to a lesser extent by the cloud of valence electrons lying above it. The XPS line of an element is found to vary over a few electron volts according to the state of chemical combination of the atom and the nature of its immediate neighbors. This is illustrated in the spectra of acetone $(CH_3)_2CO$, ethyl trifluoroacetate $CF_3CO_2C_2H_5$, and sodium azide, shown in Fig. 3. The C(1s) line of the two methyl carbons in acetone is 2–3 eV lower in binding energy than that of the more positive carbonyl carbon and is, as might be expected from the number of C atoms, twice as strong. Similar remarks apply to the two outer and the inner N(1s) electrons of the azide ion. The binding energies of the four different carbons in ethyl trifluoroacetate similarly reflect the different environments of the four carbon atoms in this molecule (note that the formula in the diagram is placed so that the lines lie below the carbon atoms with which they are associated). These "chemical shifts" have been related to the electronegativity difference of the atom from its neighbors and provide a means of estimating the effective charge on the

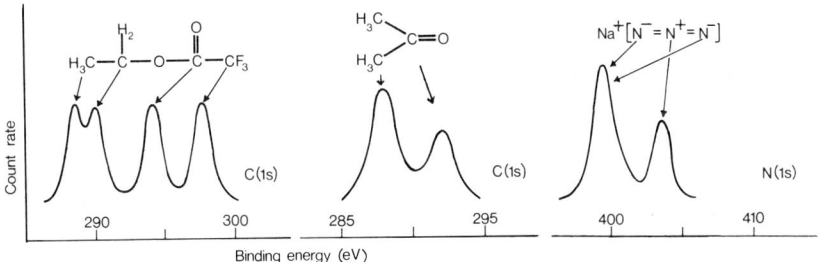

FIG. 3. Spectra of C (1s) core electrons in acetone ethyl trifluoroacetate and the N (1s) electrons in sodium azide.

core of each atom. The method has been extended to complicated molecules such as vitamin B_{12} where, using $Al(K\alpha)$, the photoelectron line of one cobalt atom could be studied in the presence of the lines of 180 other atoms. Similarly both the sulfur and the iron electron line in the anzyme cytochrome c can be recorded. It is clear that the technique is a powerful direct method of studying chemical electronic structure.

In addition to the work on mainly covalent materials, ionic and metallic solids have been studied. For metals the conduction band structure can be found and the photoelectron energy distribution curves compared with calculated densities of states. Because of the small depth of penetration of the radiation and the necessity for the photoelectrons to emerge from the material the method is clearly a surface effect. Depths down to at most 2 nm (20 Å) are involved in the process and ultrahigh vacuum techniques must be employed to eliminate the effects of surface contamination. Surface properties such as are important in catalytic and adsorption processes have received considerable attention from workers in the field. It should be mentioned that although XPS can produce some valuable information about the electrons in the valence shell, the cross sections for this radiation of valence shell orbitals are almost an order of magnitude less than those of core orbitals. For example, in hydrocarbons the cross sections of $C(2s)$ orbitals are about a tenth those of $C(1s)$ orbitals, and the cross sections of orbitals built from $C(2p)$ and $H(1s)$ are still lower by almost another order or magnitude. This is a limitation not suffered by UPS, where the cross sections particularly of 2p orbitals are generally appreciably higher than those of 2s orbitals. This fact is likely to seriously affect the prospects of XPS for detailed outer electron studies.

III. Ultraviolet Photoelectron Spectroscopy

A. General Features

Ultraviolet photoelectron spectroscopy using line sources obtained from discharges in helium or the other inert gases has produced an enormous amount of new detailed information on virtually all types of molecule and chemical species. This has been such an exciting time for chemists that nearly all the effort has so far been absorbed in studying these substances as gases. Its application to the solid phase, which involves greater experimental and interpretative difficulties, has scarcely begun though there have been some interesting studies by UPS on the conduction bands of metals such as gold in the solid and liquid phases (Eastman, 1971). Apart from the fact that the inner valence electron orbitals of diatomic and polyatomic molecules which cannot be studied by conventional spectroscopy appear directly as features in the photoelectron spectrum an additional attraction of the technique is that the UPS record itself reveals in a very convincing and readily understandable manner the orbital shell structure of the electrons in a molecule.

For monatomic gases only electronic energy is involved in photoionization and the He I photoelectron spectrum of atomic hydrogen (obtained by passing the products of a discharge in H_2 into the ion chamber) gives a single sharp peak corresponding to group of photoelectrons of energy 7.62 eV. This corresponds to the difference in energy of the helium resonance line (21.22 eV) and the ionization energy of the H(1s) electron (13.60 eV). For argon the outer electrons are in p-type orbitals and the $(3p)^5$ configuration of the ion has the two states $^2P_{3/2}$ and $^2P_{1/2}$ with statistical weights in the ratio 2 to 1 and energies differing by spin–orbit coupling. The photoelectron spectrum (see Fig. 4) corresponding to this $(3p)^{-1}$ ionized

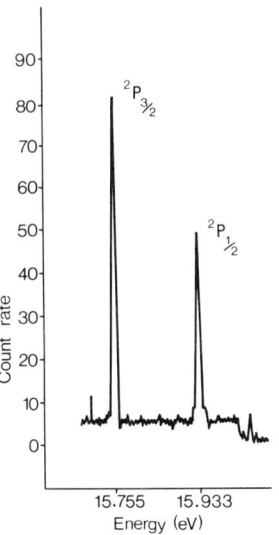

FIG. 4. Photoelectron spectrum of argon ionized by Ne (736 Å)

state is thus a doublet, one component of which is twice as strong as the other. Similar doublets are found for the other inert gases, the spin–orbit coupling increasing with their atomic weight. Narrow line spectra corresponding to ionization of atoms to the lower states of their ions (i.e., ionization from each of the outer orbitals) have also been obtained for N, O, F, Cl, Br, I, Hg, Cd, and Zn (Jonathan et al., 1972a, b; Eland, 1970; H. J. Lempka, unpublished, 1973).

For the ejection of an electron from an orbital in a molecule we have the equation

$$I_0 + E_{vib} + E_{rot} = hv - \tfrac{1}{2}mv^2$$

where I_0 is the "adiabatic" ionization energy (i.e., the pure electronic energy change) and E_{vib} and E_{rot} are the changes in vibrational and rotational energy which accompany the photoionization. To explain the nature of the information which can be obtained we shall discuss the photoelectron spectra of some simple molecules.

Figure 5b gives the potential energy curves of H_2 and H_2^+. When an electron is removed from the neutral molecule the nuclei find themselves

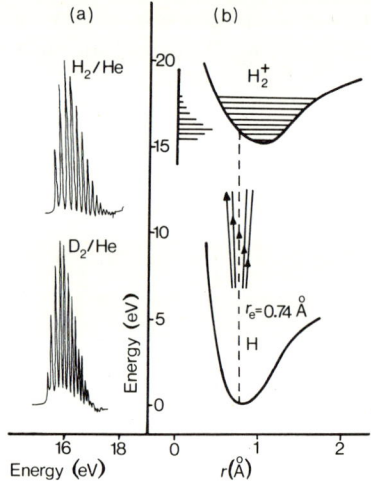

FIG. 5. (a) Photoelectron spectra of H_2 and D_2; (b) potential energy curves showing spectra plotted along ordinate.

suddenly in the potential field appropriate to the H_2^+ ion but still separated by the distance characteristic of the neutral molecule. The most probable change is thus a transition on the potential energy diagram from the internuclear separation of the ground state to a point on the potential energy curve of the ion vertically above this. This is the Franck–Condon principle and it determines to which vibrational level of the ion the most probable transition (strongest band) occurs. The energy corresponding to this change is called the vertical ionization potential I_{vert}. Transitions to vibrational levels on either side of I_{vert} are weaker, the one of lowest energy corresponding to the vibrationless state of the ion corresponding to I_{adiab}. The photoelectron spectra of H_2 and D_2 are given in Fig. 5a and that of H_2 is also plotted along the ordinate of Fig. 5b. It is worth noting that because one of the two bonding electrons is removed, the bonding energy at this internuclear distance should be halved. Thus I_{vert} should be equal to $I(H) + \frac{1}{2}D(H_2) = 13.595 + \frac{1}{2}(4.478) = 15.834$ eV. This agrees as closely as I_{vert} the maximum of the band envelope can be estimated for H_2. However, in other molecules, e.g., HCl the agreement is not so good due to compensating movements in other shells, i.e., deviations from Koopmans' theorem which assumes that the orbital binding energies of electrons are equal to the negative vertical ionization potentials (Koopmans, 1934).

It is possible to calculate the change in internuclear distance on ionization from the intensity distribution of the bands in the photoelectron spectrum. Clearly when a bonding electron is removed, the part of the photoelectron spectrum corresponding to this will show wide vibrational structure with a frequency separation that is reduced from that of the ground state vibration. The removal of relatively nonbonding electrons, on the other hand, will give rise to photoelectron spectra rather similar to those of the monatomic gases and little if any vibrational structure will accompany the main electronic band. The type of vibration associated with the pattern obtained when a bonding electron is removed can usually be identified as either a bending or a stretching mode and this can throw light on the function of the electron in the structure of the molecule, that is, either as angle forming or distance determining. From the band pattern it is frequently possible to calculate values of the changes in angle as well as changes in internuclear distance on ionization, and so the geometry of the ionic states can be found if that of the neutral molecule is known. Changes in bond lengths or angles can be calculated with about 10% accuracy for photoelectron band systems which show structure by using the semiclassical formula

$$(\Delta r)^2 = l^2(\Delta \Theta)^2 = 0.543[I_{vert} - I_{adiab}]\mu^{-1}\omega^{-2} \quad (2)$$

where r and l are in Å, Θ in radians, μ in atomic units, and ω the mean progression spacing in 1000 cm^{-1}.

B. Photoelectron Spectra of Hydrids Isoelectronic with the Inert Gases

As examples of the features mentioned above and also to illustrate how the orbitals of isoelectronic systems are related to one another we shall now discuss the photoelectron spectra of those simple hydrids for which the united atoms are the inert gases Ne, Ar, Kr, and Xe—that is, the atoms obtained by condensing all the nuclei into one central positive charge.

Hydrids such as HF, H_2O, NH_3, and CH_4 can be thought of as being formed from Ne by successively partitioning off protons from its nucleus. The solutions of the wave equation—the wave functions or orbitals—arising from the progressive subdivision of the positive charge pass continuously from the atomic to the molecular cases with a gradual splitting of the p^6-orbital degeneracy as the internal molecular fields are set up. Schematic diagrams of the orbital changes are indicated in Fig. 6. The photoelectron

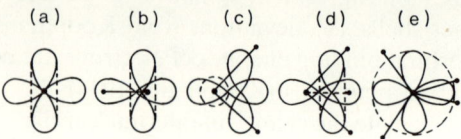

FIG. 6. Schematic diagrams of 2p-orbital structures of (a) Ne, (b) HF, (c) H_2O, (d) NH_3, and (e) CH_4.

spectra shown in Fig. 7a,b,c, and d reveal this orbital subdivision in a striking way. Figure 7a shows the spectra of the halogen acids which is the first stage in this process. The spectra of the corresponding inert gas (i.e., the united atom) is inserted on the records for comparison. The triply degenerate p^6 shell is split into a doubly degenerate π^4 shell and a singly degenerate σ^2 shell. The π^4 shell is nonbonding and represented by two bands in the photoelectron spectrum which are of equal intensity and show little vibrational structure. The σ shell is formed from the p orbital along which the proton is extracted and gives a negative cloud which binds the proton to the residual positive charge. This pσ orbital therefore gives rise in the photoelectron spectrum to a simple progression of bands with a separation corresponding to the vibration frequency of the $^2\Sigma^+$ state of the ion except where the structure is lost by predissociation.

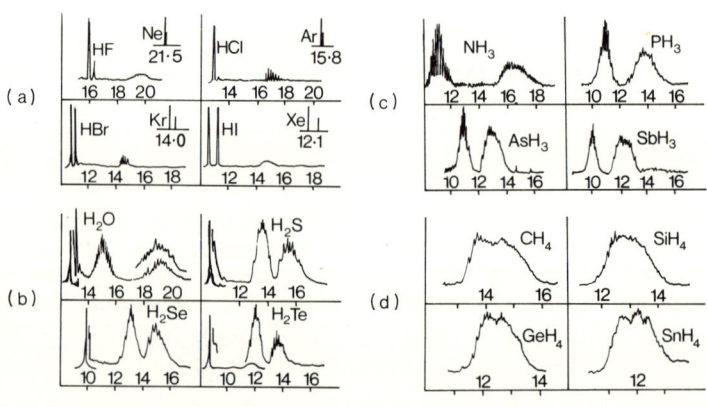

FIG. 7. Photoelectron spectra of the hydrids isoelectronic with Ne, Ar, Kr, and Xe obtained with He (21.22 eV) radiation.

The second stage of partition leads to the molecules H_2O, H_2S, H_2Se, and H_2Te. If two protons were removed colinearly from the united atom nucleus the linear triatomic molecule so formed would not have maximum stability, since only the two electrons in the p orbital lying along the line will then be effective in shielding the protons from the repulsion of the core. By moving off this line, the protons can acquire additional shelter from the central charge and from themselves through shielding by one lobe of a perpendicular p orbital (see Fig. 6c). The electrons in this orbital then become "angle determining" as distinct from the two previously mentioned p electrons which are mainly effective in determining the bond separations of the hydrogen atoms from the central atom. The remaining p orbital, because of its perpendicular orientation to the plane of the bent H_2X molecule, can affect neither the bonding nor the angle, that is, its electrons are nonbonding. These expectations are borne out by the photoelectron spectra shown in Fig. 7b. These spectra show how in all H_2X molecules the triple degeneracy of the p^6 shell of the united atom is completely split into three mutually perpendicular orbitals of different ionization energies. The lowest ionization band is sharp with little vibrational structure. The second has a wide vibrational pattern which turns out to be the bending vibration of the molecular ion. The third also has a wide vibrational pattern with band separations greater than those of the second band. These separations can be identified as the symmetrical bond stretching vibrations of the ion. The geometries of these three ionized states can be calculated from the band envelopes and pattern spacings. Only small changes are associated with the band of lowest IP. Large changes of angle accompany ionization in the second band, which in the case of H_2O causes the equilibrium configuration of this ionized state to be linear. In H_2S, H_2Se, and H_2Te the spectra show that the angles are 129°, 126°, and 124°, respectively. The third ionized state is one in which the internuclear distances are increased but little change occurs in the bond angle. The integrated intensities of all three bands are roughly equal indicating that each originates from the ionization of a single orbital.

The partitioning of three protons from the nucleus of an inert gas molecule leads to a pyramidal XH_3 molecule. One p orbital is directed along the axis of symmetry and provides the shielding which causes the molecule to have a nonplanar geometry, that is, it is angle determining. The other two p orbitals are degenerate and provide an annular cloud of negative charge passing through the three XH bonds and thus mainly determine the bond distances. The structure pattern on the first band of the photoelectron spectrum (Fig. 7c) can be assigned to bending vibrations and indicates that the molecule flies to a symmetrical planar configuration without much change in the bond distances when an electron is ionized from this orbital. The second band in the photoelectron spectrum shows Jahn–Teller splitting which is consistent with its being doubly degenerate. Although it has limited structure, such structure as can be observed corresponds to changes in bond stretching without much change in bond angle.

Finally the photoelectron spectra of the XH_4-type molecules show that the p^6 shell of the united atom has changed to another triply degenerate shell with orbitals of tetrahedral symmetry. The contour shows the presence of Jahn–Teller splitting, and the structure on the low-energy side shows that on ionization the molecule moves by contraction along one side of the enveloping cube toward a square coplanar configuration. The structure on the high-energy side indicates movement in the opposite direction toward two mutually perpendicular configurations in which opposite XH_2 angles are roughly 90°. In the heavier molecules, for example, SnH_4, the large spin–orbit splitting of the heavy atom influences the structure. The spectra also show that the movement toward coplanarity is progressively less pronounced. Further details on the spectra of these hydride molecules are given by Potts and Price (1972a,b).

It should be mentioned that in the process of partitioning off protons from the neon nucleus to form the second row hydrids, the 2s orbitals are distorted though to a lesser extent than the 2p orbitals. They are deformed from spherical symmetry in the direction of the extracted proton. They thus have "bumps" in the bond directions and in this way provide an "s" contribution to the bond. This is evident in the vibrational structure of the photoelectron bands associated with these orbitals. Figure 8 shows how the orbitals of the hydrides of F, N, O, and C are formed by proton withdrawal from neon.

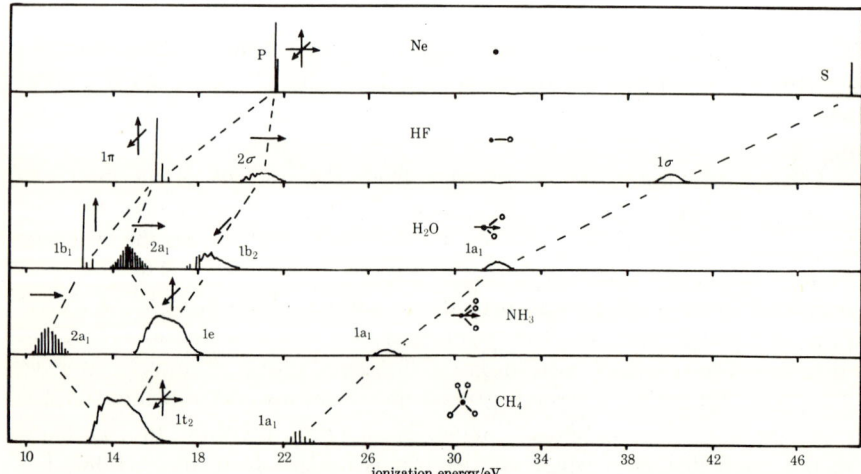

FIG. 8. Diagrammatic photoelectron spectra of hydrids formed by proton withdrawal from neon.

C. Photoelectron Spectra of the Halogen Derivatives of Methane

It is convenient to discuss at this point the photoelectron spectra of some "single bond" molecules in which some of the atoms have additional groups of nonbonding electrons. The bromomethanes, whose photoelectron spectra are illustrated in Fig. 9, are good examples of this molecular type. It can be

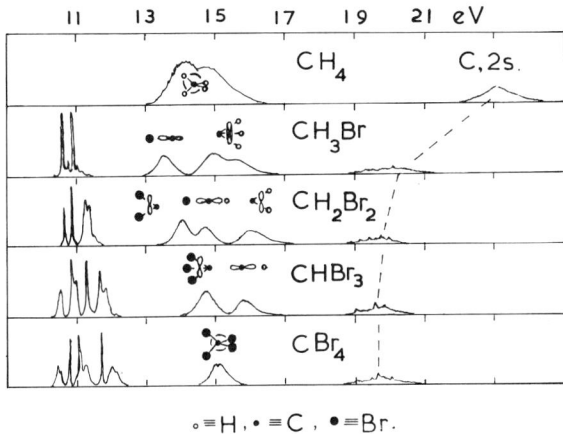

FIG. 9. Photoelectron spectra of the bromomethanes.

seen that some bands in their spectra are relatively sharp and therefore can be associated with the nonbonding electrons, while others are broad and clearly arise from electrons in strongly bonding orbitals. The orbital assignment of the latter bands are indicated schematically in the figure. A comparison of these bands with those of the hydrides of the same symmetry, e.g., CH_3Br and NH_3, CH_2Br_2, and H_2O, etc., shows that the orbitals around the carbon atom are split up in a very similar way as in the hydrids by the departure from tetrahedral symmetry brought about by the halogen substitution. This gives some support to the old concept that around an atom in a stable molecule there should be a closed (inert gas) shell of electrons. It shows further how the degeneracy of the p^6 group of these electrons is split by the fields arising from the different substituents, the splitting of the degeneracy being complete in the case of the methylene halide as it is for water in the hydrids. The sharp bands in the region of 11 eV can be readily associated with orbitals containing $4p\pi$ Br electrons which are split by spin–orbit interaction. In the case of methyl bromide two sharp bands (with very weak accompanying vibrational structure) are split by the magnitude of spin–orbit coupling constant which is 0.32 eV for bromine. Further splitting occurs as the number of bromine atoms increases and eight bands can be seen in CBr_4 corresponding to the eight nonbonding "$p\pi$" orbitals present in this molecule. A detailed account of the analysis of the photoelectron spectra of the halides of elements in groups III, IV, V, and VI of the periodic table has been given by Potts et al. (1970).

D. PHOTOELECTRON SPECTRA OF "MULTIPLE" BONDED DIATOMIC MOLECULES

To discuss "multiple" bonded systems in which molecular orbitals are formed by p atomic orbitals combining in the "broadside on" as well as in the "end on" configuration we shall take as examples the molecules N_2, NO, O_2, and F_2. Their photoelectron spectra are shown in Fig. 10. The orbitals upon which their electronic structures are built are the in-phase and out-of-phase combinations of the appropriate 2s and 2p orbitals. These are given schematically in the insert. For N_2 the electronic configuration is

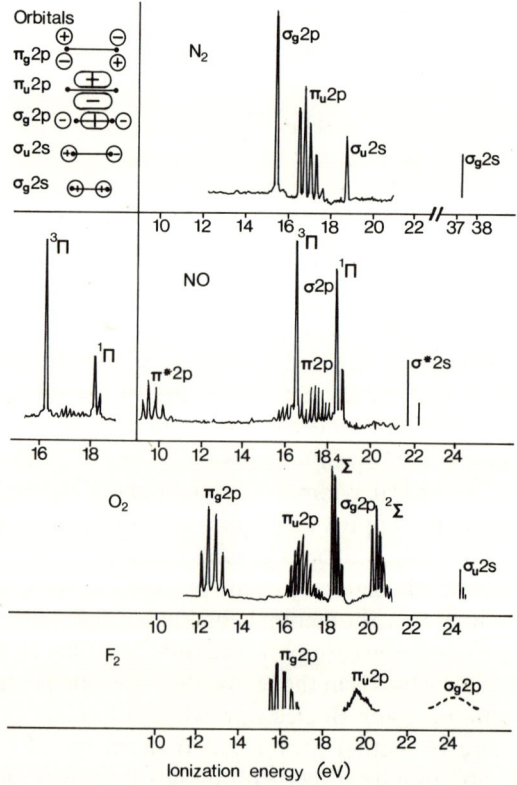

FIG. 10. Photoelectron spectra of N_2, NO, O_2, and F_2.

$(\sigma_g\,2s)^2(\sigma_u\,2s)^2(\pi_u\,2p)^4(\sigma_g\,2p)^2$, the orbitals being in order of decreasing ionization energy. With the exception of $(\sigma_u\,2s)$, all of these might be expected to provide excess negative charge density between the two nuclei and thus to account for the strength of the N_2 "triple" bond. The additional electron of NO has to go into a π^*2p orbital which is largely outside the nuclei and therefore antibonding, and so in a loose analogy N_2 is the inert gas of diatomic molecules and NO the corresponding alkali metal, since it contains one electron outside a closed shell of bonding electrons. An inspection of the photoelectron spectrum of N_2 in which the ionized states corresponding to removal of the different orbital electrons are marked, shows that by far the most vibration accompanies the removal of the π_u electron. Thus at the internuclear distance of neutral N_2, the nuclei are held mainly by the negative cloud of the $(\pi_u\,2p)^4$ electrons. The short-distance bonding character of these π electrons results in the nuclei being pulled through the $(\sigma 2p)^2$ cloud so that this σ orbital is as much outside the nuclei as between them and therefore supplies no bonding at the N–N equilibrium separation. It can be understood readily from the geometry of their overlap that the bonding of p electrons in the broadside on (π) arrangement optimizes at shorter internuclear distances than that in the end on (σ) position. In N_2 the nuclei are in fact on the inside of the minimum of the partial potential energy curve associated with the $(\sigma 2p)^2$ electrons. This orbital is thus in compression, and its electrons have both their bonding and their binding (ionization) energies reduced. In NO, O_2, and F_2 the presence of the additional electrons in antibonding $\pi 2p$ orbitals causes the internuclear distances to be relatively longer than they are in N_2, and it can be seen from Fig. 10 that the $\sigma 2p$ bands of these molecules have progressively more vibrational structure as the internuclear distance approaches more closely that separation for which the bonding of the $(\sigma 2p)^2$ orbital is optimized. The associated σ bands move through the $\pi 2p$ systems to higher ionization energies in accord with their increased effective bonding power.

The features discussed above can be illustrated by considering orbital potential energy curves as illustrated in Fig. 11. In the orbital approximation of molecular electronic structure, the complete dissociation energy curve of a molecule can be split up into the partial orbital bonding curves which give the orbital bonding and the ionization energies over the whole range of internuclear distances. Different types of orbital have their potential minima at different internuclear distances. These do not coincide with the actual equilibrium internuclear distance r_e of the molecule which is determined by the minimum of the sum of the orbital energies taken at each value of r. At any particular r the electrons in different orbitals are at different relative positions in their orbital binding energy curves. These orbital curves are not of course directly determinable but have been drawn to be as far as possible

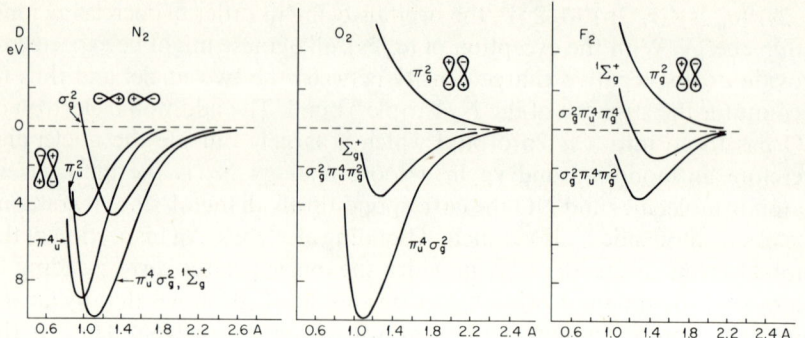

FIG. 11. Orbital contributions to the potential energy curves of the $X^1\Sigma_g^+$ states of N_2, O_2, and F_2.

consistent with the experimental facts. For instance $(\sigma_g\, 2p)^2$ binding curve has been drawn as a near Morse curve to have an r_e of about 1.4 Å a dissociation energy of 4 eV and an $\omega_e = 1200$ cm^{-1}. The curve for $(\pi_u\, 2p)^4$ can then be found for $r > r_e$ by subtracting the σ_g curve from the observed dissociation curve of N_2 on the assumption that at large distances the $(\sigma_g\, 2s)^2(\sigma_u\, 2s)^2$ orbitals do not contribute to total bonding. For the present purpose it has been assumed that this is also true at smaller values of r. The $(\pi_u\, 2p)^2$ curve can then be obtained by halving the $(\pi_u\, 2p)^4$ curve obtained by the above subtraction. The $(\pi_g\, 2p)^2$ repulsive curve can be obtained by plotting the difference between O_2, $X'\Sigma$ and $N_2 X'\Sigma$ which should check with the $(\pi_u\, 2p)^2$ curve of which it is a reflection at large r. The antibonding power of the antibonding orbital is only slightly greater than the bonding power of the bonding orbital (i.e., the former is proportional to $S(1-S)^{-1}$ and the latter to $S(1+S)^{-1}$, where S is the overlap integral). Similarly the $(\pi_g\, 2p)^2$ curve of fluorine can be obtained by plotting the difference between F_2, $X^1\Sigma_g^+$ and O_2, $X^1\Sigma_g^+$ or alternatively by halving the differences between the N_2, $X^1\Sigma_g$ and the $F_2 X^1\Sigma_g^+$ curves.

It will be noted that the optimum bonding energies of $(\sigma_g\, 2p)^2$ and $(\pi_u\, 2p)^2$ are roughly equal. This might be thought to contradict the fact that the single, double, and triple bond energies of ethane, ethylene, and acetylene are not in the ratio of 1 : 2 : 3, but in the ratio 1 : 1.76 : 2.41. However it is readily appreciated that the figures indicate that the σ bond must lose about 60% of its bonding by compression to smaller internuclear distances when forming part of the triple bond. Because of the steep slope of the repulsive part of its curve, this loss would rapidly increase with further reduction in internuclear distance. Thus in N_2 the $(\sigma_g\, 2p)$ orbital has lost nearly all its bonding power at the equilibrium internuclear distance and its photoelectron band has the features characteristic of ionization from a nonbonding orbital. The rapid increase in its bonding power with increasing r is evident from the increasing vibrational structure of the $(\sigma 2p)^{-1}$ systems in NO and O_2 (see Fig. 10). On the other hand, the $\pi_u 2p$ electrons in N_2 find themselves at internuclear distances only slightly larger than those of their potential minimum and are thus strongly bonding. As the equilibrium internuclear distance increases in passing from N_2 to NO, O_2, and F_2 as electrons are added in antibonding repulsive orbitals, the $\sigma_g 2p$ orbital acquires bonding ultimately becoming the basic single bond in F_2. The value of the dissociation energy of F_2 (1.6 eV) is less than the optimum $(\sigma_g 2p)^2$ bond energy of about 4 eV, because, as already mentioned, the bonding power of the $(\pi_u 2p)^4$ orbitals is more than offset by the antibonding power of the $(\pi_g 2p)^4$ orbitals which are filled in this molecule.

Other interesting points illustrated by Fig. 10 are that the spacing in the $\pi 2p$ antibonding bands in NO and O_2 (first systems) are larger than those of the $\pi 2p$ bonding bands (second systems). This is to be expected since the removal of an antibonding electron increases the vibration frequency while that of a bonding electron decreases this frequency. Another interesting feature is that the separation of the first and second bands, which reflects the overlap between the out-of-phase and in-phase $(\pi 2p)$ orbital, rapidly decreases with increase in internuclear distance. For NO, O_2, and F_2 these separations are 7.5, 4.5, and 3.0 eV, respectively. This indicates the reduction in "multiple" (π) bonding as the effect of the increasing number of π antibonding electrons reduces and annuls the bonding of π bonding electrons by increasing the interatomic distance and reducing the orbital overlap.

The insert of the 16 to 18-eV region of NO in Fig. 10 shows the $^3\Pi$ and $^1\Pi$ bands with an intensity ratio of 3 : 1, and thus illustrates how closely the intensities follow the statistical weights of the ionized states, agreeing with the number of channels of escape open to the electrons. Similar remarks apply to the $^4\Sigma$ and $^2\Sigma$ bands of O_2, the integrated intensities of which are in the ratio of 2 : 1. A further interesting feature which these multiplets illustrate is the greater bonding of the states of lower multiplicity. Since the lower multiplicity corresponds to the states of antisymmetric spin function, it is associated with the symmetric (summed) space coordinate wave function of the orbitals between which the spin interaction is occurring. The additional overlap to which this gives rise results in greater bonding relative to states of higher multiplicity where the total space coordinate wave function is obtained by subtracting those of the interacting states.

The photoelectron spectrum contains enough information to plot a rough potential energy diagram of the states of the molecular ion provided the internuclear distance of the neutral species is known. The shape of curve is obtained from the vibrational spacing. By placing the energy point corresponding to the maximum of the vibrational pattern at the internuclear distance of the neutral molecule its position on the horizontal scale can be fixed. That on the vertical scale is fixed by the ionization energy found for the $v' = 0$ band. Internuclear separations for excited states can then be read off the diagram. This is illustrated for NO^+ in Fig. 12 where the photoelectron spectrum is plotted along the $0Y$ axis of the potential energy diagram.

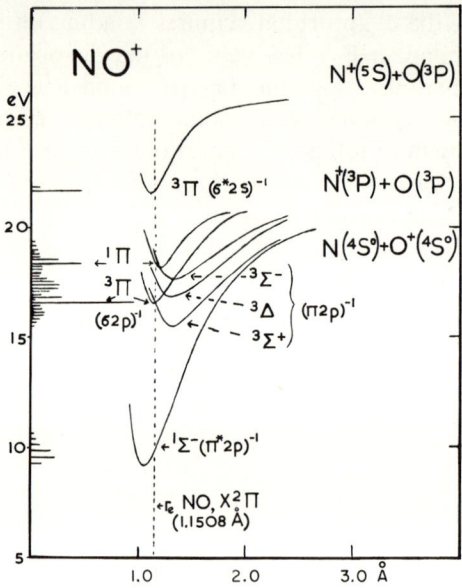

FIG. 12. Potential energy curves of NO^+ showing relation to the photoelectron spectrum plotted along $0Y$.

The "building up" of the electronic structures of molecules by adding electrons to the unfilled orbitals as an atom is changed to one with greater Z follows very closely the "aufbau" process in atoms. Although we have considered this only for diatomic molecules, it can be applied to triatomic and polyatomic systems and extends the shell concept to even the valence electrons of molecules. The difference between atoms and molecules is simply that instead of the positive charge being concentrated at one point it is split into several different centers. It might be asked whether we should now replace the bond structures of chemists by the molecular orbital description. No spectroscopist, for example, has even seen an electron in a hybridized sp^3 orbital—a concept the chemist uses constantly. No real conflict exists, since the bond between two atoms is measured by the amount of negative charge lying between the two positive centers. From the orbital point of view this is made up by contributions from two orbitals, for example, in CH_4 from the $(2pt_2)^6$ orbital and from the $(2sa_1)^2$ orbital. As pointed out in the previous discussion of this molecule the latter acquires bonding character by being distorted (by charge extraction from Ne) in the bond directions. The sp^3 hybridized notation is a shorthand way of saying this. When, however, changes in energy are involved, even, for example, in a mechanical model with certain normal vibrations being excited by a periodic force, this force must be resolved with respect to the axis of symmetry of the system and will only feed energy into a normal vibration of its own frequency. In exactly the same way the interaction of the electric vector of the light wave with the electrons in a molecule must be with the orbitals which are the irreducible representations of involved in the transition moment and the transitions can only occur between states represented by these orbitals.

E. SPECTRA AND STRUCTURE OF SOME "MULTIPLE" BONDED POLYATOMIC MOLECULES

In the same way that the spectra and orbital structure of the simple hydrids can be understood by partitioning off protons from the central charge of the inert gas atoms, the orbital structure of some simple polyatomic molecules can be followed by partitioning off protons from the nuclei of the diatomic molecules with which they are isoelectronic. This is illustrated in Fig. 13. Starting with the well-understood spectrum of N_2, we pass to HCN by displacing a unit positive charge from one of the nitrogen nuclei to a point further along the bond axis. The movement of the orbitals is indicated by the broken lines. It can be seen that the $\sigma 2p$, the $\pi 2p$, and the $\sigma_g\,2s$ orbitals have their ionization energies reduced by this lessening in the positive charge in the region in which they are mainly located. Another important effect which is apparent from comparison of the spectra is that the $\pi_u 2p$

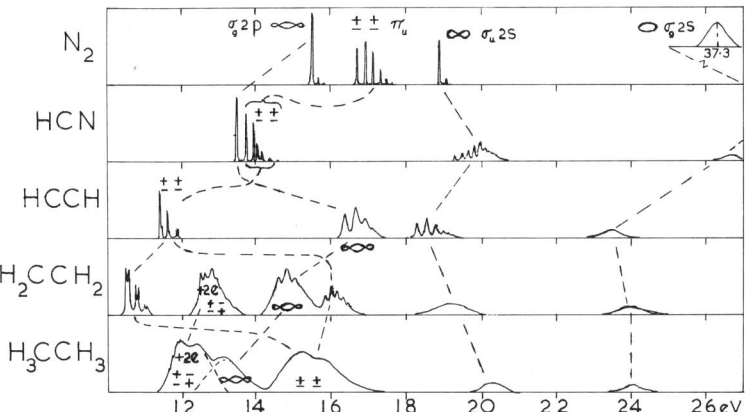

FIG. 13. Photoelectron spectra showing derivation of orbitals of HCN and C_2H_2 from the isoelectronic diatomic molecule N_2 also related orbitals in C_2H_4 and C_2H_6.

electrons lose a considerable amount of their bonding as the nuclei move further away from the position of optimum pπ orbital overlap. This can be seen from the change of the intensity distribution of the vibration band pattern (i.e., I_{max} moves toward I_{adiab}). It is also apparent from the relatively large decrease in ionization energy of this bond orbital. The change from N_2 to HCN affects the σ_u 2s orbital, which now has one of its lobes in the region between the H and the C positive centers. It consequently changes from an antibonding NN to a strongly bonding CH orbital. Its ionization energy is therefore increased and the vibrational pattern of the band changes to one which contains many vibrations of the CH bond reduced in frequency to 1690 cm^{-1} from its value in the neutral molecular ground state (3311 cm^{-1}).

A similar partitioning off of charge from the second nitrogen atom leads to the molecule acetylene, in which there is a further increase in central bond distance. The π_u 2p-orbital energy is again lowered by the positive charge reduction and also by loss of bonding due to further removal from its optimum bonding overlap position. The latter effect can be seen in the progressive lowering of the relative intensities of the vibrational bands to the first (zero) band of the system. The σ_g 2s and σ_u 2s are reduced in ionization energy by the charge removal, and the latter shows by its vibrational pattern that it is strongly CH bonding, while the former is probably CC bonding though because of its diffuseness evidence from vibrational pattern in support of this is not forthcoming. The greatest change in orbital character occurs in the σ_g 2p orbital which becomes strongly CC bonding as the internuclear separation increases toward the optimum value for pσ type bonding. This is indicated by the characteristic bonding envelope of the band pattern in which the CC frequency, reduced to about 1400 cm^{-1} from its ground state value of 1983 cm^{-1}, is strongly excited. It also now exerts some CH bonding, and both combine to raise the ionization energy by nearly 3 eV.

By bringing up two hydrogen atoms onto acetylene and suitably distorting the molecular framework one can form ethylene. The spectra show that one of the π_u CC orbitals now becomes the strongly bonding in-phase b_{2u} (CH$_2$) orbital, increasing its binding energy by about 4 eV, and the two additional electrons go into the corresponding out-of-phase b_{1g} (CH$_2$) orbital. Bringing up two more hydrogen atoms causes the remaining π-electron pair to go into the degenerate in-phase e_{1u} (CH$_3$) orbital and the two additional electrons to enter the corresponding out-of-phase orbital. In order to cover the complete energy range, these spectra have been taken from recordings with 30.4 nm (40.1 eV) radiation. It can be seen that the separations of the bands corresponding to in-phase and out-of-phase 2s C combinations (i.e., σ_g 2s and σ_u 2s, respectively) diminishes with increasing CC distances in the molecules C_2H_2, C_2H_4, and C_2H_6. This is a direct result of the reduction in 2s C orbital overlap as their separation increases.

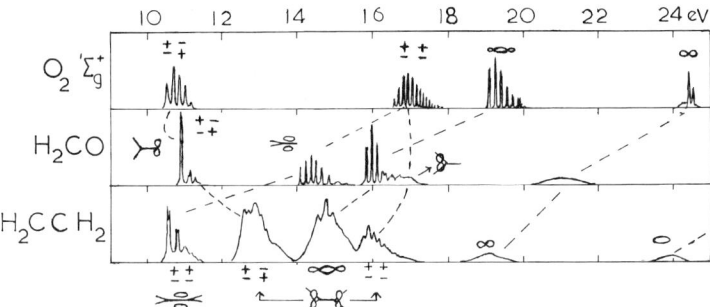

FIG. 14. Photoelectron spectra showing relation of orbitals of H_2CO and C_2H_4 to those of O_2, $X^1\Sigma_g^+$.

A similar diagram (Fig. 14) shows how the orbitals of formaldehyde and ethylene are derivable from those of oxygen. Of course in this case it is necessary to start with oxygen in its $^1\Sigma$ state, and although the relevant photoelectron spectrum has not been obtained, it can be constructed with confidence from the known spectroscopic data of the state and the photoelectron spectra of $^3\Sigma_g^-$ and $^1\Delta_g$. The spectra show how the antibonding $\pi 2p$ orbital of oxygen becomes the carbonyl " lone pair," the π_u^4 orbital splits into the carbonyl π and the b_2 (CH_2), and the $\sigma 2p$ becomes the a_1 (CO) bond orbital. Their consistent correlation with the orbitals of ethylene gives confidence to the general assignment of the orbitals.

F. SPECTRA AND STRUCTURE OF TRIATOMIC MOLECULES

Whether a triatomic molecule will be linear or bent in its ground state depends on which configuration will afford best coverage of the positive charge by the electrons in the orbitals associated with a particular geometry. The various ways in which the energies of different orbitals are affected by changing the angle from 180° to 90° were discussed by Walsh (1953) before the advent of photoelectron spectroscopy and are qualitatively described in Walsh diagrams. It is found that molecules with 16 electrons in the valence shell are linear in their ground states (e.g., CO_2, N_2O, etc.). When the number is greater than 16 (e.g., NO_2, O_3, etc.) the molecule has a bent ground state, and this situation persists until the valence shell is completely filled as in OF_2 (20 valence electrons). The structure becomes linear again for XeF_2 (22 outer electrons). For molecules formed from first row elements the orbitals can be regarded as being derived from the in-phase and out-of-phase combinations of the 2p orbitals of the outer atoms and the 2p and 2s orbitals of the central atom. The molecular orbitals so constructed for a bent molecule are shown diagramatically in Fig. 15. Those for a linear molecule are very similar and can be visualized by changing the angle to 180°.

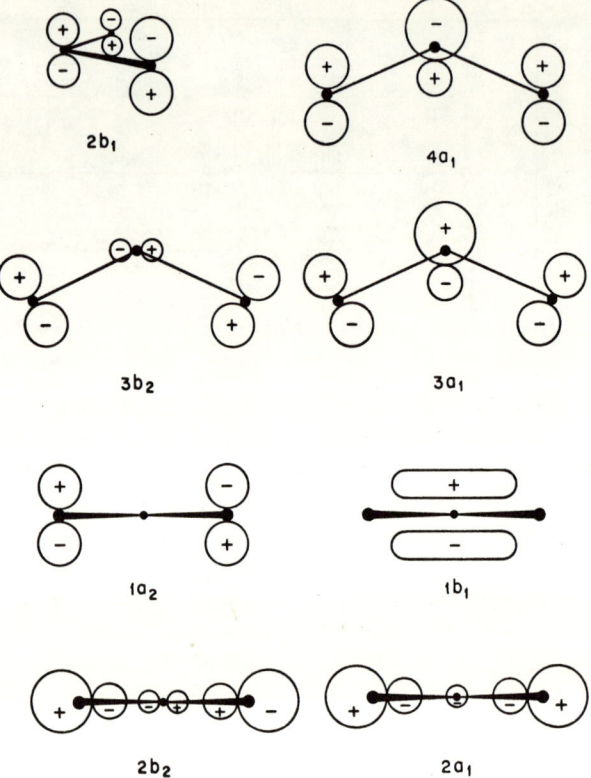

Fig. 15. The valence orbitals of bent triatomic molecules.

In carbon dioxide the orbital of lowest ionization energy corresponds to the out-of-phase combination of $2p\pi(0)$ orbitals which can be written as $(\pi_0 - \pi_0, \pi_g^4)$. This splits on bending into $3b_2$ (in-plane) and $1a_2$ (out-of-plane) orbitals. As can be seen from the photoelectron spectra shown diagrammatically in Fig. 16 this orbital is nonbonding in the linear triatomic molecules. The next filled orbital is the in-phase $(\pi_0 + \pi_C + \pi_0, \pi_u^4)$ combination which, as expected and indicated by the structure of its band system, is strongly bonding. It splits into $3a_1$ (in-plane) and $1b_1$ (out-of-plane) orbitals in the bent molecule. The next inner orbital is that corresponding to the out-of-phase $(\sigma_0 + p\sigma_C - \sigma_0, \sigma_u^2)$ orbital. It is nonbonding and followed by the in-phase $(\sigma_0 + s\sigma_C + \sigma_0, \sigma_g^2)$ bonding orbital. The similar orbital structure in the isoelectronic molecule N_2O is reflected in its similar photoelectron spectrum. The π_g and σ_u orbitals of this molecule are lower than in CO_2 because they are located near N and O rather than O and O atoms. Sim-

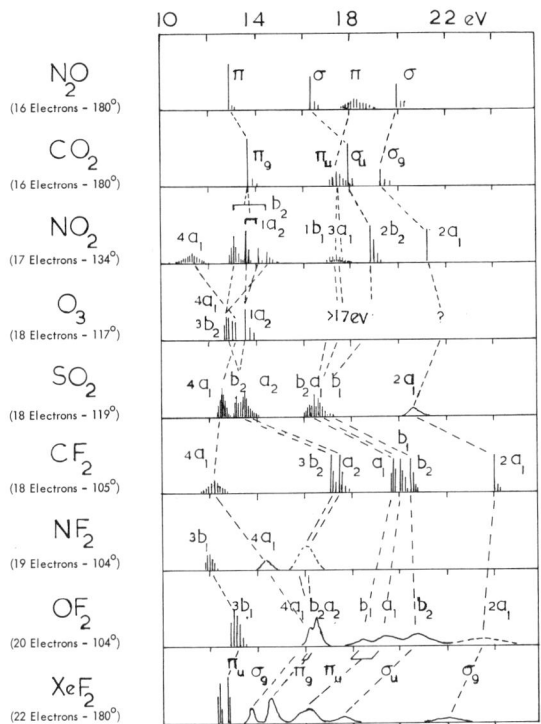

FIG. 16. Diagrammatic photoelectron spectra showing the filling of the orbitals of some triatomic molecules.

ilarly the π_u and σ_g orbitals are higher in N_2O than in CO_2 because of the higher energy contribution of the orbitals from the central atoms—N and C, respectively.

In NO_2 the additional electron has to go into the hitherto filled $4a_1$ orbital and the molecule becomes bent (ONO = 134°), finding better charge coverage in this geometry. The $(4a_1)^{-1}$ band appearing on the low-energy side of the photoelectron spectrum shows a wide vibrational pattern involving the bending motion, since, as might be expected, on ionization from this orbital the molecule moves to the linear structure associated with the 16 electrons then remaining in the valence shell. The orbitals which were doubly degenerate in CO_2 are split into in-plane and out-of-plane orbitals in bent NO_2. Further the states of NO_2^+ involving ionization from all orbitals apart from the singly occupied $4a_1$ orbital are split into singlet and triplet states because of the spin interaction of the two singly occupied orbitals, and appear in the photoelectron spectrum as similar systems with a 1 : 3 intensity ratio (Brundle et al., 1970).

The $4a_1$ shell is filled in O_3, SO_2, and CF_2 so that in NF_2 the additional electron present in its structure has to go into the $3b_1$ orbital and a new band appears in its photoelectron spectrum. This orbital becomes filled in OF_2. The spectrum of XeF_2 in which another orbital has to be provided for the two additional electrons is shown as the final spectrum in Fig. 16. The additional band corresponding to these electrons is evident in the spectrum. Electrons in this orbital are clearly responsible for causing a return to linearity. A full discussion of this spectrum is not possible here.

The building-up principle illustrated so far for di- and triatomic molecules has been extended to molecules containing four, five, six, and seven atoms, particularly for molecules of the type AX_n, where X is a halogen atom (Potts et al., 1970).

While other techniques have to be adopted for more complicated systems, it is clear that a sound understanding of what happens in the simplest cases is a very desirable preliminary to considerations of the larger molecules.

G. Methods of Orbital Assignment of Bands in Photoelectron Spectra

It would be misleading to infer from the foregoing sections that the methods described therein are generally applicable for the analysis of the spectra of more complicated molecules. This is certainly not the case. It is desirable whenever possible to assign bands from direct experimental data, as is the practice in the spectroscopy of simple molecules. However, because the resolution of photoelectron spectroscopy is not adequate to give rotational fine structural details of bands, even if these were present (which they usually are not because of the short lifetimes of the electronic excited states of polyatomic states), alternative methods must be used to find the nature of the state to be associated with a band in the photoelectron spectrum.

Obviously the vibrational pattern of a band gives a great deal of information about the orbital from which the electron has been removed, and this is greatly assisted if various isotopic species can be studied. The study of groups of compounds in which one atom is successively changed to a heavier atom in the same group of the periodic table is also valuable. Any related set of molecules, such as homologous series in which the orbitals of a chromophoric group are modified in a minor way by a substituent, can obviously help assignment. The variation in the relative intensity of the bands with photon energy, the angular dependence of photoelectron cross sections, and the possibilities of autoionization are all properties related to the orbital wave function, and if given the proper interpretation can facilitate assignment. An extension of the "aufbau" process to build the orbital structures of molecules with more atoms from molecules with fewer atoms (and electrons) is also valuable. Theoretical computations of orbital order using both *ab initio* and semiempirical approaches have been widely used to assign the bands of a photoelectron spectrum, and with improvement in molecular parameters this method has had increasing success. It has been popular with many authors who use it to justify both the assignment and the computational theory. This is obviously a dangerous procedure, and it is clearly desirable as far as possible to make assignments purely from experimental criteria. When this has been done the predictions can then be compared with the calculations, and, if the comparison is favorable, it can be used to justify the parameters employed in the theory. In this way, the experimental data can be utilized to yield molecular properties not derivable from the spectra directly and the theory can be simultaneously justified.

IV. Physical Aspects of Photoelectron Spectroscopy

A. The Intensities of Photoelectron Spectra

Apart from the quantitative data concerning orbital energies to be obtained by photoelectron spectroscopy, the intensities of the bands obviously contain much information about the nature and extent of the orbitals themselves. In this connection we must consider the transition moment of an electron from the initial orbital ϕ to the state ε in the ionization continuum into which it is ejected by the incident photon. This illustrated in Fig. 17 where it can be seen that the integral representing the transition moment depends upon the spread and nodal character of the wave function of the initial state and the wavelength and phase of the photoelectron (i.e., the final state). For the purposes of this illustration the wave functions are taken as those of a one-dimensional oscillation of an electron in a $V(1/r)$ potential field, that is, pseudo-Rydberg bound states. Plane wave functions are used for the continuum states ($\lambda[1 \text{ eV}] \sim 12 \text{ Å}$). The simplest fact that this illustrates is the fall-off of the photoionization cross section with higher energy (shorter wavelength) photoelectrons. This explains the relative transparency of electrons in valence shells to X irradiation, where the area of the graph for the transition moment is divided into small compensating positive and negative regions. It can be appreciated that the cross section appropriate to any orbital will vary with the dimensions and nodal character of this orbital and the wavelength of the photoelectron. As a rough generalization it might be expected that the photoionization cross section would maximize when $\lambda/4$ of the photoelectron is not less than the orbital dimensions. This would explain experimental observations that near the threshold of ionization the cross sections of molecular orbitals built from p atomic orbitals are greater than

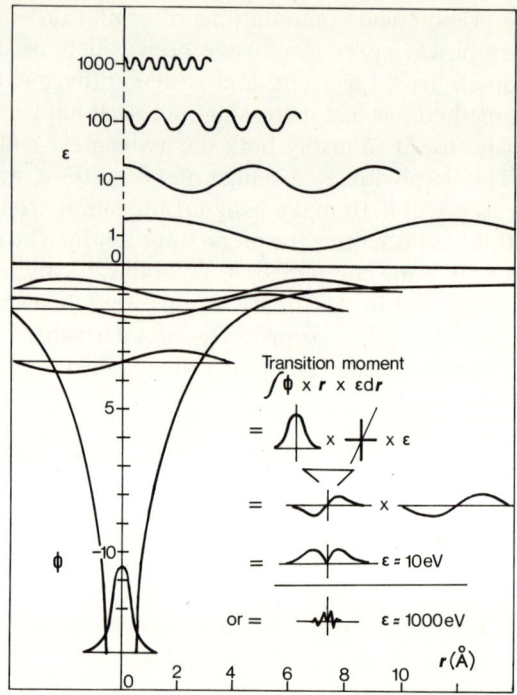

FIG. 17. Diagrammatic illustration of features of importance in the photoionization of an electron from a bound state ϕ to a continuum state ε.

those arising from s orbitals. This situation is reversed for high photoelectron energies, where, because of the smaller radial spread of s orbitals, the transition moment integral maximizes with shorter wavelength photoelectrons. The study of the variation of the relative intensity of the photoelectron spectra of different orbitals with varying irradiating wavelengths is clearly going to yield much information on the nature of orbital wave functions. A summary of the information available for gases on this topic has been given by Price et al. (1972). Photon sources of different energies derived from synchrotron radiation have already been used by Eastman and Grobman (1972) to study the band structure of gold and the density of intrinsic surface states in Si, Ge, and GeAs. The application of this source of photons of variable energy is still little exploited for gases, and it must be stated that the problem of getting adequate energy in a sufficiently narrow band is a very formidable one. It will be particularly important in the study and interpretation of autoionization phenomena and has already been used for this purpose (see Section IV,D).

B. Selection Rules

The photoionization process in the orbital approximation is a one-electron transition resulting from the interaction of the radiation with the electric dipole of the orbital. The difference of angular momentum between the neutral molecule and the molecular ion is given by

$$\Delta L = 0, \pm 1, \ldots, \pm l \qquad (3)$$

where l is the angular momentum of the ionized electron, i.e., L can change by m_l. The free electron which can be an s, p, d, or f wave requires an angular momentum of $l \pm 1$, and when two different angular momenta are possible by this rule, e.g., p → s or d, both may be utilized giving an outgoing wave of mixed character. Since one electron with spin $\pm \frac{1}{2}$ is removed the rule for the change of multiplicity is

$$\Delta S = \pm 1 \qquad (4)$$

Two electron processes are weakly allowed only when there is appreciable correlation between the motions of electrons in different orbitals, the intensities of their bands reflecting the extent of the interaction. Because all one-electron photoionizations are allowed the partial photoionization cross sections, or the probabilities of ionization to different states of the ion, are all of the same order of magnitude, i.e., integrated band intensities per orbital are roughly equal. Orbital and spin degeneracies provide multiplying factors which affect the relative intensities of related bands.

C. Dependence of Cross Section on Photoelectron Energy

As already indicated, the dependence of the transition moment on the continuum wave function implies a variation of the cross section for photoionization with the wavelength of the photoionizing radiation. The cross section is generally highest near the threshold after which it decreases with increasing photon energy. When the photon energy is very high compared with the threshold energy as, for example, in the photoionization of valence electrons by soft X rays, the cross section should be proportional to v^3. The cross sections near threshold exhibit a more complicated behavior, depending on the nature of the overlap between the initial and continuum orbitals involved. Because of experimental difficulties, few absolute measurements are available, but the different ways in which different orbitals behave with increasing photon energy can be judged from the changes in the relative intensities of the bands in spectra obtained first with low- and then with high-energy photons. Figure 18 illustrates the changes in the relative intensities of bands from p- and s-type orbitals with ionizing photons of low and high energies. It can be seen that at low energies the photoelectron bands of

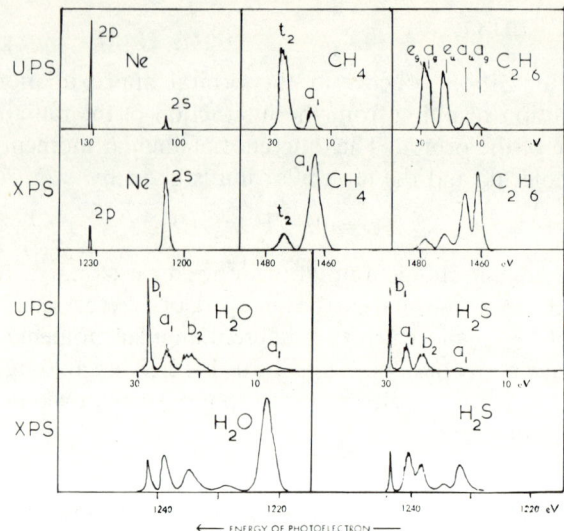

FIG. 18. Examples of change in relative intensities of bands from p- and s-type orbitals with ionizing photons of low and high energies.

p-type orbitals are generally much stronger than those of s bands from the same shell, whereas the reverse is true for their relative intensities in photoelectron spectra obtained with high energy X radiation. This behavior, which is quite general in a large number of atomic and molecular cases, reflects the greater charge density of s orbitals at smaller radii, since it is at these short distances that the major contribution to the transition moment for short wavelength photoelectrons arise. Many other examples of variation of cross section with photon energy are given by Price *et al.* (1972). Calculations of the relative partial cross sections of various orbitals from near threshold to high photoelectron energies have been carried out by many authors, the early work being referenced by Marr (1967). More recently Lohr (1972), Gelius (1972), and Iwata and Nagakura (1973) have made cross-section calculations using hydrogenlike continuum functions for the ejected electron and SCF wave functions for the orbitals of the bound electrons both in the initial and the residual ionized states. They obtained values which agree reasonably well with the experimental results particularly when the photoelectrons have high energies. Schweig and Theil (1973) by dividing the cross section into one- and two-center terms representing, respectively, contributions from atomic orbitals and their bond combinations as modified by the population coefficients, have obtained reasonable agreement for relative band intensity changes between He I and He II spectra. Their results from threshold up to about 5 eV are not expected to be significant since they use plane wave continuum functions.

D. Autoionization

The ionization continua of the outer orbitals of a system are overlapped on an energy diagram by a manifold of excited bound states associated with electrons from inner orbitals (so called super-excited states which are usually Rydberg in type). If the symmetry of the product of the wave functions of a lower ionized state and its photoelectron ($M^+ \times \varepsilon$) is the same as that of an overlapping superexcited bound state then these states mix. The theory in the case of atomic processes was given by Fano (1961) and Fano and Cooper (1968). It has been generalized for more complex systems by Mies (1968) and Bardsley (1968). The situation in the case of molecules is complicated relative to that in atoms by vibrational effects which arise from the different shapes of the potential energy curves of the three states involved in the process, *viz.*, the neutral, the super-excited, and the ionic state. In the case of a diatomic molecule, for example, absorption can take place to a vibrational level of the bound state lying vertically above the ground-state minimum followed by a swing of the system to the opposite turning point on its potential energy curve. Autoionization is most probable from values of r where ψ_{vib}^2 has maxima, i.e., particularly at the turning points of the vibration. Thus at the far turning point, autoionization can take place strongly to the vibrational state of M^+ vertically below it. Ionization can occur to vibrational levels of the M^+ state which, on Franck–Condon grounds, are not accessible by direct ionization to M^+, and one or more additional maxima may appear in the vibrational pattern. Two factors are therefore important: (1) in addition to the symmetry of the electronic wave function the nature of the autoionization is affected by the shape of the potential energy curve of the autoionizing state through its vibrational wave function, and (2) information about both can be expected from a study of the phenomenon.

One of the most interesting examples of autoionization is that of oxygen under neon irradiation. Figure 19a and b show spectra taken with constant energy bandwidth with Ne I and He I radiation respectively. Figure 19c shows a part of the spectrum taken with Ne I radiation under much higher resolution. The neon radiation consists of a strong line at 16.85 eV and a much weaker line at 16.67 eV. As can be seen from Fig. 20 the latter does not coincide with an absorption band of O_2. The $(\pi_g)^{-1}$ photoelectron spectrum due to this 16.67-eV photon corresponds to the minor peaks in the 12.2- to 13.2-eV region which have the same intensity distribution as the spectrum obtained with 584-Å irradiation and thus exhibit no autoionization. The major peaks which show a much extended system with an additional maximum are due to the 16.85-eV line which lies on top of a rather broad absorption band as shown in Fig. 20 (Geiger and Schröder, 1968). The He I and the Ne I photoelectron spectra are plotted along the ordinate of Fig. 21

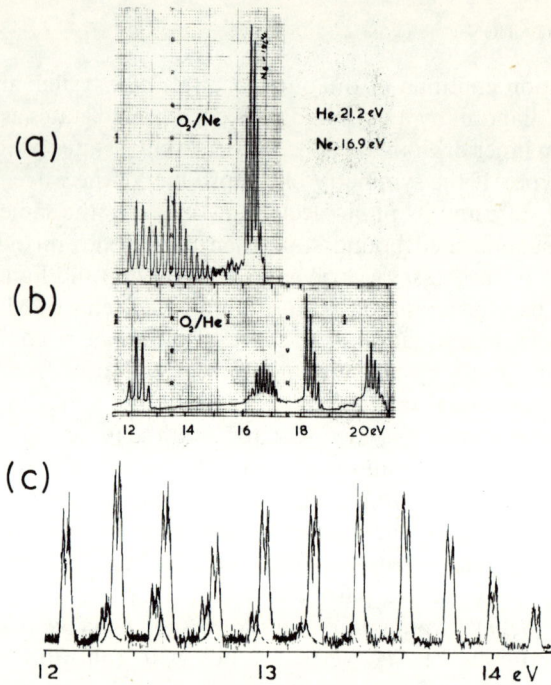

FIG. 19. (a) Photoelectron spectrum of O_2 taken with Ne I radiation. (b) Photoelectron spectrum of O_2 taken with He I radiation. (c) High resolution spectrum of O_2 using Ne I radiation showing autoionization enhancement by the 16.85-eV line (major peaks) but not by the 16.67-eV line (minor peaks).

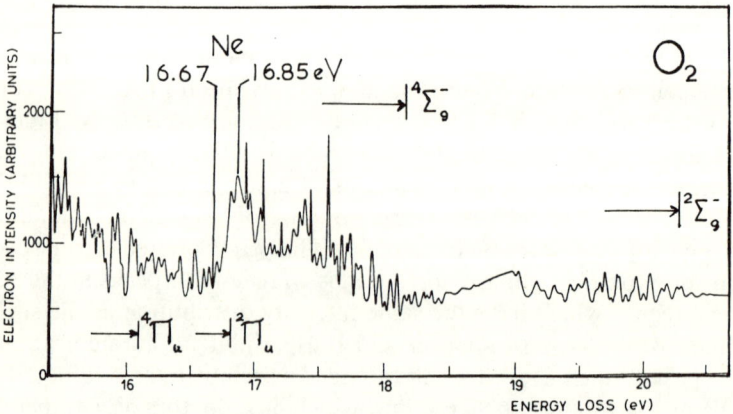

FIG. 20. Absorption energy loss spectrum of O_2 in the region of the neon lines.

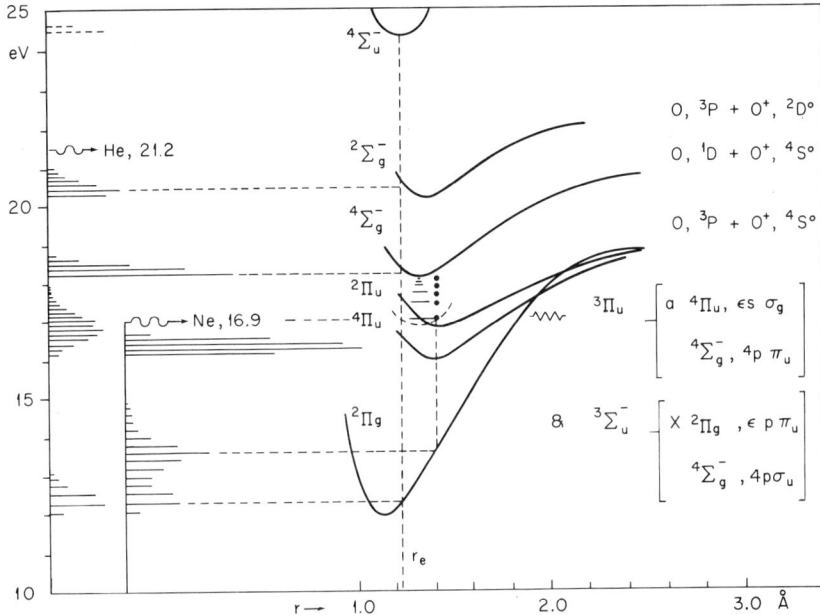

FIG. 21. Potential energy curves of O_2^+ with photoelectron spectra drawn along the ordinate indicating the nature of the autoionizing processes.

which gives also the potential curves of the states of O_2^+. It can be seen by comparison of the two spectra that both the $^2\Pi_g$ and the $^4\Pi_u$ states are affected by autoionization. While in the former an abnormal vibrational intensity pattern is produced, in the latter, as far as can be judged from the limited position of the $^4\Pi_u$ spectrum lying above 16 eV available with the neon line, autoionization enhancement occurs from the start of the system. It also appears that the autoionization in the 16-eV region produces a much stronger enhancement than that in the 12- to 15-eV region. The absorption band lying on the 16.85-eV line of neon is undoubtedly a Rydberg band going to the $(\sigma 2p)^{-1}\, ^4\Sigma_g^-$ state of O_2^+. The symmetry requirement would be met for the $^4\Pi_u$ autoionization if the bound state was $^4\Sigma_g^-$ ($4p\pi_u$): $^3\Pi_u$, this symmetry being also matched by the ionized system $^4\Pi_u\, \varepsilon s\sigma_g$: $^3\Pi_u$. [This bound state has been tentatively assigned by Lindholm (1969) to a relatively sharp band lying at 16.93 eV, but it does not appear that a state with the right symmetry to autoionize should stay sharp.] The intensities in the vibrational pattern are also compatible with this process for $v = 0$ in the bound state (see Fig. 21). The question now is whether this state could also be responsible for the autoionization of the $X^2\Pi_g$ state (in the 12- to 15-eV range). Continuum states $X^2\Pi_g$ ($\varepsilon p\sigma_u$): $^3\Pi_u$ would give the right symmetry, but according to the Franck–Condon calculations of Kissinger and Taylor (1973), it is necessary to assume that the bound state should have $v = 1$ to get the right intensity distribution in the vibration pattern of the autoionized system. This does not appear to be compatible with the absorption spectrum of O_2, and it may be that another overlying bound state is involved in autoionizing $X^2\Pi_g$ such as a $v = 4$ of a very high Rydberg state going to $A^2\Pi_u$ (note the I_{adiab} of $A^2\Pi_u$ is 16.7, i.e., less than the 16.8 eV neon photon). Autoionization through $^3\Sigma_u^-$ as well as $^3\Pi_u$ states of O_2 can occur to both $X^2\Pi_g$ and a $^4\Pi_u$ states of O_2^+.

Studies of autoionization in oxygen have engaged the attention of many workers, because oxygen is one of the major sources of free electrons in the upper atmosphere through photoionization by solar radiation. Also such studies could help in interpreting features of the O_2 absorption spectrum which, because of its diffuseness and complexity, has not yet been fully elucidated. Blake et al. (1970) observed autoionization at several different irradiating wavelength bands isolated from a helium continuum with a vacuum monochromator. The object was to obtain information about the autoionizing state by irradiating it in its successive vibrational levels, each of which will give rise to a different vibrational intensity distribution in the autoionized spectrum. The work has been carried out under higher resolution and with synchrotron radiation by Kissinger and Taylor (1973), who find reasonable agreement with the theory given by Smith (1970). This theory yields the autoionizing linewidth and the Fano line profile index of the bands of the autoionizing state which can be used to determine autoionizing rates and lifetimes of the excited states as well as assigning v' values. Autoionization bands very similar to those in O_2 have been found for NO (Kleimenov et al., 1971; Collin and Natalis, 1968; Gardner and Samson, 1973). It must be said, however, that as yet no case of the autoionization of a diatomic molecule is completely understood, particularly with reference to its effect on the energies of the superexcited state.

Since low-energy photoionizing lines are always likely to coincide with the Rydberg bands of inner orbital electrons, it is not surprising that many intensity anomalies are found in photoelectron spectra excited with hydrogen, argon, neon, and even helium resonance radiation. The spectra taken with 304 Å (He^+) radiation are free of these, since few molecules have Rydberg bands of valence electrons at this wavelength. Figure 22 shows certain striking differences in the spectra of some XF_n molecules photoionized by 584- and 304-Å radiation. It is the second and/or the third band in these spectra that are enhanced by autoionization if our explanation is correct. SF_6 has, in fact, an enormously strong second band under 584-Å excitation. It is possible that this results from interaction with the dissociative process into $F + SF_5^+$ which occurs within this band. Under 304-Å

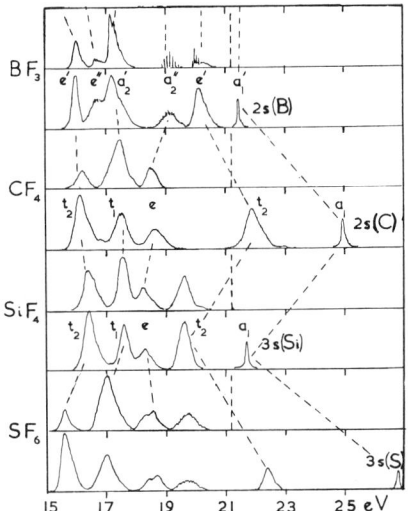

FIG. 22. Spectra of XF$_n$ molecules with 584 Å and 304 Å irradiation indicating intensity changes due to autoionization in the 584 Å systems.

excitation the intensity pattern is similar to that of UF$_6$ under 584-Å excitation. (Note: UF$_6$ has no absorption bands at 584 Å.) The enhancement in these cases is not accompanied by displacement of the maximum of the band envelope as in O$_2$ and NO, which is reasonable in view of the relatively nonbonding nature of many of the orbitals in these molecules.

A very general type of autoionization is found in the photoelectron spectra of aromatic hydrocarbons. These spectra usually are characterized by an initial set of sharp bands arising from electrons in π orbitals followed by a set of more diffuse bands due to σ-type orbitals at higher energies. The intensities of the σ bands increases substantially relative to the π bands when Ne I radiation is used instead of He I. A similar behavior is observed in angular distribution studies when the photoelectron angle is changed from 90° to 30° with the light beam. Another very interesting case has been found in the photoelectron spectra of C$_6$H$_6$ and C$_6$D$_6$ under neon (16.86 eV) irradiation (Price, 1972). A band system in the region (11.6–13 eV) was found to be selectively enhanced in C$_6$D$_6$ relative to C$_6$H$_6$. This appears to be due to the fact that one orbital of the benzene molecule has an ionization energy which in C$_6$H$_6$ is about 0.01 eV below the photon energy, whereas in C$_6$D$_6$, due to the blue zero-point energy shift, its ionization lies 0.01 eV above this energy. In the latter case the excited electron remains attached to the molecule and cannot get away. It ultimately interacts with an electron in a lower orbital thereby increasing the strength of its band by autoionization.

In C_6H_6 this electron escapes into the ionization continuum and therefore does not produce autoionization. In a simple way this may be regarded as an internal electron impact phenomenon.

Recent studies by Baer and Tsai (1973) have been made of the resonant photoelectron spectra from autoionizing states of CH_3I. It was found that Bardsley's formulation of the configuration interaction theory of autoionization was consistent with the peak intensities of the vibrational bands observed. In his theory the photon energy need not be in exact resonance with a Rydberg state in order for that state to contribute to autoionization, though its contribution is inversely proportional to the energy mismatch.

A further type of autoionization known as vibrational autoionization is possible though it has not yet been found to be very prevalent in photoelectron spectroscopy. The autoionizing state must have an electron in a Rydberg orbital with a high principal quantum number and also enough vibrational energy to bring the total energy above the minimum molecular ionization energy. Autoionization follows by interaction of the electronic and vibrational motions in which the vibrational energy of the core is converted into electronic energy and the Rydberg electron is ejected. This type of autoionization has been studied theoretically by Berry (1966) and experimentally by Berkowitz and Chupka (1969).

E. Angular Distribution of Photoelectrons

The angular distribution of the photoelectrons ejected from a molecule relative to the direction of the incident photon beam is determined by the angular momentum they must carry in order to satisfy the dipole selection rules, i.e., by the s, p, d, or f character of the outgoing electrons. In the ionization of an s electron to a p photoelectron, the axis of the p-wave function is defined by the direction of the electric vector of the light wave. Taking this as the z axis, the angular part of the electron wave function is the same as that of an atomic p_z electron and the probability of observing an electron is proportional to its square. Thus ionization of an s electron leads to a $\cos^2 \theta$ distribution about the electric vector.

$$I(\theta) \alpha Y_{10}^2 = \tfrac{3}{4}\pi \cos^2 \theta \tag{5}$$

Ionization from a p orbital leads to s and d outgoing waves. The s waves have a spherical distribution while the d waves are peaked along the direction of the electric vector. Whatever the mixture of s, p, d, etc., waves it turns out that the photoelectron angular distribution can be expressed by the formula

$$I(\theta) = (\sigma/4\pi)[1 + \beta/2(3\cos^2 \theta - 1)] \tag{6}$$

where σ is the total cross section integrated over all angles and β is an anisotropy parameter. Since in photoelectron spectrometers the light is normally unpolarized, angular variations in intensity must be measured in directions θ' away from the direction of the light beam. The distribution of intensity is then

$$I(\theta') = (\sigma/4\pi)[1 + \beta/2(\tfrac{3}{2} \sin^2 \theta' - 1)] \tag{7}$$

For a pure p wave β has the value 2 thus giving agreement with Eq. (5). It can range from -1 to $+2$. Expressions for β in terms of quantities such as l, the angular momentum of the electron before ionization; σ_{l-1} and σ_{l+1}, the partial cross sections for production of the $l-1$ and $l+1$ waves; and $(\delta_{l+1} - \delta_{l-1})$, the phase difference between the two waves have been worked out for atoms (Cooper and Zare, 1968) and diatomic molecules (Buckingham et al., 1970). However, calculations of β are impracticable for large molecules, so that hopes of assigning a photoelectron band explicitly by a measurement of its asymmetry parameter are not likely to be fulfilled in such cases. Measurements of angular distribution will help, however, to differentiate between photoelectron bands especially when these overlap, and it can be used in a semiempirical way for band assignment when the behavior of established types of bands are known. For example, experiment shows that $(\pi)^{-1}$ bands in linear or planar molecules are relatively much weaker at low angles with the light beam than $(\sigma)^{-1}$-type bands, e.g., π bands in ethylene, benzene, water, methyl, and iodide show this behavior. The variations in relative intensities for nonplanar polyatomic systems are much less and are frequently hardly within the accuracy with which the measurements can be made.

Carlson and his co-workers (Carlson, 1971a,b; Carlson et al., 1972; Carlson and McGuire, 1973) have been among the more active workers in this field. They obtain good agreement between theory and experiment for the inert gases, H_2 and N_2. In their examination of many polyatomic gases they find in addition to the features already mentioned that the angular parameter depends mostly on the initial orbital from which the electron was ejected and not on the different final states that may arise as a result of John–Teller splitting, spin–orbit splitting, or spin-coupling between two unfilled orbitals. The relative intensities of the vibrational bands of a given electronic system are usually unaltered, but in certain cases irregularities are found connected with either autoionization or some breakdown in the Born–Oppenheimer approximation involving coupling between electronic and vibrational motions.

The experimental arrangement of Carlson and others utilizing a source rotatable with respect to the electron analyzer usually involves considerable loss in photoelectron flux, which results in lower resolution spectra. This can be avoided either by using two slits and two detectors to cover two electron paths in a hemispherical analyzer (Ames, *et al.*, 1972) or by using two discharge tubes, one aligned parallel to the slit ($\theta' = 90°$) and one perpendicular to it which allows photoelectrons to be accepted by the analyzer at an angle of 30° with the beam (W. C. Price and K. Glenn, unpublished, 1971). Spectra taken by the latter method are shown in Fig. 23.

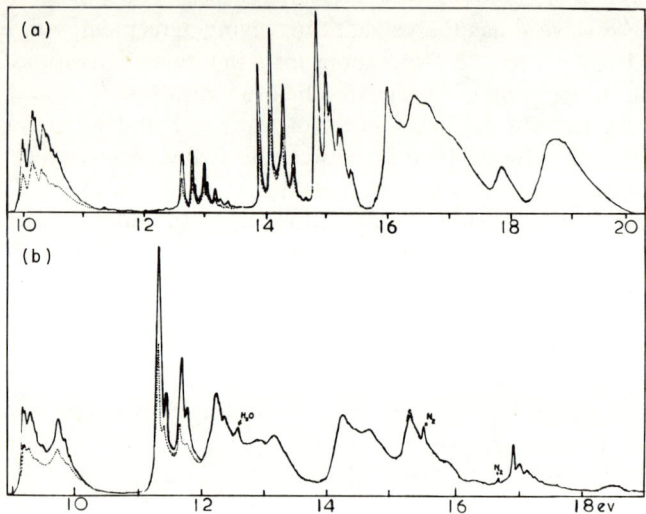

FIG. 23. Spectra showing the different relative intensities of photoemission at 90° (———) and 30° (···) with the photon beam: (a) C_6F_6, (b) C_6H_5Cl.

The theory of the partial cross sections and angular distributions for individual rotational transitions in the photoionization of diatomic molecules has been worked out by Buckingham *et al.* (1970). The only molecule for which resolved rotational photoelectron spectra can be obtained is H_2. Niehaus and Ruf (1971) using Ne I radiation measured β for the different rotational transitions, finding for $\Delta J = 0$, $\beta = 1.95$, and for $\Delta J = 2$, $\beta = 0.85$. When $\Delta J = 0$ the value of β is so close to +2 that the outgoing electron must be almost a pure p wave. For $\Delta J = 2$, theory indicates it should be a mixture of p and f waves (Sichel, 1970). The experimental value ($\beta = 0.85$) agrees well with a pure f wave for which $\beta = 0.80$. This would correspond to a simple addition of the angular momentum changes in rotational and electronic motion in determining the angular momentum of the outgoing electron. The agreement might be fortuitous since the value $\beta = 0.8$ could result from interference between p and f waves of appropriate phases and amplitude.

F. Electron–Molecule Interactions

Anomalous peaks have been observed in the photoelectron spectra of certain diatomic and triatomic molecules when the pressure in the ionization chamber is relatively high. These are attributed to collisions between emitted photoelectrons and neutral molecules leading to the formation of temporary negative ion states (Streets et al., 1972, 1973). Since the negative ion can return to the various vibrational levels of the neutral ground state, the re-emitted electron will then have its initial energy reduced by the appropriate number of vibrational quanta of the ground state. One of the most interesting examples of this occurs in the ionization of N_2 to its $(\sigma_u 2s)^{-1}$ $^2\Sigma_u^+$ state at 18.76 eV by He I (21.22 eV). The photoelectron spectra obtained at various pressures are shown in Fig. 24. As the pressure is increased

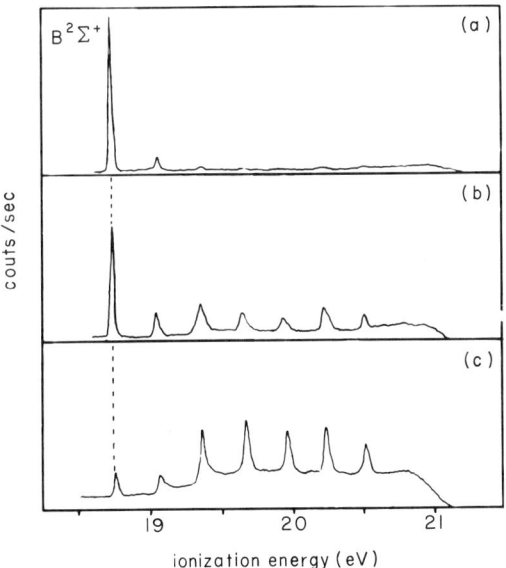

FIG. 24. Photoelectron spectra at elevated pressures of N_2 ionized by helium radiation (21.22 eV): (a) 15 μ, (b) 50 μ, (c) 110 μ.

it is seen that the photoelectrons from the main band are degraded into a series of lower energy bands with an interval which corresponds to the neutral ground-state frequency of N_2.

The resonance absorption spectrum of N_2^- can also be observed and measured with high precision by photoelectron spectroscopy. The upper curve in Fig. 25 shows the spectrum of CH_4 obtained with Ne I radiation. This of course corresponds to a well-defined band of photoelectrons having

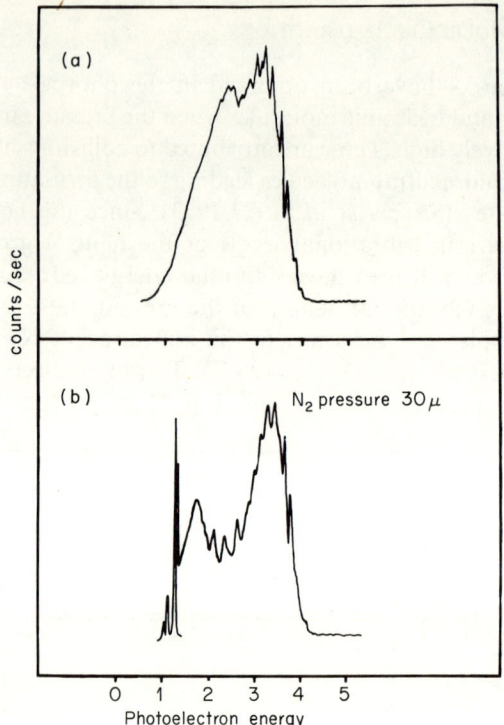

FIG. 25. Resonance absorption spectrum of N_2^- formed by the absorption of photoelectrons emitted from methane ionized by neon irradiation: (a) CH_4, (b) $CH_4 + N_2$.

energies from 0 to 4 eV. If nitrogen is now admitted to the ionization chamber, its absorption of the CH_4 photoelectrons can be observed in the photoelectron spectrum, i.e., an electron absorption spectrum can be obtained which gives the vibration pattern appropriate to the ground state of N_2^-. The halfwidths of the bands indicate the lifetimes of the negative ion states. The process is illustrated in Fig. 26. Many other negative ions can be examined in a similar way.

G. Transient Species

If a short-lived radical or free atom can be transported sufficiently quickly into the ion chamber of a photoelectron spectrometer, a spectrum can be obtained from which its orbital structure can be determined. This can be done by fast pumping from the source of the transients through the ion

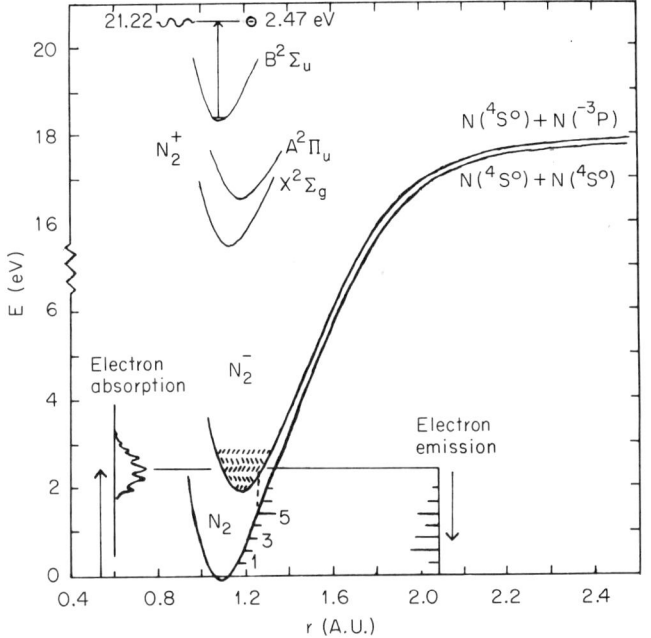

FIG. 26. Potential energy curves for the ground states of N_2 and N_2^- deduced from the excitation and absorption data.

chamber. The unstable species are produced as close as possible to the ion chamber usually by means of a high-frequency discharge, a high temperature furnace, or a gas reaction chamber (Jonathan et al., 1972 ; Cornford et al., 1972). Spectra have been obtained for atomic hydrogen, oxygen, nitrogen, and the halogens, vibrationally excited N_2, $O_2(^1\Delta_g)$, $SO(^3\Sigma^-)$, CHO^+, and CS. It is clear that much valuable information will be forthcoming as this technique is developed.

V. Conclusion

In the brief period in which it has been developed photoelectron spectroscopy has had phenomenal success in revealing the electronic structure of matter in a particularly direct way. Chemists can see the orbital structure of even fairly large molecules and no longer have to rely on the predictions of theoreticians. Future advances in the study of solids, adsorbed species, etc., can confidently be predicted to yield information on the nature and functions of the electrons involved. It is also clear that the subject will provide a happy hunting ground for the physicist for many years to come.

REFERENCES

Ames, D. L., Maier, J. P., Watt, F., and Turner, D. W. (1972). *Discuss. Faraday Soc.* **54**, 277.
Baer, T., and Tsai, B. P. (1973). *J. Electron Spectrosc.* **2**, 25.
Bardsley, J. N. (1968). *Chem. Phys. Lett.* **2**, 329.
Berkowitz, J., and Chupka, W. A. (1969). *J. Chem. Phys.* **51**, 2341.
Berry, R. S. (1966). *J. Chem. Phys.* **45**, 1228.
Blake, A. J., Bahr, J. L., Carver, J. H., and Kumar, V. (1970). *Phil. Trans. Roy. Soc. London, Ser. A* **268**, 159.
Brundle, C. R., Neumann, D., Price, W. C., Evans, D., Potts, A. W., and Streets, D. G. (1970). *J. Chem. Phys.* **53**, 705.
Buckingham, A. D., Orr, B. J., and Sichel, J. M. (1970). *Phil. Trans. Roy. Soc. London, Ser. A* **268**, 147.
Carlson, T. A. (1971a). *Chem. Phys. Lett.*, **9**, 23.
Carlson, T. A. (1971b). *J. Chem. Phys.* **55**, 4913.
Carlson, T. A., and McGuire, G. E. (1973). *J. Electron Spectrosc.* **1**, 209.
Carlson, T. A., McGuire, G. E., Jonas, A. E., Cheng, K. L., Anderson, C. P., Lu, C. C., and Pullen, B. P. (1972). *In* "Electron Spectroscopy" (D. A. Shirley, ed.), pp. 207–231. North-Holland Publ., Amsterdam.
Collin, J. E., and Natalis, P. (1968). *Chem. Phys. Lett.* **2**, 414.
Cooper, J., and Zare, R. N. (1968). *J. Chem. Phys.* **48**, 942.
Cornford, A. B., Frost, D. C., Herring, F. G., and MacDowell, C. A. (1972). *Discuss. Faraday Soc.* **54**, 56.
Eastman, D. E. (1971). *Phys. Rev. Lett.* **26**, 1108.
Eastman, D. E., and Grobman, W. D. (1972). *Phys. Rev. Lett.* **28**, 1327 and 1329.
Eland, J. D. (1970). *Int. J. Mass Spectrom. Ion Phys.* **4**, 37.
Eland, J. H. D. (1974). "Photoelectron Spectroscopy." Butterworth, London.
Fano, U. (1961). *Phys. Rev.* **124**, 1866.
Fano, U., and Cooper, J. W. (1968). *Rev. Mod. Phys.* **40**, 441.
Gardner, J. L., and Samson, J. A. R. (1973). *J. Electron Spectrosc.* **2**, 153.
Geiger, A., and Schröder, B. (1968). *J. Chem. Phys.* **49**, 740.
Gelius, U. (1972). *In* "Electron Spectroscopy" (D. A. Shirley, ed.), pp. 311–334. North-Holland Publ., Amsterdam.
Iwata, S., and Nagakura, S. (1974). *Mol. Phys.* **27**, 425.
Jonathan, N., Morris, A., Okuda, M., and Smith, D. J. (1972a). *In* "Electron Spectroscopy" (D. A. Shirley, ed.), pp. 345–350. North-Holland Publ., Amsterdam.
Jonathan, N., Morris, A., Okuda, M., Ross, K. J., and Smith, D. J. (1972b). *Discuss. Faraday Soc.* **54**, 48.
Kissinger, J. A., and Taylor, J. W. (1973). *Int. J. Mass Spectrom. Ion Phys.*, 461.
Kleimenov, V. I., Chisov, Yu. V., and Vilesov, F. I. (1971). *Opt. Spectrosc. (USSR)* **32**, 371.
Koopmans, T. (1934). *Physica* **1**, 104.
Lindholm, E. (1969). *Ark. Fys.* **40**, 117.
Lohr, L. L. (1972). *In* "Electron Spectroscopy" (D. A. Shirley, ed.), pp. 245–258. North-Holland Publ., Amsterdam.
Marr, G. V. (1967). "Photoionization Processes in Gases." Academic Press, New York.
Mies, F. H. (1968). *Phys. Rev.* **175**, 164.
Niehaus, A., and Ruf, M. W. (1971). *Chem. Phys. Lett.* **11**, 55.
Potts, A. W., and Price, W. C. (1972a). *Proc. Roy. Soc., Ser. A* **326**, 165.

Potts, A. W., and Price, W. C. (1972b). *Proc. Roy. Soc., Ser. A* **326**, 181.
Potts, A. W., Lempka, H. J., Streets, D. G., and Price, W. C. (1970). *Phil. Trans. Roy. Soc. London, Ser. A* **268**, 59.
Price, W. C. (1968). *In* "Molecular Spectroscopy" (P. W. Hepple, ed.), Vol. IV, pp. 221–236. Inst. Petroleum, London.
Price, W. C. (1972). *Discuss. Faraday Soc.* **54**, 206.
Price, W. C., and Turner, D. W. (1969/1970). A discussion on photoelectron spectroscopy. *Phil. Trans. Roy. Soc. London, Ser. A* **268**, 1–175.
Price, W. C., Potts, A. W., and Streets, D. G. (1972). *In* "Electron Spectroscopy" (D. A. Shirley, ed.), pp. 187–198. North-Holland Publ., Amsterdam.
Schweig, A., and Theil, W. (1973). *J. Electron Spectrosc.* **2**, 199.
Sichel, J. W. (1970). *Mol. Phys.* **18**, 95.
Siegbahn, K., Nordling, C., Fahlman, A., Nordberg, R., Hamrin, K., Hedman, J., Johansson, G., Bergmark, T., Karlson, S. E., Lindgren, I., and Lindberg, B. (1967). "ESCA—Atomic, Molecular and Solid State Structure Studied by Means of Electron Spectroscopy". North-Holland Publ., Amsterdam.
Siegbahn, K., Nordling, C., Johansson, G., Hedmann, J., Heden, P. F., Hamrin, K., Gelius, U., Bergmark, T., Werme, L. O., Manne, R., and Baer, Y. (1969). "ESCA—Applied to Free Molecules." North-Holland Publ., Amsterdam.
Smith, A. L. (1970). *Phil. Trans. Roy. Soc. London, Ser. A* **268**, 169.
Streets, D. G., Potts, A. W., and Price, W. C. (1972–1973). *Int. J. Mass Spectrom. Ion Phys.* **10**, 123.
Turner, D. W., and Al-Joboury, M. I. (1962). *J. Chem. Phys.* **37**, 3007.
Turner, D. W., and Al-Joboury, M. I. (1963). *J. Chem. Soc., London*, p. 5141.
Turner, D. W., Baker, A. D., Baker, C., and Brundle, C. R. (1970). "Molecular Photoelectron Spectroscopy." Wiley (Interscience), New York.
Vilesov, F. I., Kurbatov, B. C., and Terenin, A. N. (1961). *Dokl. Akad. Nauk SSSR* **138**, 1320.
Vilesov, F. I., Kurbatov, B. C., and Terenin, A. N. (1962). *Sov. Phys.—Dokl.* **8**, 883.
Walsh, A. D. (1953). *J. Chem. Soc., London*, pp. 2260 and 2266.

DYE LASERS IN ATOMIC SPECTROSCOPY

W. LANGE, J. LUTHER, and A. STEUDEL

Institut A für Experimentalphysik
Technische Universität Hannover
Hannover, Federal Republic of Germany

I. Introduction	173
II. Properties of Dye Lasers	174
A. Frequency Range	175
B. Spectral Width	176
C. Power and Energy	176
D. Pulse Properties	177
III. Applications of the High Spectral Density of Dye Lasers	177
A. Absorption Spectroscopy	179
B. Fluorescence Spectroscopy	182
C. Photoionization and Photodetachment	194
D. Two- and Multi-Photon Processes	196
IV. Applications of Tunable Dye Lasers with Extreme Narrow Bandwidth	197
A. Saturation Spectroscopy	198
B. Fluorescence Line Narrowing	205
C. Classic Absorption and Fluorescence Spectroscopy	205
D. Nonoptical Detection of Optical Resonances	209
E. Heterodyne Spectroscopy	214
F. Nonlinear Coherent Resonant Phenomena	215
References	217

I. Introduction

In the past, the application of lasers in atomic spectroscopy was essentially restricted to the study of laser transitions because only lasers of fixed frequency or very limited tuning range were available. Nevertheless many interesting investigations in molecular spectroscopy were done by means of such lasers since a large number of molecular transitions are in accidental coincidence with laser transitions. Many of these experiments have been reviewed by Demtröder (1971, 1973).

Atomic spectroscopists, doing experiments involving excitations of atoms by optical means, however, require a powerful tunable narrowband source of light. Organic dye lasers now meet this stringent demand. They offer an unrivaled versatility and simplicity in the generation of tunable coherent light over a broad range, from the near ultraviolet (340 nm) to the near infrared (1.2 μm).

Since the first description of stimulated emission from fluorescent organic dyes in liquid solution by Schäfer *et al.* (1966) and independently by Sorokin and Lankard (1966), dye lasers have been greatly developed. At the present time their narrow spectral bandwidth and their intensity per spectral interval make it possible to excite a great number of interesting levels of any atom or molecule selectively; stepwise excitation can be used for levels which cannot combine with the initial state by electric dipole radiation. The application of dye lasers could mean a great increase in the number and kinds of experiments possible in atomic spectroscopy. Problems in which conventional light sources would not work may now be solved, and experiments that were difficult with such sources will be much easier when a tunable dye laser is used. These lasers will thus greatly increase the potential of such classic spectroscopic techniques as absorption and fluorescence spectroscopy, level crossing, and optical double resonance. Moreover, tunable dye lasers will allow wide application of powerful new nonlinear techniques using saturation phenomena.

At the beginning of this new era of atomic spectroscopy we review in the present paper experiments which should clearly demonstrate what a powerful research tool tunable dye lasers in atomic spectroscopy can be. A part of the reviewed material contains experiments whose sole purpose is to check the feasibility of the methods as a whole. For this purpose the sodium D lines are particularly suited. Most experiments in which high resolution is achieved by tuning the laser frequency across the spectral structure were therefore carried out on the sodium D lines. It now seems, however, that this "sodium age" is over and that research can be extended successfully to include other subjects, especially after some of the earlier technical problems have been overcome.

The second section contains a compilation of the properties of dye lasers which are relevant to atomic spectroscopy. In the main part of the text (Sections III and IV), some of the exciting new possibilities opened up by the development of tunable dye lasers will be illustrated by a number of recent experiments, and outlines will be given for other experiments which could be done in the future.

As this review paper had to be completed in October 1973, only publications up to this date could be included in this contribution.

II. Properties of Dye Lasers

The principles of dye lasers have been treated in several excellent review articles (Snavely, 1969; Sorokin, 1969; Schäfer, 1970, 1972; Schmidt, 1970) to which the reader is referred. For convenience, several properties which are most important in applications of the laser are discussed briefly in this section. The emphasis lies on the provision of typical results, which can also be reproduced in the laboratory of the spectroscopist. Extreme (exceptional) results are only mentioned in a few cases and no completeness of references is intended. (References cited in the review articles mentioned above are not usually given in the following.) In most cases, good results are not obtained with respect to all the properties of a dye laser simultaneously; for example, dye lasers with a high output power seldom have a low beam divergence and a narrow spectral width. This means that the type and design of the dye laser must be chosen for the specific purpose to which it will be put. It should also be noted that all data given refer to rhodamine 6G as the laser medium, where not otherwise specified, because performance can be considerably poorer with other laser media.

The high power density necessary to optically pump dye lasers can be obtained either by suitable flashlamp discharges or by fixed-frequency lasers. Of the latter, giant-pulse ruby or Nd : glass lasers or N_2 lasers are the most common in use at present. The first two often have to be frequency doubled since the pumping wavelength has to be shorter than the wavelength of the dye-laser output. Xenon lasers can also be used to advantage (Hänsch et al., 1973). Continuous wave (cw) dye lasers have always, until now, been pumped by Ar^+ lasers.

A. FREQUENCY RANGE

Pulsed fixed-frequency lasers are the most universal pumping sources for dye lasers. Pulsed laser-pumped dye lasers are capable of tunable emission throughout the visible spectrum, including the near uv and the near ir region (340–1200 nm); for a compilation of laser dyes see Schäfer (1972). A somewhat smaller range is covered by fast flashlamp-pumped lasers (Warden and Gough, 1971), in particular it has been difficult to reach the uv and ir region.

Tunable cw radiation can at present be obtained in the visible region from 420 to 700 nm (Tuccio and Strome, 1972; Tuccio et al., 1973). It should be possible, however, to synthesize new laser dyes which would allow tunable cw laser action down to 375 nm, utilizing currently available Ar^+ lasers as the pumping source (Tuccio et al., 1973).

The tuning range of the dye laser depends greatly on the particular dye or dye mixture used in the laser, and this range can vary from several nanometers to over a hundred nanometers. Using dye lasers in combination with frequency-conversion systems, the range in which tunable laser radiation is available can be appreciably extended; in this way, tunable uv radiation down to 216 nm has been obtained, either by dye laser frequency doubling or by summing the dye laser and pumping laser frequencies (Dunning et al., 1973; Gabel and Hercher, 1972; Dinev et al., 1972); tunable mid-infrared radiation has been generated as the difference between the outputs of two lasers (Dewey and Hocker, 1971; Decker and Tittel, 1973).

B. Spectral Width

The linewidth of the simplest dye laser configuration is on the order of 5 nm, but this width can be dramatically reduced by adding frequency selective elements such as prisms, defraction gratings, interference filters, Fabry–Perot interferometers, and birefringent filters to the laser cavity. By means of these elements, tuning of the dye laser can also be achieved.

The smallest linewidths obtained so far with free-running dye lasers are approximately 2 MHz for a cw laser (Hartig and Walther, 1973; Schröder et al., 1973), 4 MHz for a flashlamp-pumped dye laser (Gale, 1973), and about 400 MHz for a N_2 laser-pumped dye laser (Hänsch, 1972). The frequency stability of these lasers is on the order of their bandwidth. A further improvement of linewidth and stability can be obtained by locking the dye laser to an external reference tool such as a stabilized Fabry–Perot cavity (Barger et al., 1973) or to an atomic or molecular line (Sorokin et al., 1969; Hartig and Walther, 1973); bandwidths as low as 50 kHz have been realized utilizing this technique (Barger et al., 1973).

C. Power and Energy

A typical specification of a flashlamp-pumped dye laser is a 10–100 mJ energy output in a 0.3- to 1-μsec pulse with a maximum repetition rate of 10–100 Hz. Thus the mean power output is on the order of 1 W (see, for example, Schmidt and Wittekind, 1972; Loth and Meyer, 1973). In dye lasers especially designed to give a high output, energies of 12 J (Anliker et al., 1972) and 110 J (Baltakov et al., 1973) have been obtained. In the latter case, however, the pulse length was high (~ 20 μsec). High peak powers can be obtained by mode locking (see Section II,D).

The output energy of dye lasers pumped by frequency-doubled giant-pulse Nd : glass or ruby lasers is also on the order of 10–100 mJ. The pulse length is about 20 nsec, which means that the power is high. The pulse repetition rate is typically on the order of 1 Hz. By N_2 laser pumping, output pulses with an energy of 10 μJ–1 mJ, a width of 2–8 nsec, and a repetition rate of about 100 Hz are achieved (see, for example, Hänsch, 1972).

If high output combined with narrow spectral width and low beam divergence is required, then the use of oscillator–amplifier systems can be advantageous (Huth, 1970; Itzkan and Cunningham, 1972; Carlsten and McIlrath, 1973; Curry et al., 1973).

The pumping power in cw dye lasers is on the order of a few watts. From the latest designs of dye laser cavities (Kogelnik et al., 1972; Jacobs et al., 1973), energy conversion efficiencies of up to 25% can be obtained.

D. PULSE PROPERTIES

As indicated in Section II,C, the typical pulse lengths of dye lasers are 1 μsec with flashlamp pumping, a few nanoseconds or about 20 nsec with pumping by N_2 lasers and with Q-switched Nd : glass or ruby lasers, respectively. In the techniques of active or passive mode locking (for a review, see Smith, 1970; Zel'dovich and Kuznetsova, 1972; New, 1972) the dye laser pulse is broken up into a train of very short pulses, while the total output energy is almost wholly conserved (Bradley and O'Neill, 1969; Ferrar, 1969). Using passive mode locking of a flashlamp-pumped dye laser, pulses with a length of ∼ 3 psec and an energy of 5 μJ have been obtained (Arthurs et al., 1971).

Another way of obtaining picosecond pulses is with pumping by mode-locked lasers (Lin et al., 1973). As a means of generating pulses of variable lengths in the subnanosecond region, the use of a Michelson interferometer pulse-forming network as part of the laser cavity has been proposed (Szöke et al., 1972). Short pulses adjustable between 2 and 20 nsec have been obtained from a double-cavity laser (Eranian et al., 1973).

Of course, cw dye lasers can also be mode locked (Ippen et al., 1972; O'Neill, 1972), the obtained pulse length being approximately 1 psec. The Ar^+ laser, used for pumping, can, moreover, be periodically Q-switched; in this way a train of intense pulses with constant amplitude can be produced, either in the picosecond (mode locking) or in the nanosecond region (Q-switching of the pump laser). It should be emphasized that cw dye lasers can be easily modulated by intra- and extra-cavity devices.

III. Applications of the High Spectral Density of Dye Lasers

For atoms in the field of a conventional light source the interaction with the light is usually very weak compared to the interaction with the zero point fluctuations of the vacuum. In this situation the probability of absorption for an atom in the ground state is much smaller than the probability of spontaneous emission for an atom in the excited state. (For convenience a two-level system is assumed.) Consequently, the population of the ground state is hardly influenced by the light source; in practice it is difficult to produce more than 10^{-4}–10^{-3} of the atoms in the excited state. In a (resonant) laser field, however, the probabilities can be of the same order of magnitude or the spontaneous emission can even be negligible compared with the induced processes. Under these conditions the population of the ground state is heavily influenced by the light source; the transition is said to be "saturated." Obviously it is very important in saturation phenomena to know whether the laser can strongly interact with all atoms. For this condition it is necessary that the spectral width of the laser be greater than the "inhomogeneous linewidth," i.e., the absorption width of the atomic ensemble. (The inhomogeneous linewidth includes Doppler broadening and Stark broadening by random electrical fields.) This condition is fulfilled in all experiments treated in this section. In most of them saturation is essential.

With a linewidth of 0.04 nm, which is typical for dye lasers with medium performance, a power of about 10 W/cm^2 is necessary to saturate, for example, the D_1 line of sodium, if the pulse length is long compared to the lifetime of the upper level ($\tau \sim$ 16 nsec). This is easily achieved with pulsed dye lasers. Even intercombination lines with very low oscillator strengths can be saturated by powerful dye lasers (see Section III,A).

By saturating atomic transitions it is obviously possible to prepare a large number of atoms in well-defined states for measurements of several atomic features. Because of this new technique, many experiments in atomic physics can be changed dramatically by exchanging the conventional light source with a dye laser. Moreover, many classical experimental schemes should be reconsidered to decide if a tunable laser could make any improvement. For example, Hebner and Nygaard (1971) discussed the idea of using a direct optical pumping method for a cesium maser as an alternative to the magnetic deflection scheme that is presently utilized to separate cesium atoms into groups of $F = 3$ and $F = 4$ in atomic clocks. This would be an excellent application of dye lasers. (Indeed, these authors were able to selectively depopulate the $F = 3$ hyperfine level of the Cs ground state by optical pumping with a pulsed GaAs laser.)

As was mentioned previously, in Section III experiments will be treated in which the spectral width of the laser is larger than the inhomogeneous width. Equally important in the work with saturated transitions is the opposite case, where the spectral width of the laser is smaller than the "homogeneous width," i.e., the absorption width of a single atom. In this case saturation is obtained only for a subgroup of atoms whose transition frequencies coincide with the laser frequency within the homogeneous linewidth. In the example of the D_1 line of sodium, saturation within a subgroup of atoms is obtained at power levels of a few milliwatts per square centimeter, if the spectral width of the laser is 5 MHz. (The natural linewidth of a single hyperfine component of the D_1 line is about 10 MHz.) Spectral widths on this order of magnitude can be achieved now (see Section II,B) and the power density can be obtained even with cw dye lasers. The saturation of subgroups of atoms gives rise to interesting phenomena. Some of them are very useful in spectroscopy; they are treated in Section IV.

While saturation effects occur at relatively modest power levels, the high power of pulsed dye lasers can induce several other kinds of nonlinear effects which are not so familiar to the atomic spectroscopist. One of these is multiquantum transitions, which are briefly discussed in Section III,D.

A. ABSORPTION SPECTROSCOPY

Some possibilities for the application of dye lasers to atomic spectroscopy were shown in the early work of McIlrath (1969) and of Bradley and associates (1970a). McIlrath used a frequency-doubled Nd : glass laser (300 mJ pulses of 40-nsec duration at 530 nm) to pump a dye laser system. The resultant output at 657.278 nm, the wavelength of the spin forbidden $4s^2$ $^1S_0$–4s 4p $^3P_1$ transition in calcium, was 30 mJ (10^{17} photons) with a linewidth of about 0.04 nm. Even with this large emission width, the laser power was sufficient to saturate this transition. The beam was directed into a resonance vessel containing calcium vapor held at 800°C with 40 Torr He as buffer gas. Immediately after the laser pulse, which excited the $^3P_1$ state, a fast rise flashlamp was fired; this served as light source for absorption spectroscopy. By comparison of the absorption spectra obtained with and without dye laser pumping, the transitions originating from the ^3P term can easily be found. In this case they were of the type 4s 4p ^3P–4s ns ^3S and 4s 4p ^3P–4s n'd ^3D. In an improved experiment, transitions with n up to 14 and n' up to 22 were observed (McIlrath and Carlsten, 1973). In addition, absorption lines corresponding to the autoionizing transitions 4s 4p ^3P–3d 4d ^3D and 4s 4p ^3P–3d 4d ^3S in the continuum, and transitions to the nonautoionizing levels 3d 4d ^3P and 3d 5d ^3P also lying in the continuum have been observed.

The experiment of Bradley et al. (1970a) was similar; the 5 $^2P_{3/2}$ state of rubidium was populated using a dye laser of narrow bandwidth, pumped by a giant-pulse ruby laser. Because of the shorter lifetime of the 5 $^2P_{3/2}$ state of rubidium compared with the 4s 4p $^3P_1$ state of calcium, absorption spectroscopy using a flashlamp would not be suitable in this case. Instead, the ruby laser was used to pump simultaneously a second (broadband) dye laser, producing a quasicontinuum automatically synchronized with the pulse exciting the 5 $^2P_{3/2}$ state. As a check of this method, the absorption line 5 $^2P_{3/2}$–5 $^2D_{5/2}$ was recorded. The disadvantage of this technique, however, is that only a small part of the absorption spectrum can be recorded at one time. Air sparks, produced by focusing the output of the pump laser, have been proposed as an alternative absorption background source in experiments probing short-lived states (McIlrath, 1969). Another possibility is

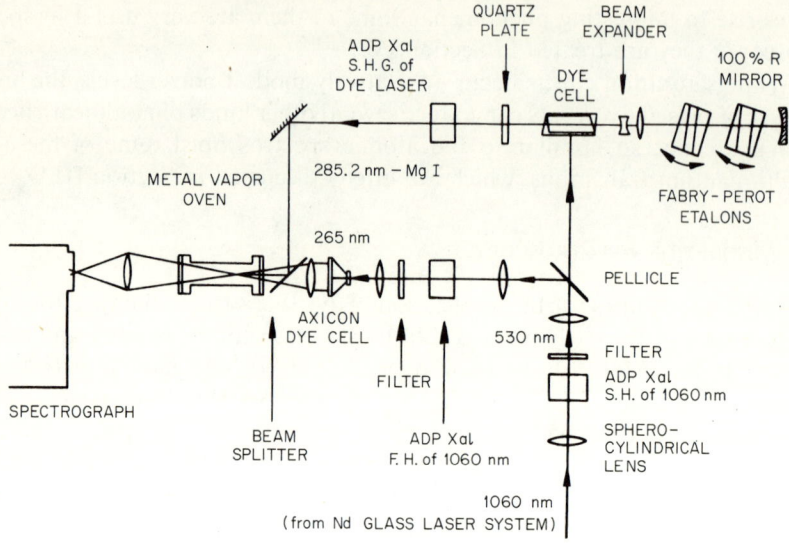

FIG. 1. General experimental arrangement for absorption experiments. From Bradley *et al.* (1973a).

employed in the work of Bradley *et al.* (1973a). The experimental setup is shown in Fig. 1. They use the frequency-doubled output of a Nd : glass oscillator-amplifier system (1 J at 530 nm) to pump a dye laser populating the excited 3s 3p $^1P_1$ state of magnesium. Part of the pumping beam is further frequency doubled to the fourth harmonic of the Nd laser. This 265-nm radiation is used to excite fluorescence in scintillator dyes which then serves as a background continuum. Absorption spectra were recorded from 400 to 290 nm. In this experiment the series 3s 3p $^1P_1$–3s nd $^1D_2$ was obtained in absorption for the first time, up to $n = 24$ (Fig. 2). In emission the series has been seen up to $n = 13$ only. Furthermore, the absorption due to the autoionizing transition 3s 3p $^1P_1$–3p^2 $^1S_0$ could be clearly seen and its width determined by microdensitometry. Very similar results have been obtained in barium (Bradley *et al.*, 1973b).

These types of experiment seem extremely useful in the analysis of spectra because of the new capability of observing absorption transitions which cannot be recorded from the ground state or low-lying thermally excited states. There is a special advantage that levels above the normal ionization limit can now be investigated without employing vacuum-ultraviolet techniques. In addition there is the possibility that resonance fluorescence might be observed from a highly excited level, selectively populated by a narrowband dye laser.

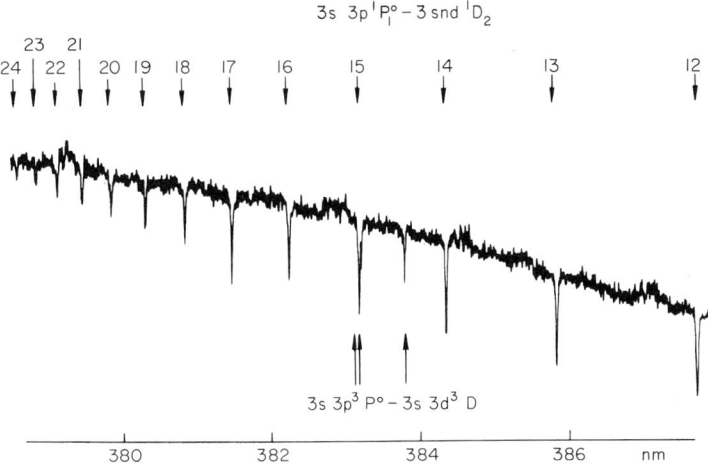

FIG. 2. Microdensitometer trace showing higher terms of the 3s 3p $^1P_1$–3s nd $^1D_2$ absorption series obtained from selectively excited Mg I. Triplet absorption lines due to collision-induced ^1P–^3P transitions are also present. From Bradley *et al.* (1973a).

In all the absorption experiments mentioned above a few torr of buffer gas were used in the metal-vapor oven. Thus, collisions occurred and transitions starting from other levels besides the one originally populated were observed. This might be used as a tool in the study of collision processes, if, for example, the absorption spectra are recorded with different time delays after the exciting laser pulse.

Of more practical interest are absorption experiments using dye lasers to detect low concentrations of atoms or molecules. Added sensitivity has resulted from laboratory experiments in which the absorbing medium is placed within the dye laser cavity (Peterson *et al.*, 1971; Thrash *et al.*, 1971). In these experiments performed on Na and I_2, and on Ba^+ and Sr, respectively, the dye laser runs without frequency-determining elements and consequently emits a broad continuum. At the absorbed wavelengths the laser action is substantially decreased or completely quenched. In this laser-enhanced absorption method (absorption cell inside cavity) the detectability limit is lowered by about two orders of magnitude compared with the conventional type of experiment (cell outside cavity). This technique is also useful for kinetic studies on transient chemical species (Atkinson *et al.*, 1973).

Within the different schemes for optical detection of pollutants in the atmosphere, absorption experiments are the most sensitive. The application of an absorption technique presents practical difficulties, however, if remote detection of pollutants and determination of their spatial distributions are desired. These may be overcome by detecting the absorption in the Mie backscattering of a dye-laser beam, as discussed by Ahmed (1973). Recently this scheme has been successfully employed by Walther (1973b) and associates. It seems to offer a much higher sensitivity than other schemes applied in LIDAR (laser radar) studies.

B. Fluorescence Spectroscopy

1. Lifetimes

The most obvious fluorescence experiment that can be performed using a pulsed dye laser is the direct determination of lifetimes by observing the exponential decay of selectively excited levels. For this purpose it is necessary to switch off the laser source in a time T which is shorter than the lifetime τ to be measured. In flashlamp-pumped dye lasers with their relatively slow time behavior, this condition can be fulfilled by the use of electrooptical switches or similar devices. [For example, Erdmann et al. (1972) employed switching of the wavelength-determining intracavity Lyot filter in their remeasurement of the $3\ ^2P_{1/2}$ lifetime of sodium.] The condition $T < \tau$ will usually be fulfilled in atomic lifetime studies if a nitrogen laser-pumped dye laser is employed (pulse length 2–5 nsec); a disadvantage, however, is the small output energy of these lasers (typically $\sim 50\ \mu J$), which is insufficient to saturate transitions with low f-values. By pumping dyes with Q-switched, frequency multiplied ruby or Nd : glass lasers, short pulses with high output energy can be obtained. In principle, the short duration, high-energy pulses from mode-locked flashlamp-pumped dye lasers could also be employed. Since only one level will be excited by a narrowband dye laser, lifetime measurements can be made even in those cases where the level structure is fairly complex.

The great advantage of using dye lasers in lifetime studies is that by stepwise excitation direct measurements can be made on levels which do not combine with the ground state or with levels which are thermally populated at reasonable temperatures. As an example a part of the level scheme of sodium is shown in Fig. 3. Only the $n\ ^2P$ levels can be excited by electric dipole radiation without special means, since under normal conditions only the $3\ ^2S_{1/2}$ ground state is populated. In order to excite other levels, electron bombardment or beam foil excitation are generally employed. Both methods suffer from a lack of selectivity, as cascading from higher excited states tends to complicate the results. Thus, very few reliable measurements have been performed even in the simple case of sodium. It has been shown that this problem can be overcome by the use of dye lasers (Gornik et al., 1973a). The experimental scheme employed by Gornik et al. is shown in Fig. 4. A

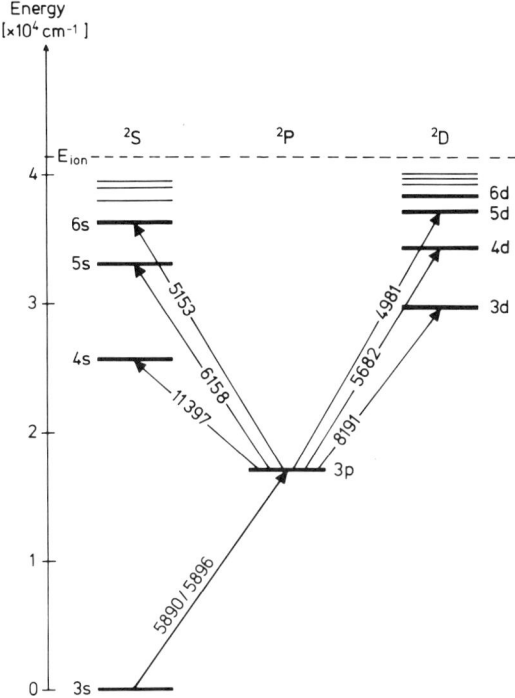

FIG. 3. Partial term diagram of Na. (All wavelengths are given in angstroms.)

flashlamp-pumped dye laser (pulse duration 500 nsec) tuned to the sodium D_2 line was used to populate the $3\ ^2P_{3/2}$ level. The upper transition (cf. Fig. 3) starting from this intermediate level was induced by the 5-nsec pulse of a N_2 laser-pumped dye laser. In both lasers the linewidth was reduced to the same order as the Doppler width and the wavelengths were carefully adjusted to resonance. The two laser beams intersected in an atomic beam of sodium. The N_2 laser was triggered by the light of the flashlamp via an adjustable delay line, thus providing time overlap of the two laser pulses. The fluorescence radiation was detected by a photomultiplier, which was protected against strong radiation from the intermediate state by means of a monochromator or an interference filter together with suitably absorbing solutions. Immediately after the end of the second laser pulse the decay of the $n\ ^2D$ or the $n\ ^2S$ fluorescence intensity was registered in the 20 channels of a fast transient recorder (Fig. 5). The data were accumulated and processed by a small computer which also operated as the time control of the experiment. Thus the lifetime of the $4\ ^2D$ state was deduced. Recently the measurements have been extended up to the $9\ ^2S$ and $7\ ^2D$ levels (Gornik *et*

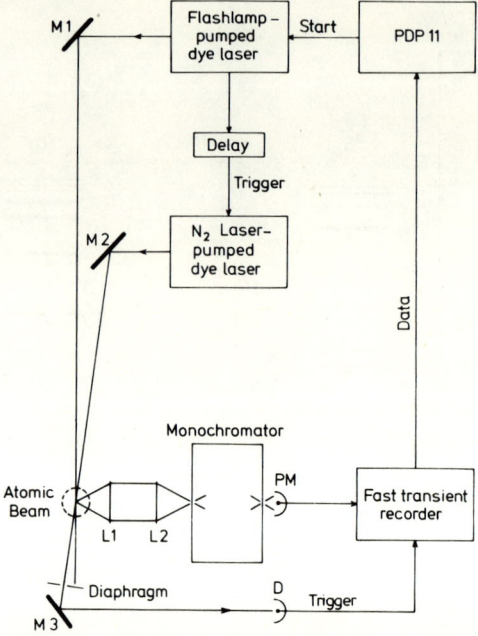

Fig. 4. Schematic drawing of the apparatus for lifetime measurements (M1, M2, M3, mirrors, PM, photomultiplier, D, *PIN* photodiode).

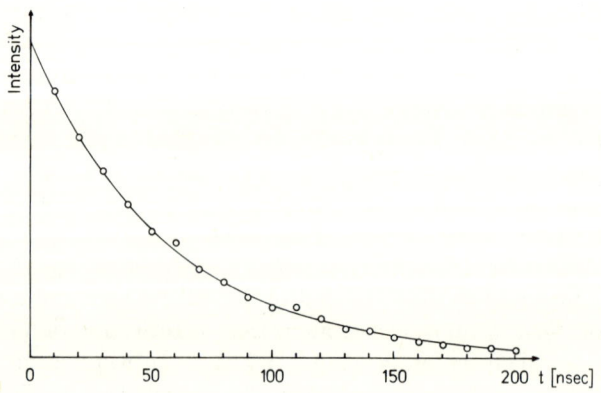

Fig. 5. Decay of the 4 ^{2}D fluorescence intensity (250 single decay curves have been averaged; circles are experimental points).

al., 1973b). An accuracy of 1% should be obtainable. Moreover, it should be possible to employ other experimental techniques, e.g., level crossing, once the atoms are produced in the excited states. Care must be taken in these experiments to avoid effects induced by the coherence of the exciting light, which may disturb the measurements. This has been thoroughly discussed by McIlrath and Carlsten (1972). It can be assumed, however, that the degree of coherence is rather low in most nitrogen-pumped dye lasers.

For a given power of the exciting light, the maximum population of the intermediate state is nearly reached after one natural lifetime of this state. Thus, stepwise excitation processes involving short-lived intermediate states can be performed quite favorably by the use of two dye lasers, pumped separately by a single N_2 laser. Since the wavelength region of such a dye laser extends from 340 to 1100 nm, the method offers extreme versatility. In particular, levels up to 7.2 eV can be reached; this limit is extended to more than 9.5 eV if both lasers are frequency doubled. It is not necessary, of course, to employ two lasers to achieve the advantage of selectivity in exciting higher levels. It can be more advantageous to utilize an electric discharge for the first step of excitation if the energy of the intermediate state is high or the oscillator strength very low. Collins *et al.* (1972) started with the metastable $2\ ^3S$ state of helium present in a flowing afterglow and induced the step from the $2\ ^3S$ to the $5\ ^3P$ state by a frequency-doubled dye laser. Gornik *et al.* (1973c) populated the metastable $4s\ 4p\ ^3P_0$ and the $4s\ 4p\ ^3P_2$ state of calcium by a Penning discharge and, in a second step, excited the $4s\ 5s\ ^3S_1$ and $4s\ 4d\ ^3D_1$ states with a N_2 laser-pumped narrowband dye laser.

2. Quantum-Beat and Modulation Experiments

Under certain experimental conditions, pulsed optical excitation of an atomic ensemble results in a modulation, called quantum beats, of the decaying fluorescence intensity (Corney and Series, 1964a; Dodd *et al.*, 1964, 1967; Alexandrov, 1964). The frequencies of this modulation are proportional to the energy separation of the coherently excited levels, and the damping of the oscillations is determined by the relaxation of the coherence in the excited ensemble. Thus the time-resolved observation of the fluorescent light following a pulsed coherent excitation makes it possible to measure, simultaneously, energy splittings in the excited states and the relaxation times of population and coherence.

Quantum beats are only observed if the exciting light pulse is kept short compared to the times that are characteristic for the time evolution of the excited ensemble, that is, the light pulse has to be appreciably shorter than the decay times and the modulation periods of the fluorescent light. For atomic states, this usually means that pulse lengths longer than several nanoseconds are not tolerable. Since it is extremely difficult to obtain, simultaneously, short pulses and sufficiently high power from classical light source systems, the method of quantum beats has only been used in a very few cases in the past. Now, using pulsed tunable dye lasers as light sources this method seems to be becoming a favorable tool in atomic spectroscopy.

In the following, we describe such a quantum-beat experiment involving pulsed dye laser excitation (Gornik *et al.*, 1972). The principle of this experiment is shown in Fig. 6. Ytterbium atoms in a homogeneous magnetic field

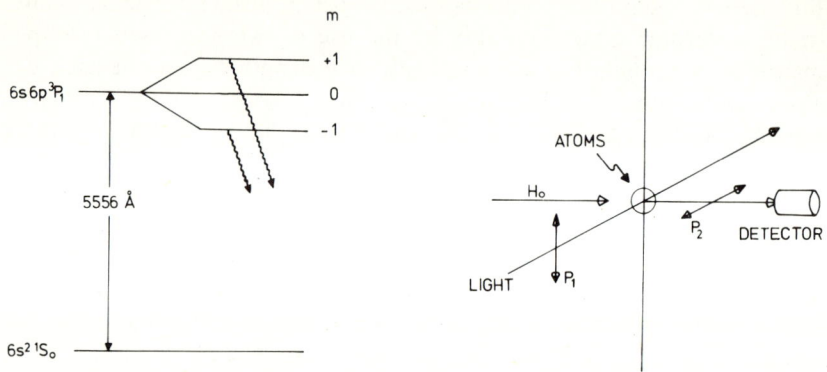

FIG. 6. Level scheme and experimental configuration for quantum-beat observation in the triplet resonance line of Yb. (P_1, P_2, polarizers).

of about 1 Oe are excited to the 6s 6p $^3P_1$ state (lifetime $\sim$ 1 μsec) by the short pulse (5 nsec) of an N_2 laser-pumped dye laser. Since the polarization of the exciting light was chosen to be perpendicular to the magnetic field, the $m = \pm 1$ sublevels of the $^3P_1$ are excited coherently. If the fluorescent light in the direction of the magnetic field is detected through a second polarizer a modulation of the decaying intensity is observed. The modulation of a few megahertz is equal to the energy separation of the $m = \pm 1$ Zeeman sublevels. Figure 7 shows a quantum-beat signal produced by a single shot of excitation. The photo was taken from a fast oscilloscope which was directly connected to a 50-ohm-terminated photomultiplier. The great advantage of laser excitation becomes evident if one considers that the information displayed in the photo was obtained in only 5 μsec. In experiments using conventional light sources, an integration time of several hours is necessary to obtain a similar signal-to-noise ratio (Dodd *et al.*, 1967). This dramatic change in sampling time now makes it possible to apply the quantum-beat method at a large variety of atomic levels.

In order to deduce precise relaxation times and splitting factors from quantum-beat signals, it is appropriate to average the signals over several

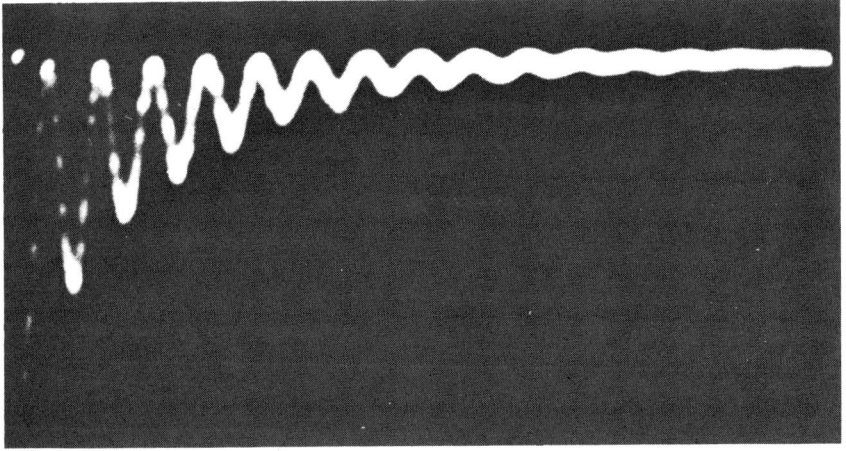

FIG. 7. Quantum-beat signal from a single pulse of excitation (full scale 5 μsec).

individual shots of excitation; moreover, the signals should be prepared in a digital form. Figure 8 shows an experimental setup which satisfies such requirements. Using this setup, signals such as those in Fig. 9 are obtained. From such a measurement the lifetime and the magnetic splitting factor g_J of the 6s 6p $^3P_1$ of Yb could be determined with an error of 5 and 0.5%, respectively.

Because of the good signal-to-noise ratio obtained in quantum-beat experiments under laser excitation, this method seems to present a favorable way to investigate the influence of collisions (Omont, 1965; D'Yakonov and Perel, 1965a) and multiple scattering (Holstein, 1947; Pringsheim, 1948; Barrat, 1959; D'Yakonov and Perel, 1965b) on the relaxation times of the excited atomic ensemble. As mentioned above, the decay constant of both population and coherence can be simultaneously deduced from quantum-beat signals. Examples of such measurements are given in Figs. 10 and 11. At considerably higher atomic densities than in Fig. 9, a slower decay of the fluorescence intensity due to radiation trapping is observed; moreover, coherent multiple scattering can be seen which results in smaller amplitude and a longer relaxation time of the modulation (Barrat, 1959). Collisions with noble gas atoms modify the quantum-beat signals in a way shown in Fig. 11; the relaxation time of coherence T_c is appreciably shortened, but on the other hand, almost no influence of the collisions on the relaxation time of population T_p can be observed.

Evidently the quantum-beat technique is not restricted to the case of Zeeman splitting. The method can also be used to measure, for example, Stark splittings or hyperfine structures, provided that the modulation

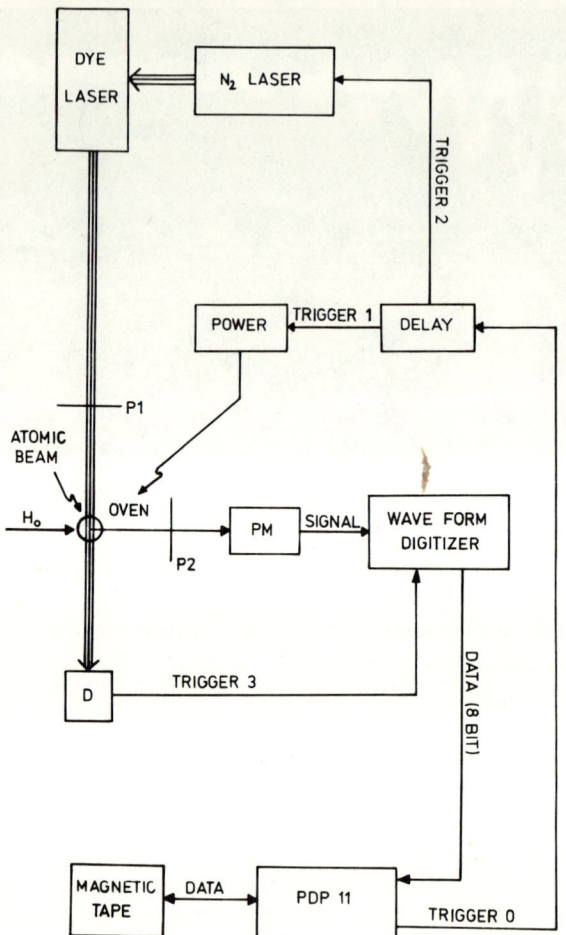

FIG. 8. Experimental setup for quantum-beat observation and analysis. (D, photodiode; PM, photomultiplier; P1, P2, polarizers).

frequencies are low enough to be detected electronically. Haroche et al. (1973) observed hyperfine quantum beats under pulsed dye laser excitation in the 444.5 nm resonance line of Cs (6 $^2S_{1/2}$–7 $^2P_{3/2}$). From the beat signals (Fig. 12) of the coherently excited hyperfine levels they were able to deduce the three hyperfine intervals (82.9, 66.5, 49.9 MHz) of ^{133}Cs to an accuracy of about 1%.

If the single pulse excitation used in quantum-beat experiments is changed into an excitation by a pulse train, the fluorescent light will display a steady-state modulation (Alexandrov, 1963; Dodd et al., 1964; Corney and Series,

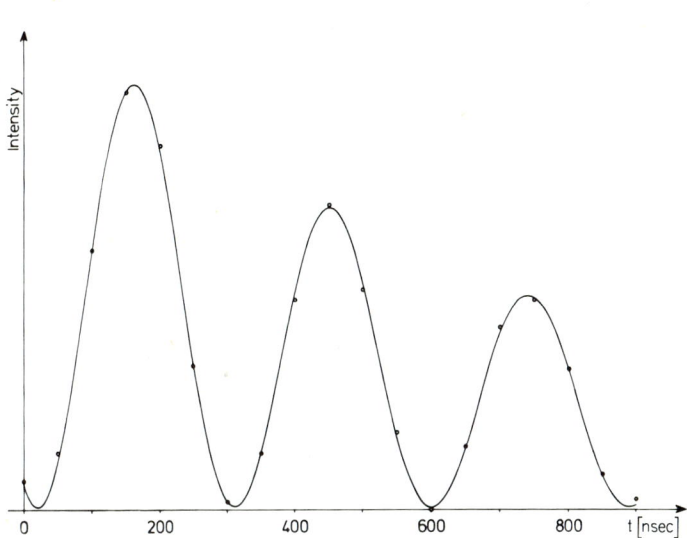

FIG. 9. Computer analysis of a quantum-beat signal obtained by means of the apparatus as shown in Fig. 8. The signals from 100 individual shots have been averaged, total measuring time 2 sec, low density in the atomic beam. (Circles are experimental points.)

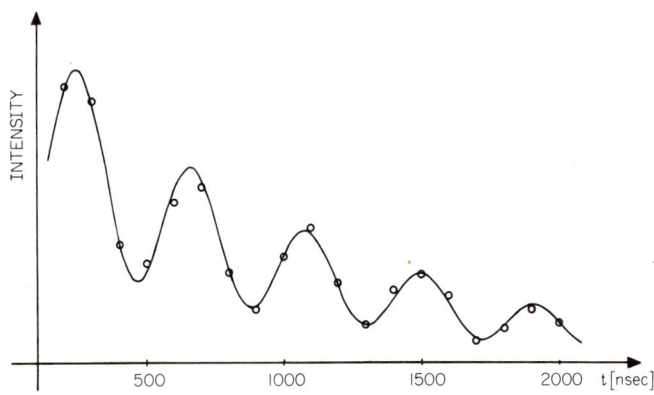

FIG. 10. Quantum-beat signal recorded with a substantially higher particle density in the atomic beam than in Fig. 9. The change of the signal is partly due to multiple scattering and partly to coherent multiple scattering.

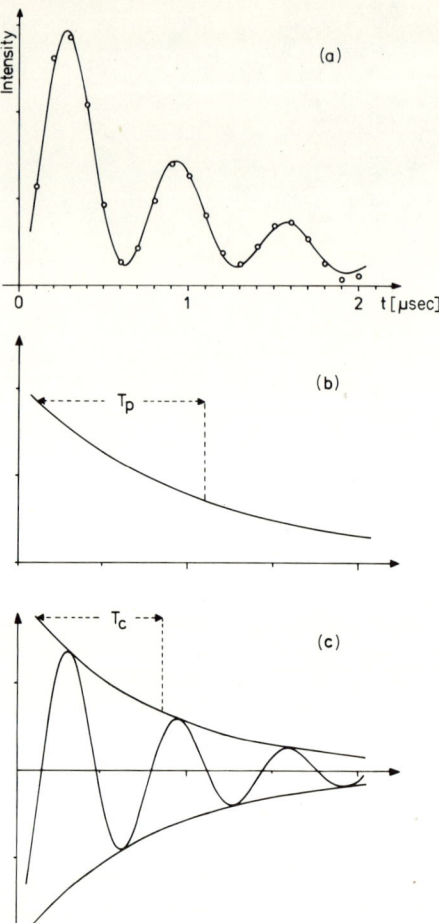

FIG. 11. Quantum-beat signal and analysis showing the influence of collisions (Ar with Yb) on the relaxation times of the sample (Yb). (a) The quantum-beat signal; (b) the unmodulated, (c) the modulated component of the fitted curve. (T_p, T_c relaxation times of population and coherence, respectively; circles are experimental points.)

1964a,b). Obviously its amplitude will show a maximum if the separation of two pulses is equal to the period of a quantum beat. Thus by simply measuring the amplitude, but not the frequency, of the modulation in the fluorescent light as a function of the pulse repetition frequency, it is possible to determine the energy separation between coherently excited states. Evidently such an experiment is more conveniently performed with a modulated light beam rather than with a train of pulses. This experimental technique is especially advantageous in the case of short-lived states, where the double-

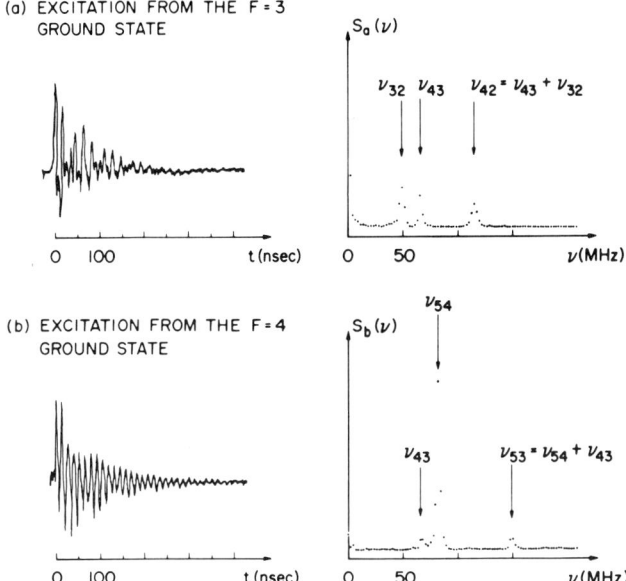

FIG. 12. Modulated component and frequency analysis of the hyperfine quantum-beat signal in the 455.5-nm resonance line ($6\,^2S_{1/2}$–$7\,^2P_{3/2}$) of ^{133}Cs. ν_{ij} is the frequency difference between the hyperfine levels $F = i$ and $F = j$ of the $7\,^2P_{3/2}$. From Haroche et al. (1973).

resonance technique requires high rf fields and time resolution becomes critical in quantum-beat experiments. Until now dye lasers have not been used in such modulation experiments, though they should offer a great advantage compared to classical illumination systems which cannot produce intense rf-modulated light beams.

3. Level Crossing and Optical-Radiofrequency Double Resonance

Closely related to the modulation and beat effects described above are pulsed level-crossing experiments (Copley et al., 1968). These experiments—which are easily performed using pulsed dye lasers—may in principle offer some advantages with respect to the energy resolution of the method. If the fluorescent light following the pulse excitation is not taken from all atoms but from a sample that has been at least a certain time in the excited state, an undulation is superimposed on the normal level-crossing signal. The central peak of this pattern can be considerably narrower than a signal found under cw excitation (Copley et al., 1968; zu Putlitz, 1969) (see Fig. 13). Technically this type of level-crossing signal is obtained by means of a gated detection

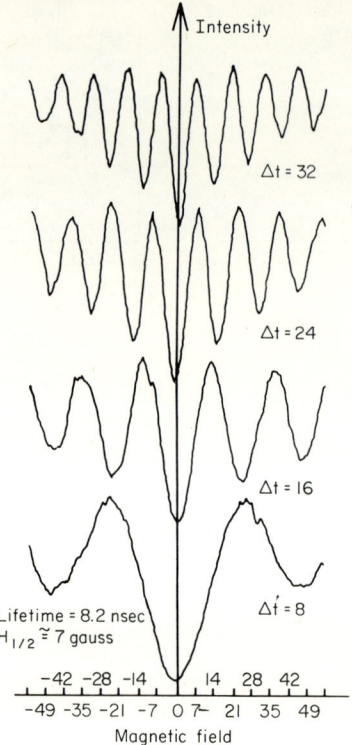

FIG. 13. Signals from a time-delayed level-crossing experiment in the Ba 553.5 nm line ($6s^2\ ^1S_0$–$6s\ 6p\ ^3P_1$). Δt, delay time in nsec. From Schenck et al. (1973).

system, the typical delay times between excitation and opening of the gate being several natural lifetimes of the level under study. In the case of narrow spaced level crossings, the spurious signals, that is, the undulations just mentioned, produce difficulties in the interpretation of the measurements. If, however, the time function of the exciting pulse and the time response of the gating system is known, these spurious signals can, in principle, be eliminated during the evaluation of the measurements.

Delayed level-crossing experiments using pulsed dye laser excitation have been performed by Schenck et al. (1973) in the Ba and Ca resonance lines and by Walther (1973a) in the Na D_2 line. The widths of the central peaks observed in these experiments are by a factor of 2 to 6 smaller than those determined by the natural lifetime of the levels concerned.

Methods such as quantum beats or modulated excitation, though known for many years, have never become a standard technique in the past since no adequate light sources have been available. The invention of the dye laser will most probably change these methods into widely used spectroscopic tools. Also, in standard spectroscopic techniques such as double resonance or level crossing, the dye laser will increase greatly their range of applicability. As discussed earlier, high lying atomic levels may become considerably populated by dye laser excitation and hence investigated with the approved methods just mentioned. An experiment of this kind has been performed by Svanberg et al. (1973). A two-step optical excitation method involving an rf lamp and a cw dye laser was used for the population of highly excited D levels in Cs and Rb. Using this technique, it was possible to determine the hyperfine splitting constants of the 8, 9, and 10 ^{2}D levels of ^{133}Cs and the 6 and 7 ^{2}D levels of ^{87}Rb by means of the level-crossing technique and the optical double-resonance method. Moreover, a double-resonance signal was obtained from the 5 ^{2}F of ^{133}Cs that was populated by cascading from the highly excited ^{2}D levels.

Due to the broad linewidth of thermal light sources, a large number of hyperfine or Zeeman levels are usually excited in level-crossing, double-resonance, or related experiments. On the other hand, a remarkable increase in signal-to-noise ratio and in resolution may be obtained if a particular subgroup of such a level system were populated (Kretzen, 1973). This has been shown in a few classical double-resonance experiments using Zeeman-scanned and/or monoisotopic light sources (for a review, see zu Putlitz, 1965). Obviously this method of selective excitation of Zeeman or hyperfine levels in the type of experiments just mentioned should be applicable in many cases, using narrowband tunable dye lasers.

4. Other Experiments

Collisional quenching of resonance radiation and excitation-transfer have been studied using the advantages of dye laser sources to produce high populations of the atomic levels involved. Examples are given in the work of Collins et al. (1972) mentioned above, and by Burell and Kunze (1972) on helium, and by Grandin and Husson (1973), who observed time-resolved sensitized fluorescence arising from energy transfer in rare gases which were excited by a pulsed dye laser. Sorokin and Lankard (1969b) observed infrared laser action in helium-buffered strontium vapor, resulting from the combined effects of optical pumping by a dye laser and inelastic collisions. Infrared laser action was also observed in vapors of potassium, rubidium, and cesium, photoexcited by intense fixed frequency lasers or by dye lasers (Sorokin and Lankard, 1969a, 1971). As an explanation for this phenomenon the authors suggested that the radiation of the primary laser is absorbed by alkali-metal molecules, which subsequently dissociate into excited atomic states. In the case of cesium the laser emission was strongly changed when helium buffer gas was present.

The strong resonance fluorescence occurring when an atomic transition is saturated or nearly saturated has been successfully utilized for the detection of low concentrations of atoms. For example, flame-fluorescence work has been performed on barium (Denton and Malmstadt, 1971), sodium (Kuhl and Marowsky, 1971) and, with a frequency-doubled dye laser, on magnesium, nickel, and lead (Kuhl and Spitschan, 1973). In this work the detection limits were 0.0003 ppm for magnesium, 0.1 ppm for nickel, and 0.03 ppm for lead.

Resonance scattering lends itself favorably to use in LIDAR studies (see Section III,A), and a particularly spectacular example is the work in the detection of sodium in the upper atmosphere. The concentration of sodium was measured up to heights of 100 km by use of a flashlamp-pumped dye laser (Bowman et al., 1969; Gibson, 1969). The spatial resolution was 2.5 km in this experiment.

C. Photoionization and Photodetachment

Tunable lasers may be very helpful in photoionization and photodetachment experiments, especially when cross sections of these processes are to be studied as a function of wavelength.

Very recently, the first measurement of photoionization from a rare gas metastable state was reported by Stebbings et al. (1973). In the first step, helium atoms in an atomic beam were excited to the $2\,^1S$ and $2\,^3S$ metastable states by electron impact. By irradiating the beam with the light from a helium discharge lamp, the helium ($2\,^1S$) metastables could be quenched so that, switching the lamp on and off, data appropriate to each metastable species could be derived from the observations. The ultraviolet radiation for the second step, the photoionization process, was obtained by frequency doubling the output of a tunable dye laser. The ultraviolet beam obtained in this way had a linewidth of 0.05 nm, a pulse power of several thousand watts, and a pulse length of 5 nsec. As is usually done in photoionization experiments, the produced ions were detected by particle multipliers. From these measurements, cross sections for the photoionization of the $2\,^1S$ and $2\,^3S$ were derived from threshold (312 nm for ^1S and 260 nm for ^3S) to 240 nm. Interestingly, the result turns out to be that the ratio of singlet-to-triplet photoionization cross section is not in agreement with theoretical predictions.

Other experiments using selective two-step ionization of atoms were reported on Rb by Ambartzumian and Letokhov (1972) and on Mg by

Bradley *et al.* (1973a). For the first step in these experiments (selective excitation of an atomic level), the light of a tunable dye laser was used; the second step was performed by means of a frequency-doubled ruby laser and a continuous light source, respectively.

In photodetachment studies, the application of continuously tunable dye lasers instead of conventional light sources offers the possibility of increasing considerably the signal-to-noise ratio and consequently the resolution obtained in such experiments. Lineberger and Woodward (1970), Hotop and Lineberger (1973), and Hotop *et al.* (1973) studied the threshold behavior of the photodetachment cross sections for various negative ions in cross-beam experiments. The negative ion beam, accelerated to 1 or 2 keV and mass analyzed, was crossed at right angles with the light of a pulsed (5-Hz), flashlamp-pumped tunable dye laser ($\Delta \lambda = 0.1$–0.2 nm, 0.5–5 mJ/pulse). The fast neutrals resulting from the photon–negative ion interaction, the ion beam current, and the laser beam intensity were measured and the cross sections derived. As an example, Fig. 14 shows the relative cross section for the $S^- + h\nu \rightarrow S + e^-$ reaction in the first 1200 cm^{-1} above threshold. Four thresholds are indicated by arrows. The known $3p^4\ ^3P_2$–$3p^4\ ^3P_1$ splitting of the neutral S corresponds to the separation between the first two and between the second two arrows. Thus the distance between the first and the

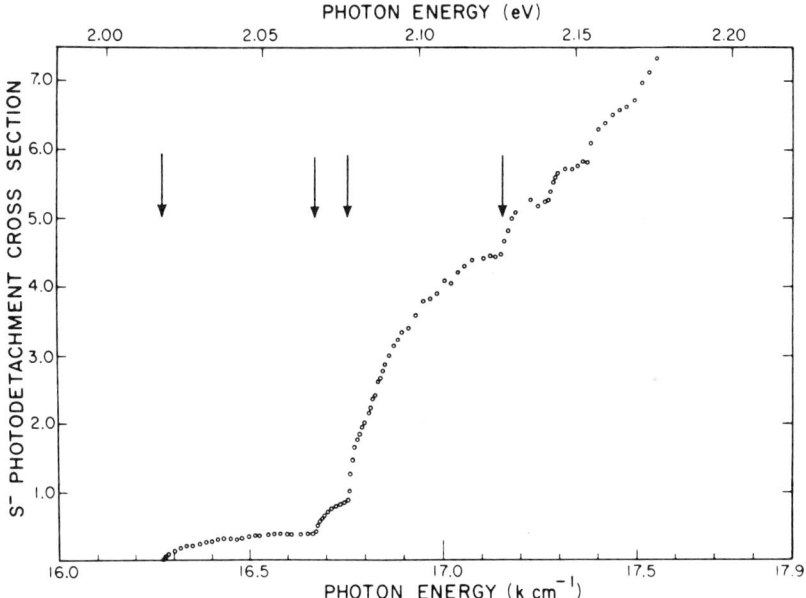

FIG. 14. Relative cross section for photodetachment of S^-. Arrows indicate the thresholds, which are mentioned in the text. From Lineberger and Woodward (1970).

third indicated threshold has to be attributed to the $3p^5\ ^2P_{1/2}$–$3p^5\ ^2P_{3/2}$ splitting of S^-. This was the first photodetachment measurement of a negative ion fine structure. Obviously from the third threshold (transition $3p^5\ ^2P_{3/2} \to 3p^4\ ^3P_2$) the electron affinity of S^- can be determined.

D. Two- and Multi-Photon Processes

The most important two-photon process is Raman scattering. The tunability of dye lasers allows advantage to be made of the increase in cross section of the resonant Raman effect as compared with the usual Raman scattering. This may be very useful in work on molecules and in studying effects in the solid state, but it has found no application in atomic spectroscopy. There has been reported, however, simultaneous stimulated Stokes and anti-Stokes electronic Raman scattering by selectively excited potassium and rubidium atoms (Bradley et al., 1971). Stimulated Raman scattering and four-photon parametric interaction induced by a dye laser have also been studied in rubidium by other authors (Korolev et al., 1970, 1971).

The main application of stimulated Raman scattering in atoms, however, seems to be in the generation of short-wavelength radiation. For example, Vinogradov and Yukov (1972) have proposed to use the higher order anti-Stokes components of stimulated Raman scattering in atomic iodine for a very efficient conversion of high power laser pulses into the ultraviolet. This should be feasible, since the population of the atomic iodine levels $5\ ^2P_{1/2}$ and $5\ ^2P_{3/2}$ can be highly inverted by photodissociation of CF_3I and C_3F_7I (Henzel et al., 1971).

At high intensities of light, two-photon absorption with frequency $\omega_1 + \omega_2$ (where ω_1 and ω_2 are individual photon frequencies) is no longer improbable. The obvious cause is that the probability of this process is proportional to $n_1 n_2$, where n_1 and n_2 are the occupation numbers in the light fields of frequencies ω_1 and ω_2, respectively. In atomic spectroscopy the effect has been observed in potassium (Tan-no et al., 1973). The $6\ ^2S_{1/2}$ level of potassium was excited by a two-photon process with red light, and the violet emission corresponding to the transition $5\ ^2P_{3/2}$–$4\ ^2S_{1/2}$ was observed with some delay; the emission process was interpreted as a spontaneous cascade transition within the $6\ ^2S_{1/2}$, $5\ ^2P_{3/2}$, and $4\ ^2S_{1/2}$ array.

For spectroscopic purposes two-photon processes should be interesting, if $\omega_1 \sim \omega_2$. If the two-photon interaction takes place within a standing wave of monochromatic light, it should give a sharp resonance, if the frequencies ω_l and ω_r of the left and right travelling waves seen by an atom, where $\omega_l = \omega_0 - \omega_D$ and $\omega_r = \omega_0 + \omega_D$ are adjusted to give $\omega_l + \omega_r = 2\omega_0 = \omega_{tp}$. Here ω_D is the Doppler shift, ω_{tp} the frequency of the two-photon absorption, and ω_0 the laser frequency. In this situation the resonance would not be affected by Doppler broadening (see Cagnac et al., 1973).

Higher order processes induced by a dye laser have also been reported, e.g., three-photon excitation and ionization of potassium vapor (Agostini et al., 1972); the observed emission was interpreted as being due to a stimulated three-photon process (two absorbed, one emitted).

Closely related to these phenomena is the third harmonic frequency generation by metal vapors in phase-matching gases (Ward and New, 1969; Harris and Miles, 1971; Harris, 1973; Miles and Harris, 1973; Young et al., 1971; Kung et al., 1973). This process utilizing high-power mode-locked dye lasers promises to yield intense coherent vacuum ultraviolet radiation which is tunable to some extent.

Note added in proof: In the meantime D. Pritchard, J. Apt, and T. W. Ducas [*Phys. Rev. Lett.* **32**, 641 (1974)] measured the fine structure splitting of the 4d ^{2}D state of sodium using high-resolution, cw, simultaneous two-photon excitation from the 3s ^{2}S ground state in an atomic beam. T. W. Hänsch, K. C. Harvey, G. Meisel, and A. L. Schawlow at Stanford were able to obtain high-resolution spectra of the same transition without Doppler broadening with two counter-propagating cw laser beams, as discussed in the text. A similar experiment was performed by M. Levenson and N. Bloembergen at Harvard, who observed the Na 3s–5s transition without Doppler broadening, using a nitrogen-laser-pumped pulsed dye laser.

IV. Applications of Tunable Dye Lasers with Extreme Narrow Bandwidth

In this section, we will discuss laser applications in which the narrow bandwidth of the tunable laser output plays the essential role. Thus, the main part of this section is devoted to high-resolution optical spectroscopy.

In this field, the resolution is usually limited by the Doppler effect when conventional light sources are used; the natural linewidth can only be approached in few particular cases. Therefore the study of narrow spectral features is, in general, impossible using conventional purely optical spectroscopy. Besides insufficient resolution, lack of intensity also often prevents the investigation of finer details in atomic spectra. Both difficulties may be overcome by the application of tunable lasers; moreover, the narrow spectral bandwidth of the laser output and the tunability can together eliminate the need for costly and bulky spectrometers (see Sections IV,A and C).

In the methods of so-called Doppler-free spectroscopy described in this section, the laser frequency is often tuned across atomic resonances. To see narrow spectral structures the laser light must have a bandwidth much narrower than the inhomogeneous linewidth, a condition which can be fulfilled with the dye lasers, which are now available.

As lasers can be many orders of magnitude superior to conventional light sources in their intensity per spectral interval, new nonlinear spectroscopic techniques in the optical frequency range could be developed. Among these, saturation spectroscopy (Section IV,A) has now been applied in high-resolution atomic spectroscopy using tunable dye lasers.

Saturation spectroscopy and the closely related fluorescence line narrowing (see Section IV,B) require only saturation. There are, however, other nonlinear effects which make use of the coherence of the laser light. Some of these nonlinear effects are obtained under steady-state conditions; others appear in the transient regime. Those coherent effects which are likely to be applied successfully in atomic spectroscopy along with tunable dye lasers are briefly discussed at the end of this section (Section IV,E).

It is not only the application of nonlinear effects which characterizes the latest development in high-resolution atomic spectroscopy, but also, in traditional absorption and fluorescence spectroscopy, tunable single-mode lasers will more and more replace conventional light sources. In these cases the Doppler effect is nearly eliminated by using a highly collimated atomic beam which is illuminated normal to its direction as an absorber. Successful applications of this technique are given in Section IV,C; several modifications using nonoptical detection of the optical resonances are described in Section IV,D. This kind of detection may be advantageous when radioactive isotopes are to be studied.

As is well known, Doppler-free spectroscopy can also be done with conventional incoherent light sources when rf spectroscopy or level-crossing methods are used. These methods are now complemented by the purely optical high-resolution techniques described in this section.

A. SATURATION SPECTROSCOPY

Lasers produce an extremely high intensity per unit bandwidth; because of this, new nonlinear high-resolution phenomena such as saturation spectroscopy and fluorescence line narrowing become accessible. In saturation spectroscopy, the narrow laser spectral width is used to produce velocity-resolved changes in the population distribution of atoms or molecules.

The inhomogeneously broadened absorption profile (Doppler profile) of a gaseous absorber can be described as a superimposition of many narrow subprofiles, whose width $\delta\omega_h$ is due to such homogeneous broadening processes as spontaneous emission and collisions. Thus $\delta\omega_h$ characterizes the bandwidth of the interaction of an absorbing atom with a monochromatic electromagnetic field. Consequently a laser beam with frequency ω passing through an absorber in the z direction can only be absorbed by atoms having a velocity component within the interval $v_z \pm \delta v_z/2$, where $v_z = (\omega - \omega_0)c/\omega_0$ (ω_0 being the center frequency of the Doppler absorption profile) and where δv_z corresponds to $\delta \omega_h$. If the narrowband laser light is sufficiently intense so as to induce an appreciable saturation (or nonlinearity), the velocity selective absorption process results in a "hole burning" in the population distribution of the ground state atoms (Bennett, 1962), as illustrated in Fig. 15b. A profile with the linewidth $\delta \omega_h$ can of course only be

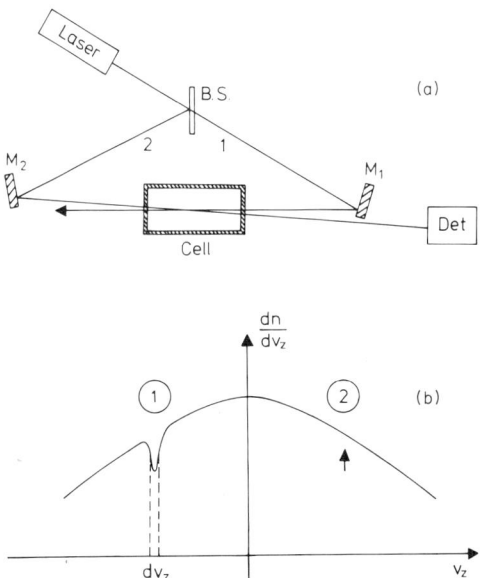

FIG. 15. Principle of saturation spectroscopy: (a) Experimental setup; (b) saturation within Doppler profile.

obtained if the bandwidth of the laser beam is much smaller than $\delta \omega_h$, and if the output power of the laser is not too high; otherwise power broadening would occur.

The hole profile with its sub-Doppler linewidth can be detected by probing the entire Doppler profile with a second, tunable (weak) laser beam. In many cases this procedure allows a hole to be detected, even if only 1% or less of the atoms in the velocity interval have been excited by the first "saturating" laser beam. A convenient technique for saturation spectroscopy was described by Hänsch et al. (1971a) and applied to a study of the hyperfine structure of a molecular iodine line. The principle is illustrated in

Fig. 15a. The output of a tunable laser is divided by a beam splitter into a saturating beam (labeled 1 in the figure) and a weaker probe beam (2 in the figure) which are sent in nearly opposite directions through a cell containing the absorbing gas sample. Generally the two beams will excite different species of atoms, having velocities of $v_z \pm \delta v_z/2$ and $-(v_z \pm \delta v_z/2)$, respectively; consequently, the intensity of both beams will be diminished when passing the absorber. If, however, the laser frequency coincides with the center of the Doppler-broadened absorption line, only one group of atoms with a velocity component of $0 \pm \delta v_z/2$ is available for absorption. The probe beam then sees the hole burned out by the saturating beam and consequently will pass through relatively undiminished. Hence, if the laser frequency is tuned across one of the absorption frequencies of the gas, the intensity of the probe beam shows a sharp peak at the resonance frequency. As absorption is detected only for atoms moving within a small velocity interval in one direction, rather than in all directions, the Doppler effect is nearly eliminated, and consequently the width of the hole is given by the width of the lower and upper level of the absorbing transition which are due to the radiative lifetimes of the levels and collision effects. If the pressure in the gaseous sample is sufficiently low and if power broadening is avoided, the width of the peak is caused essentially by the natural linewidth.

Saturation spectroscopy, which is both a sensitive and convenient technique, is based on a similar effect, already known for many years, which occurs in the amplifying medium inside a laser as light is reflected back and forth between the two end mirrors of the cavity (Lamb, 1964). The spectroscopic technique using the resulting saturation effect inside the laser cavity is often named Lamb-dip spectroscopy. The first Lamb-dip studies were by Szöke and Javan (1963) and McFarlane et al. (1963).

Earlier saturation spectroscopy measurements had been restricted essentially to either gas laser transitions (for example, Lee and Skolnick, 1967; Letokhov, 1967; Lisitsyn and Chebotayev, 1968) or to molecular transitions in accidental coincidence with gas laser lines (for example, CH_4: Barger and Hall, 1969; NH_2D: Brewer et al., 1969; Hanes and Dahlstrom, 1969; SF_6: Shimizu, 1969; Rabinowitz et al., 1969; Basov et al., 1969). A significant improvement was achieved by Hänsch et al. (1971b), who took advantage of the broad spectral range of tunable dye lasers to apply, for the first time, high resolving saturation spectroscopy to atomic resonance lines. A schematic diagram of their experimental setup is shown in Fig. 16. A pulsed (80-Hz) tunable dye laser was used in the experiment (Hänsch, 1972). By means of a piezoelectrically tunable confocal Fabry–Perot interferometer outside the laser cavity, the linewidth (FWHM) was reduced from 300 to 7 MHz and the pulse length thereby stretched from 5 to 30 nsec. Continuous scanning was accomplished by tilting an intracavity etalon while simultaneously tuning

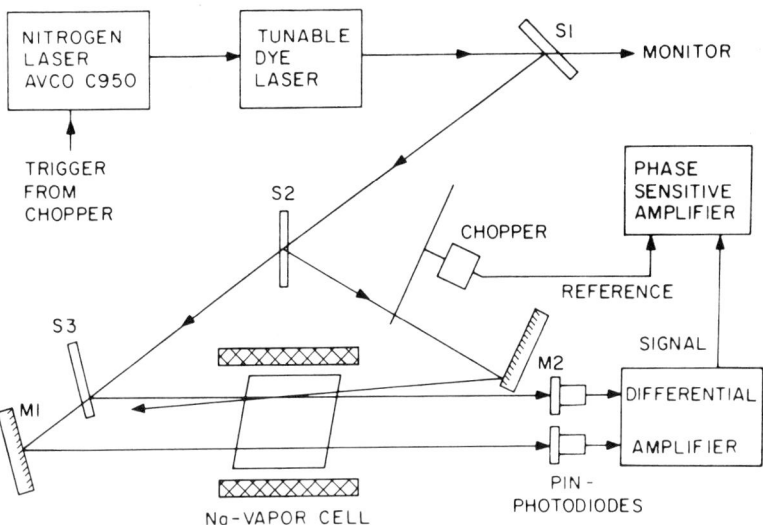

FIG. 16. Scheme of a laser saturation spectrometer. From Hänsch et al. (1971b).

the external passband filter by applying a suitable voltage. Lock-in techniques were used to detect small bleaching effects. To reduce the noise due to random laser fluctuations, a differential detection scheme was used in which the intensity of the probing beam after absorption was compared with the intensity of a reference beam passing the cell outside the bleached region. It was possible to register changes of the probe intensity down to 10^{-3} with a time constant of 1 sec.

With this experimental arrangement, the sodium resonance lines D_1 at 589.6 nm and D_2 at 589.0 nm were studied. The smallest observed linewidth of a single hyperfine component in the D_1 line was 40 ± 4 MHz, which is about four times the natural linewidth caused by the radiative lifetime. The difference between observed and natural linewidth is ascribed to the finite laser bandwidth and to residual Doppler broadening due to the finite crossing angle of saturating and probing beam. It was not possible, of course, to resolve the narrow hyperfine splitting of the $^2P_{3/2}$ level in the D_2 line (see Fig. 21), with a linewidth of 40 MHz, and hence only a line broadening was observed. (In Section IV,C, fluorescence experiments are described in which higher resolution of the hyperfine splitting of the sodium D lines was achieved.)

Essentially the same experimental technique was used by the authors in an investigation of the red Balmer line H_α of atomic hydrogen at 656.3 nm (Hänsch et al., 1972). The study of the fine structure of hydrogen and hydrogenlike spectra has always been of particular interest because, owing to the simplicity of the hydrogen atom, corresponding theoretical calculations can be subjected to crucial tests by comparing the calculations with the experimental results.

The spectroscopic investigation of the H_α line is especially interesting for two reasons. (i) From a measurement of the wavelength of H_α the Rydberg constant can be accurately determined (see, for example, Kibble et al., 1973). (ii) The determination of the energy difference of the levels $2\,^2S_{1/2}$ and $2\,^2P_{1/2}$ (the Lamb shift) provides a crucial test for the correctness of quantum electrodynamics. Unfortunately, in conventional optical spectroscopy the small mass of hydrogen causes the lines to have a large Doppler width which masks all finer details of the line structure, and although there have been many efforts to study, with an increasing high resolution, the Balmer lines in the visible, accurate knowledge of the fine structure is at present still mainly derived from rf spectroscopy and level-crossing techniques (for example, Series, 1957; Taylor et al., 1969; Kibble et al., 1973). Using saturation spectroscopy, however, it was possible for the first time to resolve H_α into single fine structure components and to observe the Lamb shift at visible frequencies. After the success of these experiments, great progress can be expected in the near future, especially in the more accurate determination of the Rydberg constant.

Since the observation of the H_α line in an absorption experiment requires the population of the $n = 2$ states (see Fig. 17), the hydrogen was excited in a

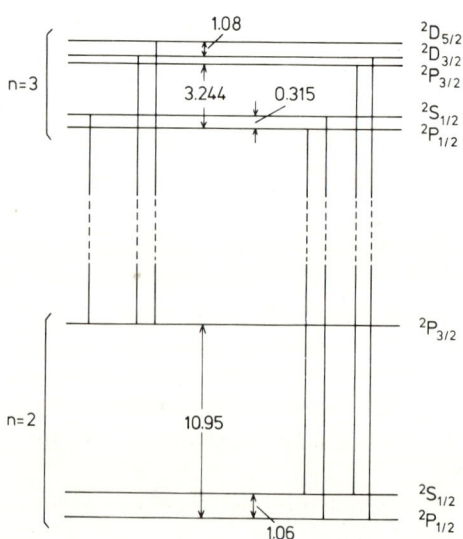

FIG. 17. Energy level scheme of H_α, fine structure splitting in gigahertz. The separation of the levels $^2P_{3/2}$ and $^2D_{3/2}$ in the figure does not indicate the real energy difference.

simple Wood discharge tube, yielding an estimated population of 10^{10} cm^{-3} in the $n = 2$ state. The saturation spectrum was observed in the afterglow, 1 μsec after termination of the discharge, in order to minimize the influence of Stark broadening and shifts on the measurements.

Figure 18 shows a saturation spectrum of H_α in which the four strongest components are clearly resolved. The splitting of the doublet at the extreme right is essentially due to the term difference $2\,^2P_{1/2}$–$2\,^2S_{1/2}$ (Lamb shift). The width of the narrowest transition observed was about 300 MHz, that is, 20 times smaller than the Doppler width in hydrogen, which is nearly 6000 MHz at room temperature. A resolution of 5 parts in 10^7 was achieved, an order of magnitude better than the most recent value obtained by conventional emission spectroscopy from a liquid-helium-cooled deuterium discharge (Kibble *et al.*, 1973).

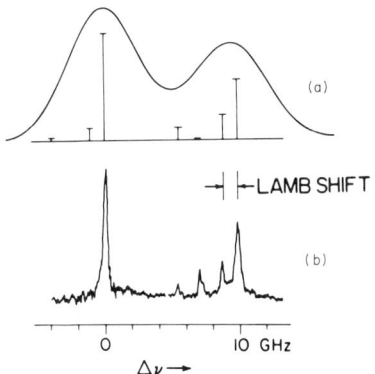

FIG. 18. Hydrogen line H_α. (a) Theoretical fine structure (intensities according to relative oscillator strengths) and Doppler profile at room temperature, (b) saturation spectrum (Hänsch *et al.*, 1972; Hänsch, 1973).

Besides the components expected according to the level scheme (Fig. 17), additional signals halfway between two lines that share a common upper or lower level may occur in saturation experiments (Schlossberg and Javan, 1966b). The weak component belonging to the transition $2\,^2P_{1/2}$–$3\,^2S_{1/2}$ almost coincides with a "crossover" signal due to the common lower $2\,^2S_{1/2}$ level of the two neighboring transitions (see Figs. 17 and 18).

The observed intensities indicate that the $2\,^2S$ and $2\,^2P$ levels are nearly equally populated, although one would expect a higher population for the metastable $2\,^2S$ level compared with the short-living $2\,^2P$ levels (1.6 nsec). Time-resolved measurements in the afterglow reveal, however, that the lifetime of the $2\,^2P$ states is lengthened by about a factor 10^3, caused by resonance trapping of the emitted ultraviolet Lyman-alpha radiation. By this process the velocity distribution of the $2\ ^2P$ states is rapidly thermalized compared with the $2\ ^2S$ state. It was possible to show this by means of measurements using time-delayed probe pulses (see below).

The observed linewidth can be explained by unresolved hyperfine splittings, by radiative lifetimes, by the finite laser bandwidth, and by residual Doppler broadening due to the finite crossing angle between saturating and probing beam; therefore large contributions from pressure broadening and Stark effect can be excluded.

In recent experiments (Hänsch, 1973) with reduced laser bandwidth and with a smaller crossing angle between the two beams, it was even possible to perceive the hyperfine splitting (170 MHz) of the $2\ ^2S_{1/2}$ level in the $2\ ^2S_{1/2}-3\ ^2P_{3/2}$ component (second component from the right in Fig. 18). By means of experiments of the type described, it should be possible to measure the wavelength of one of the components of H_α in saturated absorption and thereby the Rydberg constant, with an accuracy of 2 parts in 10^8. This is the accuracy of the present standard of length.

By means of saturation spectroscopy, relaxation phenomena and collision processes can also be studied as was first demonstrated in the work on the sodium D lines by Hänsch et al. (1971b). Owing to the use of a pulsed laser, the required time resolution can be achieved by delaying the probe pulse with respect to the saturating pulse. This is done by sending the probe pulse through a folded optical delay line. Such experiments revealed, for example, that the bleaching at the resonance frequencies remained observable despite a delay of the probe beam of several radiation lifetimes. This remanent hole in the velocity distribution of the two ground-state levels can be explained in view of a velocity-selective optical-pumping cycle. Only some of the atoms excited by the bleaching beam return to their original level, others end in the other ground-state level. The holes (and peaks) in the velocity distribution of the ground-state levels remain for several radiation lifetimes after the bleaching.

If the bleaching intensities are high enough, the linewidth will decrease when the probe is delayed by more than the laser pulse length. The line broadening observed in the presence of the saturating light field cannot be explained in terms of atomic level population changes and is instead ascribed to a line splitting due to optical nutations of the absorbing sodium atoms in the saturating light field (Shahin and Hänsch, 1973).

The line broadening in the presence of buffer gases can be used to determine collision cross sections. By varying the delay time of the probe beam the line broadening is observed as a function of time. Thus a rather direct observation of collisional velocity changes is possible, and the rate at which equilibrium is re-established by such collisions can be determined. This may give information on the nature of the collision processes.

B. Fluorescence Line Narrowing

The application of saturation spectroscopy is limited to lines that can be studied in absorption. There exists, however, a simple complementary technique which permits Doppler-free spectroscopy in fluorescence lines: the laser-induced fluorescence line narrowing (Szabo and Erickson, 1972). In this technique, as in saturation spectroscopy, a highly monochromatic radiation interacts with the inhomogeneously broadened ensemble under study (see Fig. 19). This results in a group of excited atoms or molecules which all

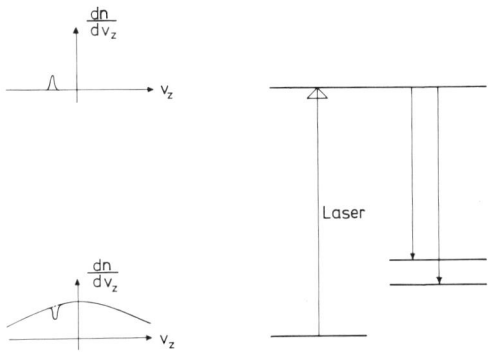

FIG. 19. Laser-induced fluorescence line narrowing: population profiles and level diagram.

have nearly the same velocity component in the direction of the laser beam. Obviously the fluorescence radiation emitted by such an ensemble shows a linewidth which is unaffected by Doppler broadening as long as the direction of observation is collinear to the laser beam. Thus, if a structure of the fluorescence line is studied by means of a spectrograph the final linewidth will in most cases be determined only by the resolution of the spectrograph. Up to now, the fluorescence line narrowing has not been used in atomic spectroscopy; this situation will most likely change, especially if high quality tunable uv lasers become available, since a narrowband excitation of a highly excited state is the basis for interesting applications of this new spectroscopic method.

C. Classic Absorption and Fluorescence Spectroscopy

The unique properties of laser light opened the way to powerful new nonlinear spectroscopic techniques such as saturation spectroscopy and fluorescence line narrowing, which allow the removal of the Doppler broadening and thus provide a tool for high-resolution spectroscopy. These methods can now be widely applied since lasers with a sufficiently small linewidth, good tunability, and high frequency stability are becoming available. On the other hand, the recent advances in the development of narrowband tunable dye lasers also renewed the interest in the old-fashioned methods of fluorescence and absorption spectroscopy which may have some advantages when compared to nonlinear spectroscopy. Among these advantages are the following. (i) A resolution of one natural linewidth can be achieved if the Doppler broadening is sufficiently reduced, which can be done, for example, with the use of a highly collimated atomic or molecular beam; this is in contrast to the fluorescence line narrowing, which is limited in its resolution principally to two natural linewidths (as are optical double resonance and level crossing), whereas, in saturation spectroscopy, the resolution of narrow spaced lines suffers from crossover signals and power broadening. (ii) Detailed measurements of line shapes by fluorescence spectroscopy can easily and unambiguously be compared with theory. Moreover, the experimental setup, even including the atomic beam, is not more complicated than that used in high-resolution saturation spectroscopy.

The feasibility of fluorescence spectroscopy by scanning an optical spectrum with a narrowband tunable single-mode cw dye laser, and observing the scattered fluorescent light was demonstrated independently by three groups (Schuda et al., 1973; Hartig and Walther, 1973; Lange et al., 1973). All studied the D_2 line ($3\ ^2P_{3/2}$–$3\ ^2S_{1/2}$) in sodium and used cw dye laser configurations similar to that described by Hercher and Pike (1971). The results obtained are generally similar; we report in the following on the work by Lange et al. (1973) because these authors have achieved the highest resolution. Figure 20 shows schematically the experimental setup. Details of the cw dye laser are given by Schröder et al. (1973). In the experiments the laser had a bandwidth of 7 MHz which, under certain conditions, could be reduced to 2 MHz. The attenuated laser light was scattered on a sodium atomic beam with collimation ratio 1 : 400 resulting in a Doppler width of

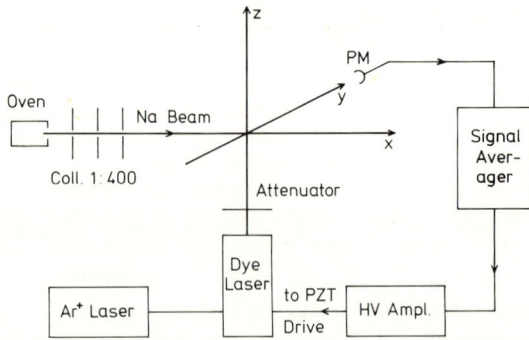

FIG. 20. Experimental setup of a high-resolution fluorescence experiment.

less than 5 MHz with the atoms viewed at right angles to the beam. This width is sufficiently small when compared with the natural width of 10 MHz of the D_2 line. The scattered fluorescent light was observed as the laser was

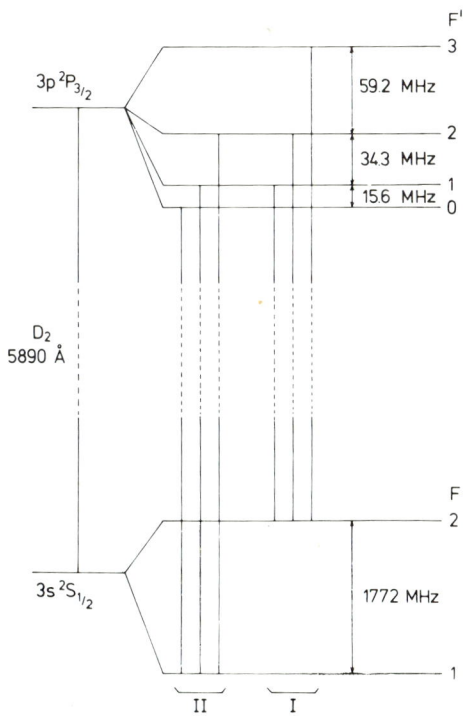

FIG. 21. Hyperfine level scheme of the Na D_2 line (different scales in upper and lower state).

tuned across the hyperfine structure. For this purpose the signal averager in Fig. 20 was synchronized with the piezoelectric drive providing the frequency variation of the dye laser. The time constant of the detection channel was kept rather low (1 msec) so that possible short time variations of the dye laser were able to be seen. As is demonstrated in Figs. 22b and 23b, all hyperfine components of the D_2 line were resolved, the linewidth was about 15 MHz, and for comparison, ideal signal curves are shown in Figs. 22a and 23a. These curves were calculated for atoms at rest excited by a tunable monochromatic light source using the known hyperfine splitting factors, the lifetime of 3 $^2P_{3/2}$, and the theoretical relative transition probabilities. The linewidth of the theoretical curves is 10 MHz. In very recent experiments the observed linewidth could be reduced to less than 12 MHz.

It should be mentioned, however, that the excellent resolution shown in Figs. 22b and 23b can only be observed at low intensities of the exciting

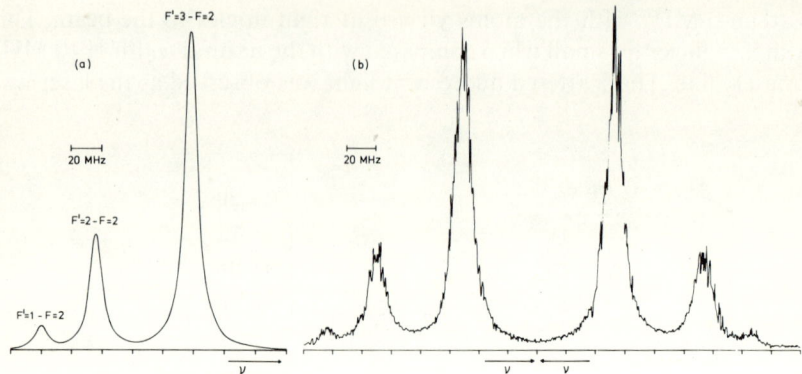

FIG. 22. Part I of the hyperfine structure of the Na D_2 line (cf. Fig. 21) (a) Calculated signals with natural linewidth, (b) experiment; triangular sweep mode was used in the averaging system, thus the right part of the experimental curves should be the mirror-image of the left part; sweep time 8 sec, 3 sweeps have been averaged.

light. At higher power densities (a few milliwatts per square millimeter) optical hyperfine pumping becomes noticeable; this is caused by the frequency-selective excitation process. In the $F = 2$–$F' = 3$ transition, all excited atoms have to return to the $F = 2$ ground level, whereas when the laser is tuned to the $F = 2$–$F' = 2$ transition, the excited atoms can decay to either the $F = 2$ or the $F = 1$ ground level. This results in a gradual depletion of the population of the $F = 2$ ground level and the radiation which is absorbed and scattered at the $F = 2$–$F' = 2$ transition frequency is diminished. The same happens in the group of components starting in the $F = 1$ ground level; in this case the $F = 1$–$F' = 0$ transition is not affected by optical hyperfine pumping and becomes dominant. Similar effects occur in saturation spectroscopy.

The results of the hyperfine splitting of the $3\ ^2P_{3/2}$ level of ^{23}Na obtained by Lange et al. (1973) are given in Fig. 21. They agree quite closely with earlier optical double-resonance measurements by Ackermann (1966) and with values derived from recent time-delayed level-crossing experiments by

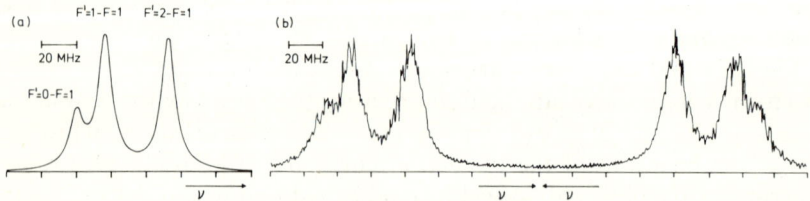

FIG. 23. Part II of the hyperfine structure of the Na D_2 line. (a) Calculated signals with natural linewidth, (b) experiment; sweep time 4 sec; 4 sweeps have been averaged (cf. Fig. 22).

Cha et al. (1973) (see Section III,B,3). The extreme accuracy which can now be achieved by the application of tunable dye lasers enables one to check earlier measurements made with rf spectroscopy methods. It then turns out that, for example, the results of a level-crossing experiment on the Na 3 $^2P_{3/2}$ level by Mashinskii (1970) do not agree with the above-mentioned more recent investigations by Lange et al. and Cha et al.

D. Nonoptical Detection of Optical Resonances

One of the existing classes of Doppler-free optical spectroscopy is the special class which uses nonoptical detection of optical resonances. In these kinds of techniques, as in Doppler-free fluorescence spectroscopy (see Section IV,C), the Doppler effect is almost completely eliminated by illuminating a highly collimated atomic beam normal to its direction. Three methods have been experienced so far; magnetic deflection of the atomic beam (Rabi-type experiment), deflection of the atomic beam by recoil of resonant atoms, and photoionization from the upper state. In all three cases, the optical resonance is detected through the atoms themselves. This gives these methods an important advantage over methods using detection of the light in the cases when radioactive atoms are to be studied. In these cases the radioactivity of the atoms makes it possible to use very sensitive detection systems, as is well known from the successful application of the atomic-beam magnetic resonance method used to study the hyperfine structure of several hundred radioisotopes. In the process of producing the radioactive sample, a very small number of radioactive atoms is created in the presence of a large number of carrier atoms. These stable atoms are of no significance if the radioactivity is detected; they may, however, cause considerable trouble in the optical detection of the resonances. In this case, time-consuming separation processes have usually to be performed, which then sets lower limits on the halflives of the radioisotopes that can be studied. It should be possible, however, to use the atomic-beam methods described below, on line with an accelerator producing the radioisotopes so that very short-lived nuclides also can be investigated. The halflife has at least to be sufficient for the atom to get down the length of the atomic-beam machine, that is $\tau_{1/2} > 10^{-4}$ sec. Moreover, methods using atomic beams do not suffer from the wall problems which may arise with methods using resonance cells because the interaction of the microscopic number of radioactive atoms with the walls of their container can quickly destroy the usefulness of the sample. As is well known, the ground-state hyperfine splitting of radioisotopes can be studied by means of the atomic-beam magnetic-resonance method with good results. The methods to be discussed below may, however, be advantageously applied in (i) investigations of the hyperfine structure of excited levels and in

(ii) isotope shift measurements. The second point deserves particular attention because it offers the possibility of measuring isotope shifts of short-lived isotopes. Isotope shift measurements are mostly made with conventional, purely optical techniques. These do not allow the study of isotopes with halflives of less than about 1 hour. On the other hand, measurements of long chains of isotopes including isotopes which are far from the valley of nuclear stability and which therefore have short lifetimes may reveal interesting features in the change of the mean square nuclear charge radius on the addition of neutrons as Bonn et al. (1973a,b) have demonstrated recently in the example of the isotopic chain ^{181}Hg–^{205}Hg. Owing to the large splittings in the mercury intercombination line, these authors did not have to take care of Doppler broadening and could therefore use a conventional light source for their optical-pumping experiments on the radioactive isotopes. The three methods which will now be described, however, allow an almost Doppler-free optical spectroscopy and use as light source a narrowband tunable dye laser.

1. Magnetic Deflection of an Atomic Beam

This method is, in the optical range, the counterpart of the Rabi magnetic resonance method (see, for example, Ramsey, 1956). Figure 24 shows a

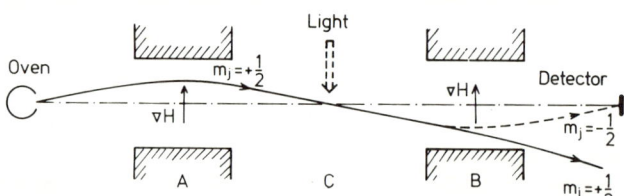

FIG. 24. Scheme of the experimental setup for the Rabi-type magnetic deflection method.

schematic diagram of the apparatus. The gradients of the inhomogeneous magnetic fields produced by the magnets A and B have the same direction; thus an atom, flying through the A field along the path shown in the figure, can reach the detector only if it undergoes, in the intermediate C region, a transition into another magnetic substate having an effective magnetic moment of opposite sign, so that, in the B field, the atom will be deflected into the desired opposite direction. Since the field strengths of the A and B magnets are, in general, in the Paschen–Back region the quantum number m_J has to be changed to achieve a detectable change in the effective magnetic moment. In the magnetic resonance method, this change is achieved by means of a rf field which, being in a homogeneous magnetic C field, induces magnetic dipole transitions between appropriate magnetic sublevels of the atomic ground state or of a metastable state, and information on the splitting of these states is thus obtained. In the optical analog the atoms are illuminated in the C region by resonance radiation, and an atom excited in this way from some magnetic sublevel may fall by spontaneous emission to another magnetic sublevel with an effective magnetic moment of opposite sign; this would result in an increase of the signal at the detector.

There were two earlier approaches to this method, both of which used conventional light sources and consequently needed an additional device to scan the structure to be studied. Perl *et al.* (1955) illuminated an atomic sodium beam in the C region with resonance radiation from a sodium discharge lamp and applied simultaneously a rf field at an appropriate frequency which caused magnetic dipole transitions among the excited state hyperfine levels. In this way the hyperfine splitting of $3\ ^2P_{1/2}$ and $3\ ^2P_{3/2}$ could be measured. Marrus and McColm (1965) excited atoms of a cesium atomic beam in a similar way, but they applied a static electric field in the C region. They were then able to study the optical Stark effect (Marrus *et al.*, 1966; Marrus and Yellin, 1969) and, using different isotopes in the light source and in the atomic beam and by tuning of the Stark levels of the beam atoms, measured the isotope shift in the D_1 line of several radioactive cesium isotopes against the stable ^{133}Cs isotope (Marrus *et al.*, 1969). Moreover, the determination of the hyperfine splitting of the excited level was also possible (Marrus *et al.*, 1967).

A more elegant analog in the optical range of the Rabi magnetic resonance method is, however, the application of a narrowband tunable dye laser for the illumination of the atomic beam in the C region. This method offers the greatest versatility and at the same time a high resolution, which is, in principle, only limited by the natural linewidth.

This method was studied recently by Duong *et al.* (1973a). The sodium D lines were used to check the method. The configuration of the tunable dye laser was an improved version of the type described by Hercher and Pike (1971). On single-mode operation, the frequency of the dye laser is servo-controlled by means of a very stable external Fabry–Perot interferometer. A signal at the detector is observed when the frequency of the laser comes into resonance with the frequency of a hyperfine transition and "optical pumping" takes place from a substate of $3\ ^2S_{1/2}$ with $m_j = +1/2$ $(m_j = -1/2)$ to a substate of $3\ ^2S_{1/2}$ with $m_j = -1/2$ $(m_j = +1/2)$ (see Fig. 25). All hyperfine components of the D lines were observable in this way, except the transition $3\ ^2P_{3/2}$ $(F' = 0)$–$3\ ^2S_{1/2}$ $(F = 1)$ which, in principle, cannot be detected with the method being described because the $F' = 0$ level can only combine with the $F = 1$ ground state level which has only $m_j = -1/2$ sublevels. Thus the smallest hyperfine splitting of the $3\ ^2P_{3/2}$ is not attainable for the test. The

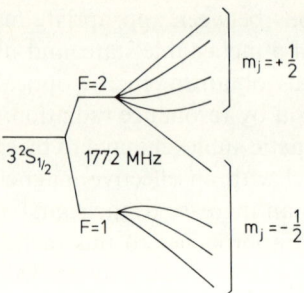

FIG. 25. Magnetic hyperfine splitting of the sodium 3 $^2S_{1/2}$ ground state.

halfwidth of the components was about 20 MHz, which made possible the resolution of the three hyperfine components $^2P_{3/2}$ ($F' = 3, 2, 1$)–$^2S_{1/2}$ ($F = 2$).

Since the method uses magnetic deflection of an atomic beam, its application can only be to atoms with a paramagnetic ground state. This is one reason to discuss another method which does not suffer from this restriction. This method will be covered in the next section.

2. Deflection of an Atomic Beam by Recoil

Magnetic deflection of an atomic beam is used in the detection of optical resonance; however, atomic-beam deflection based on the transfer of linear momentum between photons and atoms during resonant interaction can also be applied to high-resolution optical spectroscopy. The recoil effect was first studied experimentally many years ago by Frisch (1933), who showed that a highly collimated sodium atomic beam is deflected if the atoms are transversely illuminated by a sodium discharge lamp. Two recent experiments using as the light source, either the whole spectral profile of the resonance lines emitted by discharge lamps (Picqué and Vialle, 1972) or a laser (Schieder et al., 1972), demonstrated the effect more clearly.

With the absorption of a photon with frequency v, a linear momentum $p = hv/c$ in the direction of this photon is transferred to the atom. There are two different ways in which a new photon can be re-emitted. (i) If stimulated emission takes place, every atom recoil following emission cancels exactly the recoil momentum acquired upon absorption, because the induced photons are radiated precisely in the same mode as the photons being absorbed, that is, with the same frequency, direction, and polarization. (ii) In spontaneous emission, however, the recoil momentum p is scattered in random directions, leaving from the absorption process a net, directed, recoil.

Thus, only absorption followed by spontaneous emission contributes to the deflection of the atomic beam. If the beam is illuminated normal to its direction, the transfer of the transverse momentum p to an atom of mass m and velocity v leads to a deflection angle $\alpha = p/mv$, which is typically of the order of 10^{-5} radiants. Because of the smallness of this value the linear momentum transfer is not found to play a role in the usual resonance experiment. A well-collimated atomic beam and a very narrow detector are required for the observation of the effect.

Much larger deflections of the atomic beam may be achieved when a high power laser beam is used, the excited levels then being saturated. The interaction time between the atomic beam and the light beam then determines the number of excitations of each atom, and consequently the resulting angular deflection of the atomic beam. Moreover, the high value of radiation pressure in a strong, resonant laser field offers many other possible applications, as was pointed out by Ashkin (1970). It was suggested, for example, that different isotopes of the same atom may be separated by means of resonance radiation pressure by virtue of the isotope shift in the resonant frequency.

A high-resolution-spectroscopy application of atomic-beam deflection by resonant light was given by Jacquinot et al. (1973). Again, sodium was used to test the method. An atomic beam with a collimation ratio greater than 10^4 was irradiated normal to the beam axis by a tunable single-mode dye laser. A hot-wire detector with a width of 10 μm was fixed at the center of the undeflected beam profile. One excitation of a sodium atom of average velocity resulted in a deflection of 23 μm with the geometry of the atomic beam apparatus as used in the experiment. When the laser is tuned to a hyperfine transition of the sodium D lines, a reduction of the number of detected atoms is observed and in this way, flop-out signals of the order of 10^7 atoms per second were obtained. The recorded linewidth was 30 MHz, which does not allow a complete resolution of the hyperfine splitting of the $3\ ^2P_{3/2}$ level. The recorded intensity distribution of the hyperfine pattern indicates that the experiment was carried out near the limiting situation in which each atom undergoes a single excitation. With higher intensities of the exciting laser beam, optical pumping may become noticeable, and the observed intensity distribution would be changed because an atom may fall into a hyperfine level different from the original one and so being lost for further excitation. This may occur when the excited levels $F' = 1$ or 2 are involved in the transition.

In contrast to the Rabi-type method previously described, the method of the atomic-beam deflection by recoil can also be used for atoms with a diamagnetic ground state. This method is especially suitable for such atoms because, owing to the lack of a ground-state hyperfine splitting, no optical pumping can occur and consequently higher laser beam intensities can be used, which in turn allows a lower atomic beam collimation resulting finally in a stronger signal.

3. Detection by Photoionization

In this method, the atomic beam is simultaneously illuminated with two radiation fields: the first is the light of a single-mode tunable dye laser which induces the transition to be studied from a lower level A which is initially populated, to a higher unpopulated level B; the second is the light of either a laser or a conventional source, with a frequency which allows only the atoms in the higher level B to be ionized. The ions are deflected away from the atomic beam by means of an accelerating electric field and observed, for example, by an electron multiplier. In this way, the optical transition from A to B is detected via the photoionization of level B. Level A can be the ground state or an excited state.

A check on the feasibility of this method was made by Duong et al. (1973b). In their preliminary experiment on sodium a satisfactory resolution of the hyperfine structure of the D_1 line was achieved.

A disadvantage of this method are the low cross sections for photoionization; however, it should be possible to overcome this difficulty using high-power pulsed lasers which have now become available. The method seems to be particularly suitable for the detection of optical resonances between two excited states, because in this case the Rabi resonance method and the recoil deflection technique may suffer from a large background created by the atoms in the ground level. Such experiments, however, should also be possible with the fluorescence method (see Section IV,C).

E. HETERODYNE SPECTROSCOPY

In experiments where the dye laser frequency is scanned, the precise calibration of the abscissa presents problems. An elegant possibility for their solution is the application of methods of heterodyne spectroscopy. In the simplest version of this technique, the beat frequency between two lasers is observed; these are in turn stabilized at different atomic or molecular transitions. The beat frequency can be measured with very high accuracy; it gives, directly, the energy difference between the two transitions. With gas lasers the method is especially successful. A recent example is given in the work of Petersen et al. (1973), in which the beat frequency between two CO_2 lasers stabilized on the Lamb dip (see Section IV,A) of different vib–rot transitions of CO_2 ($\lambda \sim 10$ μm) was measured. The beat frequency was between 30 and 60 GHz, the halfwidth of the signal was 20–30 kHz, and the uncertainty of the result about 3 kHz.

Many variations and improvements of the idea of heterodyne spectroscopy are possible. One of the most promising is the "Frequency Offset Locked Laser Spectroscopy" (see Hall, 1973). In this method, a laser is stabilized on a frequency $v_{ref} + v_{off}$ by means of beat techniques, where v_{ref} is the frequency of a stabilized reference laser and v_{off} a well-known variable frequency in the rf or microwave region; the laser frequency can thus be scanned very precisely.

It is to be expected that these techniques will be applied in dye laser spectroscopy in the near future. As a first step, Walther (1973b) stabilized a dye laser on a hyperfine component of the Na D_2 line using a highly collimated atomic beam.

Heterodyne methods can also be used for the investigation of incoherent light sources. If the light from a small thermal source is mixed at a photodetector with the radiation of a monochromatic laser, a beat signal which contains information about the source spectrum in the region of the laser frequency occurs in the rf region. The analysis of this beat spectrum should give a higher signal-to-noise ratio than is reached in classic high-resolution spectroscopy when the source has a small angular diameter. The application of this technique to astronomy has been discussed by Nieuwenhuijzen (1970); Hinkley and Kelley (1971) proposed its application in pollution control. The use of a tunable laser is obviously an essential in this method because the beat spectrum has to be in a region reached by rf equipment.

F. Nonlinear Coherent Resonant Phenomena

In radio-frequency spectroscopy, which in contrast to optical spectroscopy is always performed with coherent fields, phenomena such as spin echoes (Hahn, 1950), transient nutation (Torrey, 1949), and free induction decay (Bloch, 1946; Hahn, 1950) are already familiar. [They are reviewed in the book of Abragam (1961), for example.] These phenomena are resonant, that is, in these experiments the rf field is almost in resonance with a transition frequency of a quantum mechanical system, and since the populations of the levels involved change, in times which are short compared with the dephasing times, then the interaction between field and sample is nonlinear. These effects have their precise optical counterparts, which can be observed with lasers as light sources, that is, if coherent optical fields are applied. A treatment of these effects is beyond the scope of this paper, but the reader is referred to the review given by Courtens (1972). In the above-mentioned effects, the coherent field has to be almost in resonance with the system under study; therefore dye lasers should find applications in this field. Up to now, however, no such experiments with dye lasers have been reported.[1]

---

[1] Other interesting nonlinear coherent resonant phenomena which seem, however, to be of less importance to spectroscopy, have been observed with dye lasers. Among these are self-induced transparency (Bradley et al., 1970b), self-defocusing of light (Grischkowsky and Armstrong, 1972), and slow propagation (Grischkowsky, 1973) and self-steepening (Grischkowsky et al., 1973) of optical pulses.

This is most probably due to the stringent demand for monochromaticity in these experiments. The effects might nevertheless sometimes influence other experiments as unwanted effects (see Section III,B,1).

Photon echoes (Abella *et al.*, 1966), which are similar to spin echoes, mainly provide information on relaxation times (cf. Wang, 1970). It has been pointed out, however, that since they contain information on level structures, which are usually hidden within the inhomogeneous width of a probe, hyperfine spectroscopy should be possible (Alekseyev, 1970).

The transient nutation experiment has its counterpart in the optical nutation experiment (Tang and Silverman, 1966). The principle is as follows: if a monochromatic light beam is switched on in the way of a step function, a sinusodial modulation of the populations (in a two-level system) is then produced, resulting in a sinusodial change in macroscopic polarization, which in turn results in a modulation of the transmitted light intensity. The modulation is damped by relaxation processes until a steady-state population is reached. This effect may also give information on relaxation times and, in more complicated systems, on level structures hidden under the inhomogeneous linewidth. The counterpart of the third transient phenomenon known in rf spectroscopy is the free induction decay; it has also been recently observed by Cheo and Wang (1970) and Brewer and Shoemaker (1972). This phenomenon arises if a coherent field is suddenly switched off (see also Hopf *et al.*, 1973). All these effects might become useful spectroscopic tools, especially in connection with dye lasers.

The most important nonlinear coherent effect in spectroscopy has, until now, been the laser-induced line narrowing in coupled Doppler-broadened transitions, often referred to as optical–optical double resonance (Schlossberg and Javan, 1966a; Hänsch and Toschek, 1968; Feld and Javan, 1969). The principles involved in a typical experiment of this kind are shown in Fig.26. Two optical transitions 0–1 and 0–2 sharing a common level 0, interact simultaneously with two laser fields, ω_1 and ω_2, which propagate in the same direction. If one laser frequency, say ω_1, is tuned across the Doppler-broadened profile of the transition 0–1, then a nonlinear response will occur, but only if both radiation fields interact with the same narrow velocity subgroup of the ensemble, that is, if $\omega_2 - \omega_1 = \omega_{02} - \omega_{01}$. The width of a signal due to this coupling of two Doppler-broadened transitions depends only on the radiative lifetimes of the levels 1 and 2 (if power broadening is neglected); thus the energy separation between levels 1 and 2 can be measured with a linewidth that is affected neither by a finite lifetime of the level 0 nor by inhomogeneous broadening effects. The maximum energy separation that can be measured by means of this method is limited only by the laser wavelengths available, while in scanning techniques such as saturation spectroscopy, the limiting factor is the tuning range of the laser.

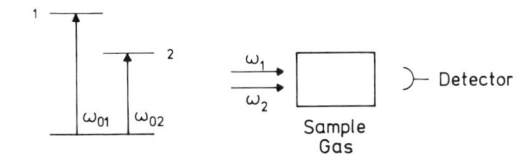
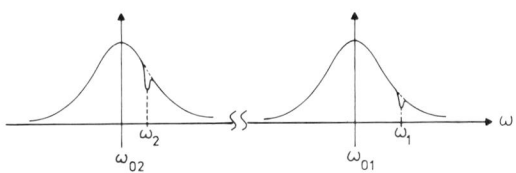

Fig. 26. Optical–optical double resonance: level diagram, simplified experimental arrangement and the absorption line shapes as influenced by the laser fields.

A technical modification of the optical–optical double resonance is mode crossing (Schlossberg and Javan, 1966b). In this technique the frequencies ω_1 and ω_2 are fixed (for example, two modes of one laser), whereas the levels 1 and 2 are tuned by means of magnetic or electric fields; thus using this technique, Zeeman and Stark splittings can be measured with great precision.

Until now, the optical–optical double resonance has only been realized in experiments utilizing fixed-frequency lasers. Most work of this kind has been restricted to the laser medium itself, because for such experiments two laser frequencies have to be available which coincide with the two coupled transitions of the sample. The introduction of tunable lasers to this technique will, however, extend its applicability to a large number of spectroscopic problems.

REFERENCES

Abella, I. D., Kurnit, N. A., and Hartmann, S. R. (1966). *Phys. Rev.* **141**, 391.
Abragam, A. (1961). "The Principles of Nuclear Magnetism." Oxford Univ. Press (Clarendon), London and New York.
Ackermann, H. (1966). *Z. Phys.* **194**, 253.
Agostini, P., Bensoussan, P., and Boulassier, J. C. (1972). *Opt. Commun.* **5**, 293.
Ahmed, S. A. (1973). *Appl. Opt.* **12**, 901.
Alekseyev, A. I. (1970). *Phys. Lett. A* **31**, 495.
Alexandrov, E. B. (1963). *Opt. Spectrosc. (USSR)* **15**, 232.
Alexandrov, E. B. (1964). *Opt. Spectrosc. (USSR)* **17**, 522.
Ambartzumian, R. V., and Letokhov, V. S. (1972). *Appl. Opt.* **11**, 354.
Anliker, P., Gassmann, M., and Weber, H. (1972). *Opt. Commun.* **5**, 137.

Arthurs, E. G., Bradley, D. J., and Roddie, A. G. (1971). *Appl. Phys. Lett.* **19**, 480.
Ashkin, A. (1970). *Phys. Rev. Lett.* **24**, 156; **25**, 1321.
Atkinson, G. H., Laufer, A. H., and Kurylo, M. J. (1973). *J. Chem. Phys.* **59**, 350.
Baltakov, F. N., Barikhin, B. A., Kornilov, V. G., Mikhnov, S. A., Rubinov, A. N., and Sukhanov, L. V. (1973). *Sov. Phys.—Tech. Phys.* **17**, 1161.
Barger, R. L., and Hall, J. L. (1969). *Phys. Rev. Lett.* **22**, 4.
Barger, R. L., Sorem, M. S., and Hall, J. L. (1973). *Appl. Phys. Lett.* **22**, 573.
Barrat, J. P. (1959). *J. Phys. Radium* **20**, 541.
Basov, N. G., Kompanets, I. N., Kompanets, O. N., Letokhov, V. S., and Nikitin, V. V. (1969). *Sov. Phys.—JETP Lett.* **9**, 345.
Bennett, W. R., Jr. (1962). *Phys. Rev.* **126**, 580.
Bloch, F. (1946). *Phys. Rev.* **70**, 460.
Bonn, J., Huber, G., Kluge, H.-J., Köpf, U., Kugler, L., Otten, E.-W., and Rodriguez, J. (1973a). *In* "Atomic Physics 3" (S. J. Smith and G. K. Walters, eds.), p. 471. Plenum, New York.
Bonn, J., Huber, G., Kluge, H.-J., and Otten, E.-W. (1973b). *In* "Summaries of the Vth EGAS Conference." Lund.
Bowman, M. R., Gibson, A. J., and Sandford, M. C. W. (1969) *Nature (London)* **221**, 456.
Bradley, D. J., and O'Neill, F. (1969). *Opto-Electron.* **1**, 69.
Bradley, D. J., Gale, G. M., and Smith, P. D. (1970a). *Proc. Phys. Soc., London (At. Mol. Phys.)* [2] **3**, L11.
Bradley, D. J., Gale, G. M., and Smith, P. D. (1970b). *Nature (London)* **225**, 719.
Bradley, D. J., Gale, G. M., and Smith, P. D. (1971). *Proc. Phys. Soc., London (At. Mol. Phys.)* **4**, 1349.
Bradley, D. J., Ewart, P., Nicholas, J. V., Shaw, J. R. D., and Thomson, G. D. (1973a). *Phys. Rev. Lett.* **31**, 263.
Bradley, D. J., Ewart, P., Nicholas, J. V., Shaw, J. R. D., and Thomson, G. D. (1973b). *In* "Summaries of the Vth EGAS Conference." Lund.
Brewer, R. G., and Shoemaker, R. L. (1972). *Phys. Rev. A* **6**, 2001.
Brewer, R. G., Kelly, M. J., and Javan, A. (1969). *Phys. Rev. Lett.* **23**, 559.
Burrell, C. F., and Kunze, H.-J. (1972). *Phys. Rev. Lett.* **38**, 1.
Cagnac, B., Grynberg, G., and Biraben, F. (1973). *J. Phys. (Paris)* **34**, 845.
Carlsten, J. L., and McIlrath, T. J. (1973). *Opt. Commun.* **8**, 52.
Cha, S. C., Figger, H., and Walther, H. (1973). *In* "Summaries of the Vth EGAS Conference." Lund.
Cheo, P. K., and Wang, C. H. (1970). *Phys. Rev. A* **1**, 255.
Collins, C. B., Johnson, B. W., and Shaw, M. J. (1972). *J. Chem. Phys.* **57**, 5310.
Copley, G., Kibble, B. P., and Series, G. W. (1968). *Proc. Phys. Soc., London (At. Mol. Phys.)* [2] **1**, 724.
Corney, A., and Series, G. W. (1964a). *Proc. Phys. Soc., London* **83**, 207.
Corney, A., and Series, G. W. (1964b). *Proc. Phys. Soc., London* **83**, 213.
Courtens, E. (1972). *In* "Laser Handbook" (F. T. Arecchi and E. O. Schulz-Dubois, eds.), Vol. 2, pp. 1259-1322. North-Holland Publ., Amsterdam.
Curry, S. M., Cubeddu, R., and Hänsch, T. W. (1973). *Appl. Phys.* **1**, 153.
Decker, C. D., and Tittel, F. K. (1973). *Opt. Commun.* **8**, 244.
Demtröder, W. (1971). *In* "Topics in Current Chemistry," Vol. 17. Springer-Verlag, Berlin and New York.
Demtröder, W. (1973). *Phys. Rep.* **7**, 223.
Denton, M. B., and Malmstadt, H. V. (1971). *Appl. Phys. Lett.* **18**, 485.
Dewey, C. F., and Hocker, L. O. (1971). *Appl. Phys. Lett.* **18**, 247.
Dinev, S. G., Stamenov, K. V., and Tomov, I. V. (1972). *Opt. Commun.* **5**, 419.

Dodd, J. N., Kaul, R. D., and Warrington, D. (1964). *Proc. Phys. Soc., London* **84**, 176.
Dodd, J. N., Sandle, W. J., and Zissermann, D. (1967). *Proc. Phys. Soc., London* **92**, 497.
Dunning, F. B., Tittel, F. K., and Stebbings, R. F. (1973). *Opt. Commun.* **7**, 181.
Duong, H. T., Jacquinot, P., Liberman, S., Picqué, J.-L., Pinard, J., and Vialle, J.-L. (1973a). *Opt. Commun.* **7**, 371.
Duong, H. T., Jacquinot, P., Liberman, S., Pinard, J., and Vialle, J.-L. (1973b). *C.R. Acad. Sci., Ser. B* **276**, 909.
D'Yakonov, M. I., and Perel, V. I. (1965a). *Sov. Phys.—JETP* **20**, 227.
D'Yakonov, M. I., and Perel, V. I. (1965b). *Sov. Phys.—JETP* **20**, 997.
Eranian, A., Dezauzier, P., and de Witte, O. (1973). *Opt. Commun.* **7**, 150.
Erdmann, T. A., Figger, H., and Walther, H. (1972). *Opt. Commun.* **6**, 166.
Feld, M. S., and Javan, A. (1969). *Phys. Rev.* **177**, 540.
Ferrar, C. M. (1969). *IEEE J. Quantum Electron.* **5**, 550.
Frisch, R. (1933). *Z. Phys.* **86**, 42.
Gabel, C., and Hercher, M. (1972). *IEEE J. Quantum Electron.* **8**, 850.
Gale, C. M. (1973). *Opt. Commun.* **7**, 86.
Gibson, A. J. (1969). *J. Sci. Instrum.* [2] **2**, 802.
Gornik, W., Kaiser, D., Lange, W., Luther, J., and Schulz, H. H. (1972). *Opt. Commun.* **6**, 327.
Gornik, W., Kaiser, D., Lange, W., Luther, J., Radloff, H.-H., and Schulz, H. H. (1973a). *Appl. Phys.* **1**, 285.
Gornik, W., Kaiser, D., Lange, W., Luther, J., Radloff, H.-H., and Schulz, H. H. (1973b). *In* "Summaries of the Vth EGAS Conference." Lund.
Gornik, W., Kaiser, D., Lange, W., Luther, J., Meier, K., Radloff, H.-H., and Schulz, H. H. (1973c). *Phys. Lett. A* **45**, 219.
Grandin, J.-P., and Husson, X. (1973). *In* "Summaries of the Vth EGAS Conference." Lund.
Grischkowsky, D. (1973). *Phys. Rev. A* **7**, 2096.
Grischkowsky, D., and Armstrong, J. A. (1972). *Phys. Rev. A* **6**, 6.
Grischkowsky, D., Courtens, E., and Armstrong, J. A. (1973). *Phys. Rev. Lett.* **31**, 422.
Hahn, E. L. (1950). *Phys. Rev.* **80**, 580.
Hall, J. (1973). *In* "Atomic Physics 3" (S. J. Smith and G. K. Walters, eds.), p. 615. Plenum, New York.
Hanes, G. R., and Dahlstrom, C. E. (1969). *Appl. Phys. Lett.* **14**, 362.
Hänsch, T. W. (1972). *Appl. Opt.* **11**, 895.
Hänsch, T. W. (1973). *In* "Atomic Physics 3" (S. J. Smith and G. K. Walters, eds.), p. 579. Plenum, New York.
Hänsch, T. W., and Toschek, P. (1968). *IEEE J. Quantum Electron.* **4**, 467.
Hänsch, T. W., Levenson, M. D., and Schawlow, A. L. (1971a). *Phys. Rev. Lett.* **26**, 946.
Hänsch, T. W., Shahin, I. S., and Schawlow, A. L. (1971b). *Phys. Rev. Lett.* **27**, 707.
Hänsch, T. W., Shahin, I. S., and Schawlow, A. L. (1972). *Nature (London), Phys. Sci.* **235**, 63.
Hänsch, T. W., Schawlow, A. L., and Toschek, P. (1973). *IEEE J. Quantum Electron.* **9**, 553.
Haroche, S., Paisner, J. A., and Schawlow, A. L. (1973). *Phys. Rev. Lett.* **30**, 948.
Harris, S. E. (1973). *Phys. Rev. Lett.* **31**, 341.
Harris, S. E., and Miles, R. B. (1971). *Appl. Phys. Lett.* **19**, 385.
Hartig, W., and Walther, H. (1973). *Appl. Phys.* **1**, 171.
Hebner, R. E., Jr., and Nygaard, K. J. (1971). *J. Opt. Soc. Amer.* **61**, 1455.
Henzel, P., Hohla, K., and Kompa, K. L. (1971). *Appl. Phys. Lett.* **18**, 48.
Hercher, M., and Pike, H. A. (1971). *Opt. Commun.* **3**, 346.
Hinkley, E. D., and Kelley, P. L. (1971). *Science* **171**, 635.
Holstein, T. (1947). *Phys. Rev.* **72**, 1212.
Hopf, F. A., Shea, R. F., and Scully, M. O. (1973). *Phys. Rev. A* **7**, 2105.

Hotop, H., and Lineberger, W. C. (1973). *J. Chem. Phys.* **58**, 2379.
Hotop, H., Patterson, T. A., and Lineberger, W. C. (1973). *Phys. Rev. A* **8**, 763.
Huth, B. G. (1970). *Appl. Phys. Lett.* **16**, 185.
Ippen, E. P., Shank, C. V., and Dienes, A. (1972). *Appl. Phys. Lett.* **21**, 348.
Itzkan, I., and Cunningham, F. W. (1972). *IEEE J. Quantum Electron.* **8**, 101.
Jacobs, R. R., Lempicki, A., and Samelson, H. (1973). *J. Appl. Phys.* **44**, 2775.
Jacquinot, P., Liberman, S., Picqué, J.-L., and Pinard, J. (1973). *Opt. Commun.* **8**, 163.
Kibble, B. P., Rowley, W. R. C., Shawyer, R. E., and Series, G. W. (1973). *Proc. Phys. Soc., London (At. Mol. Phys.)* [2] **6**, 1079.
Kogelnik, H. W., Ippen, E. P., Dienes, A., and Shank, C. V. (1972). *IEEE J. Quantum Electron.* **8**, 373.
Korolev, F. A., Bakhramov, S. A., and Odintsov, V. I. (1970). *JETP Lett.* **12**, 90.
Korolev, F. A., Bakhramov, S. A., and Odintsov, V. I. (1971). *Opt. Spectrosc. (USSR)* **30**, 425.
Kretzen, H. H. (1973). *Phys. Lett. A* **44**, 79.
Kuhl, J., and Marowsky, G. (1971). *Opt. Commun.* **4**, 125.
Kuhl, J., and Spitschan, H. (1973). *Opt. Commun.* **7**, 256.
Kung, A. H., Young, J. F., and Harris, S. E. (1973). *Appl. Phys. Lett.* **22**, 301.
Lamb, W. E., Jr. (1964). *Phys. Rev. A* **134**, 1429.
Lange, W., Luther, J., Nottbeck, B., and Schröder, H. W. (1973). *Opt. Commun.* **8**, 157.
Lee, P. H., and Skolnick, M. L. (1967). *Appl. Phys. Lett.* **10**, 303.
Letokhov, V. S. (1967). *JETP Lett.* **6**, 101.
Lin, C., Gustafson, T. K., and Dienes, A. (1973). *Opt. Commun.* **8**, 210.
Lineberger, W. C., and Woodward, W. (1970). *Phys. Rev. Lett.* **25**, 424.
Lisitsyn, V. N., and Chebotaev, V. P. (1968). *Sov. Phys.—JETP* **27**, 227.
Loth, C., and Meyer, Y. H. (1973). *Appl. Opt.* **12**, 123.
McFarlane, R. A., Bennett, W. R., Jr., and Lamb, W. E., Jr., (1963). *Appl. Phys. Lett.* **2**, 189.
McIlrath, T. J. (1969). *Appl. Phys. Lett.* **15**, 41.
McIlrath, T. J., and Carlsten, J. L. (1972). *Phys. Rev. A* **6**, 1091.
McIlrath, T. J., and Carlsten, J. L. (1973). *Proc. Phys. Soc., London (At. Mol. Phys.)* [2] **6**, 697.
Marrus, R., and McColm, D. (1965). *Phys. Rev. Lett.* **15**, 813.
Marrus, R., and Yellin, J. (1969). *Phys. Rev.* **177**, 127.
Marrus, R., McColm, D., and Yellin, J. (1966). *Phys. Rev.* **147**, 55.
Marrus, R., Wang, E., and Yellin, J. (1967). *Phys. Rev. Lett.* **19**, 1.
Marrus, R., Wang, E., and Yellin, J. (1969). *Phys. Rev.* **177**, 122.
Mashinskii, A. L. (1970). *Opt. Spectrosc. (USSR)* **28**, 1.
Miles, R. B., and Harris, S. E. (1973). *IEEE J. Quantum Electron.* **9**, 470.
New, G. H. C. (1972). *Alta Freq.* **41**, 711.
Nieuwenhuijzen, H. (1970). *Opt. Technol.* **2**, 13.
Omont, A. (1965). *J. Phys. Radium* **26**, 26.
O'Neill, F. (1972). *Opt. Commun.* **6**, 360.
Perl, M. L., Rabi, I. I., and Senitzky, B. (1955). *Phys. Rev.* **97**, 838; **98**, 611.
Petersen, F. R., McDonald, D. G., Cupp, J. D., and Danielson, B. L. (1973). *Phys. Rev. Lett.* **31**, 573.
Peterson, N. C., Kurylo, M. J., Braun, W., Bass, A. M., and Keller, R. A. (1971). *J. Opt. Soc. Amer.* **61**, 746.
Picqué, J.-L., and Vialle, J.-L. (1972). *Opt. Commun.* **5**, 402.
Pringsheim, P. (1948). "Fluorescence and Phosphorescence," 1st ed. Wiley (Interscience), New York. (2nd ed., 1963 p. 61).
Rabinowitz, P., Keller, R., and LaTourrette, J. T. (1969). *Appl. Phys. Lett.* **14**, 376.
Ramsey, N. F. (1956). "Molecular Beams." Oxford Univ. Press, London and New York.

Schäfer, F. P. (1970). *Angew. Chem.* **82**, 25.
Schäfer, F. P. (1972). *In* " Laser Handbook " (F. T. Arecchi, and E. O. Schulz-Dubois, eds.), Vol. 1, pp. 369–424. North-Holland Publ., Amsterdam.
Schäfer, F. P., Schmidt, W., and Volze, J. (1966). *Appl. Phys. Lett.* **9**, 306.
Schenck, P., Hilborn, C. H., and Metcalf, H. (1973). *Phys. Rev. Lett.* **31**, 189.
Schieder, R., Walther, H., and Wöste, L. (1972). *Opt. Commun.* **5**, 337.
Schlossberg, H. R., and Javan, A. (1966a). *Phys. Rev. Lett.* **17**, 1242.
Schlossberg, H. R., and Javan, A. (1966b). *Phys. Rev.* **150**, 267.
Schmidt, W. (1970). *Lasers* **4**, 47.
Schmidt, W., and Wittekindt, N. (1972). *Appl. Phys. Lett.* **20**, 71.
Schröder, H. W., Welling, H., and Wellegehausen, B. (1973). *Appl. Phys.* **1**, 343.
Schuda, F., Hercher, M., and Stroud, C. R., Jr. (1973). *Appl. Phys. Lett.* **22**, 360.
Series, G. W. (1957). " Spectrum of Atomic Hydrogen." Oxford Univ. Press, London and New York.
Shahin, I. S., and Hänsch, T. W. (1973). *Opt. Commun.* **8**, 312.
Shimizu, F. (1969). *Appl. Phys. Lett.* **14**, 378.
Smith, P. W. (1970). *Proc. IEEE* **58**, 1342.
Snavely, B. B. (1969). *Proc. IEEE* **57**, 1374.
Sorokin, P. (1969). *Sci. Amer.* **220**, 30.
Sorokin, P. P., and Lankard, J. R. (1966). *IBM J. Res. Develop.* **10**, 162.
Sorokin, P. P., and Lankard, J. R. (1969a). *J. Chem. Phys.* **51**, 2929.
Sorokin, P. P., and Lankard, J. R. (1969b). *Phys. Rev.* **186**, 342.
Sorokin, P. P., and Lankard, J. R. (1971). *J. Chem. Phys.* **54**, 2184.
Sorokin, P. P., Lankard, J. R., Moruzzi, V. L., and Lurio, A. (1969). *Appl. Phys. Lett.* **15**, 179.
Stebbings, R. F., Dunning, F. B., Tittel, F. K., and Rundel, R. D. (1973). *Phys. Rev. Lett.* **30**, 815.
Svanberg, S., Tsekeris, P., and Happer, W. (1973). *Phys. Rev. Lett.* **30**, 817.
Szabo, A., and Erickson, L. E. (1972). *Opt. Commun.* **5**, 287.
Szöke, A., and Javan, A. (1963). *Phys. Rev. Lett.* **10**, 521.
Szöke, A., Goldhar, J., Grieneisen, H. P., and Kurnit, N. A. (1972). *Opt. Commun.* **6**, 131.
Tang, C. L., and Silverman, B. D. (1966). *In* " Physics of Quantum Electronics " (P. Kelley, B. Lax, and P. Tannenwald, eds.), p. 280. McGraw-Hill, New York.
Tan-no, N., Kan-no, K., Yokoto, K., and Inaba, H. (1973). *IEEE J. Quantum Electron.* **9**, 423.
Taylor, B. N., Parker, W. H., and Langenberg, D. N. (1969). *Rev. Mod. Phys.* **41**, 375.
Thrash, R. J., von Weyssenhoff, H., and Shink, J. S. (1971). *J. Chem. Phys.* **55**, 4659.
Torrey, H. C. (1949). *Phys. Rev.* **76**, 1059.
Tuccio, S. A., and Strome, F. C. (1972). *Appl. Opt.* **11**, 64.
Tuccio, S. A., Drexhage, K. H., and Reynolds, G. A. (1973). *Opt. Commun.* **7**, 248.
Vinogradov, A. V., and Yukov, E. A. (1972). *JETP Lett.* **16**, 447.
Walther, H. (1973a). *Proc. Conf. Laser Spectrosc., 1973* Plenum, New York, in press.
Walther, H. (1973b). Lecture given at the 4th Annual Meeting of the European Quantum Electronics Division, Stresa (unpublished).
Wang, C. H. (1970). *Phys. Rev. B* **1**, 156.
Ward, J. F., and New, G. H. C. (1969). *Phys. Rev.* **185**, 57.
Warden, J. T., and Gough, L. (1971). *Appl. Phys. Lett.* **19**, 345.
Young, J. F., Bjorklund, G. C., Kung, A. H., Miles, R. B., and Harris, S. E. (1971). *Phys. Rev. Lett.* **27**, 1551.
Zel'dovich, B. Y., and Kuznetsova, T. I. (1972). *Sov. Phys.—Usp.* **15**, 25.
zu Putlitz, G. (1965). *Ergeb. Exakten Naturwiss.* **37**, 105.
zu Putlitz, G. (1969). *Comments At. Mol. Phys.* **1**, 74.

RECENT PROGRESS IN THE CLASSIFICATION OF THE SPECTRA OF HIGHLY IONIZED ATOMS

B. C. FAWCETT

Science Research Council
Astrophysics Research Division of the Appleton Laboratory
Culham Laboratory, near Abingdon
Berkshire, England

I. Introduction	224
II. Laboratory Light Sources	225
A. The Theta Pinch and Other Plasma Sources	226
B. Laser-Produced Plasmas	228
C. The High-Voltage Vacuum Spark	230
D. The Low-Inductance High Capacity Spark	231
E. The Beam Foil Source	231
III. Measuring and Experimental Techniques	232
A. Measuring Techniques	232
B. Experimental Techniques for Recording Time-Resolved Spectrograms	234
C. Separation of Ionization Stages	234
D. Separation of Lines of the Same Ion Due to Variation of Their Intensity with Source Density	235
IV. Theoretical Calculations	236
A. The Hartree-X Method	236
B. The Multiconfiguration Hartree–Fock Method	237
V. Line Classifications of Highly Ionized Systems	239
A. The $2s^2 2p^n$–$2s 2p^{n+1}$ and $2s 2p^n$–$2p^{n+1}$ Transition Arrays	239
B. The $3s^2 3p^n$–$3s 3p^{n+1}$ and $3s^2 3p^n$–$3s^2 3p^{n-1} 3d$ Transition Arrays	241
C. Transitions to Levels of Principal Quantum Number $n = 2$ from Those with Higher Principal Quantum Numbers	247
D. Helium-like Spectra and Their Satellites	250
E. Transitions between Levels of Principal Quantum Number 3 and 4	254
F. The First Long Period	255
G. The Spectra of Heavier Elements	257
H. References to Published Data on Line Classification	258
VI. Identification of Emission Lines of Highly Ionized Atoms in the Solar Spectrum	262
A. History	262
B. Solar Spectra between 20 and 2000 Å	262
C. Eclipse Solar Ultraviolet Spectra	281
D. Solar Spectra between 1 and 25 Å	283
VII. Discussion	284
References	285

I. Introduction

Since 1960 there has been rapid progress and renewed enthusiasm in the classification of the spectra of highly ionized atoms. Several factors contributed to this renewal of interest, which had waned since the early days of atomic spectroscopy in the late 1920's and early 1930's when physicists eagerly studied spectra to support new theories in atomic physics and quantum mechanics. One event which marked the beginning of this contemporary research was the launching of a rocket in January 1960 (Hinteregger, 1960) under the auspices of the Air Force Cambridge Research Laboratory hereafter abbreviated AFCRL, USA. The rocket payload was a scanning grazing incidence spectrometer fitted with a photomultiplier insensitive to radiation longer than 1300 Å which had previously (Violett and Rense, 1959) masked solar rocket spectra below 300 Å. Solar spectra were obtained which could not be explained using current tabulations of spectral lines or atomic energy levels. Improved records which will be referred to later (see Table IV) were obtained between 1960 and 1963 at AFCRL and the Naval Research Laboratory (NRL), USA and Goddard Space Flight Center (GSFC), USA. The quantity and intensities of the unknown spectral lines recorded were such as to account for the major part of the Sun's energy in the far ultraviolet falling on the Earth's outer atmosphere. The quest for knowledge of the nature of this high energy radiation, which affects the Earth's ionosphere and hence such mundane things as radio communications, not only had a stimulating effect on spectral line classification but also radically improved knowledge of the physics of the solar corona. More surprises were in store as rocket experiments from these and other groups such as the Astrophysics Research Unit (ARU), Culham, Leicester University, England, The Aerospace Corporation, USA, and the Lebedev Institute in Moscow brought back a wealth of revealing solar spectra. These included solar flare helium-like spectra of iron XXV which few astrophysicists anticipated and excellent records of solar forbidden lines between 1000 and 2000 Å—observations which have considerable impact on atomic and solar physics.

Apart from the solar rocket program, another adventure furthered advances in the classification of the spectra of highly ionized atoms. Since the explosion of the hydrogen bomb, man has desired to harness its power for peaceful purposes. This motivation brought heavy investments in effort and expenditure in plasma physics aimed at attaining the requirements of controlled fusion. To attempt to fulfill these needs gases were heated to electron temperatures of over ten million degrees and supported by magnetic fields. This coincided with the criteria for a spectroscopic light source able to emit spectra of highly ionized atoms.

Yet another new development brought a bountiful harvest of new line classifications. The giant pulse laser with its focused beam generated an excellent spectroscopic light source which emitted impurity-free spectra of elements from highly ionized atoms.

What follows is a review of these recent contributions to the understanding of multiply ionized systems and their solar applications. For earlier work on atomic spectra, which will not be dealt with here, the reader is referred to the listing of atomic energy levels (Moore, 1949, 1952, 1958) which also includes full references to line classification papers. This article will also omit those relevant aspects so thoroughly dealt with in the authoritative treatment of atomic spectra by Edlén (1964) in *Handbuch der Physik*. Another article (Edlén, 1963) covers spectrograph design, earlier light sources, and wavelength standards. Before moving attention away from this earlier work it is fitting to pay tribute to its quality. Research workers of the last decade have had the benefit of building on firm foundations in atomic spectroscopy laid after painstaking and accurate work. Outstanding are the prolific and highly reliable contributions of Professor Edlén and other Swedish researchers at Lund and Uppsala Universities. One of the highlights of his analysis is the classification of the visible forbidden lines (Edlén, 1942) which later provided a clue to the identity of solar lines in the extreme ultraviolet.

The laboratory sources cited above are discussed in Section II along with other light sources which are still of service. A small selection of experimental and measuring techniques which are of special benefit are described in Section III. Theoretical calculations which are becoming increasingly important as more complex spectra are tackled are referred to in Section IV. Section V includes a comprehensive account of new line classifications of spectra of highly ionized atoms with a full indexed reference list to recent work. Section VI deals with their application to the solar corona and includes an up-to-date list of line identifications in the solar spectrum. The sections, although related, can to some extent be studied separately.

II. Laboratory Light Sources

Light sources which provide clear observations of the spectra under investigation are a prerequisite to spectral analysis. The high voltage vacuum spark was the main laboratory source of highly ionized spectra during the past 40 years but the new technology of the past 15 years (Gabriel, 1970) has given birth to light sources which generate highly ionized spectra under conditions which permit new, interesting features to be seen on the spectrograms. The most productive of these new sources are the theta pinch, laser-produced plasma, and low inductance spark, which will be discussed along with other sources both from the point of view of their construction and their advantages for spectral identification in the next subsections.

A. The Theta Pinch and Other Plasma Sources

The high temperature plasmas of fusion research contrast with spark discharges and provide conditions which are more favorable to the study of some highly ionized spectra. The main difference is that these plasmas are formed from an initial volume of gas restricted by a magnetic field and heated so that atoms are progressively ionized to higher stages of ionization and finally lost or recombine, whereas in spark discharges injection and loss of material controls their physical conditions. It is possible, therefore to select plasmas for line classification purposes which radiate spectra from successive stages of ionization sequentially. Emission lines belonging to particular ionization stages can therefore be identified by recording their intensity variation as a function of time. The intensities can furthermore be related to theory since the parameters of the plasma such as electron and ion temperature and density are meaningful and can be measured. Such discrimination of ionization stages was achieved in the first line classification project (Fawcett et al., 1961) on a high temperature plasma which was generated in Zeta: a toroidal pinch of 3-msec pulse duration and 10^{14} cm^{-3} electron density. Spectra were recorded photographically and their time history studied photographically (see Section III,B) or photoelectrically (see Fig.1). Another similar toroidal device SCEPTRE was subsequently applied to line classification (Bockasten et al., 1963). The use of these machines for atomic spectroscopy has been restricted because of their large running cost, but a moderate-sized theta pinch can be maintained by a small group with a limited budget. Furthermore, the theta pinch has no electrodes to contaminate the discharge and generates a plasma with high electron temperatures of up to 350 eV and densities of about 10^{16} cm^{-3}, which radiates intense spectra. It has been built in sizes ranging from 1 kJ to 1 MJ and has become the most favored spectroscopic plasma light source. The first theta pinch used for this purpose was Scylla at Los Alamos (Sawyer et al., 1962) yielding identifications of Ne IX and Si XIII. There followed contributions to the analysis of neon and argon (Fawcett et al., 1964) at Culham, of iron at Los Alamos (Peacock et al., 1965), and an extensive analysis of highly ionized sulfur, chlorine, and argon on Scylla I at the High Altitude Observatory, Boulder, Colorado (Deutschman and House, 1966, 1967). At Uppsala University comprehensive studies of N IV, N V (Hallin, 1966a,b) and O IV, O V (Bromander, 1969; Bockasten and Johansson, 1968) utilized a 2-kJ theta pinch. There are other papers dealing with neon based on a theta-pinch light source (Hermansdorfer, 1972; Druetta et al., 1972; Tondello and Paget, 1970).

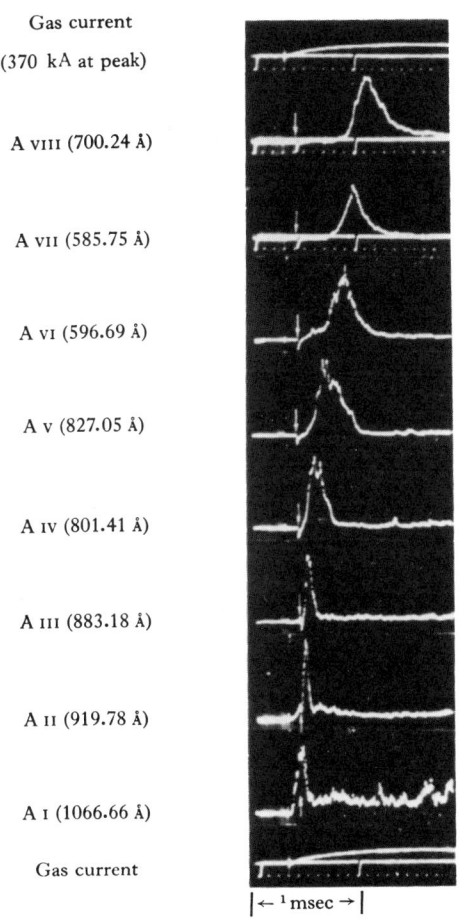

FIG. 1. Temporal variation of the intensities of spectral lines due to Ar I to Ar VIII inclusive showing the successive appearance of the ionization stages in the ZETA plasma. The argon is introduced as a 5% impurity in deuterium with a total gas pressure of 0.5 mTorr. From Fawcett et al. (1961).

The last reference is to the first of a number of line classification projects (Fawcett et al., 1971, 1972a,b; Fawcett, 1971a,b,d) undertaken at ARU, Culham with a theta pinch specially constructed for spectroscopic studies of interest to astrophysics (Gabriel and Paget, 1972). It has produced some of the purest highly ionized theta pinch spectra and will therefore be described here. It consists of a 40-kV capacitor bank connected via switches to a single turn coil of 12.5 cm diameter and 30 cm long which surrounds a quartz tube containing hydrogen at 40 mTorr. To the hydrogen is added a small percentage of the gas or volatile compound under investigation. A reverse bias magnetic field of up to 900 gauss is applied before and throughout the discharge. The gas is preionized with a short pulse of 10 MHz applied capacitively to the outside of the quartz tube near the end of the coil. Ten microseconds later a preheater discharge of 1 mF at 40 kV through the main coil heats the plasma which remains fully ionized when the main bank is discharged 20 msec later. The main bank is typically operated at 30 kJ and 37 kV. Electron temperatures of 300 eV and densities of 8.6×10^{16} cm^{-3} are attained and measured by laser scattering. The parameters chosen result in a plasma relatively free from injected impurities when compared with most other theta pinches. The quartz tube has a side arm fitted through the center of the coil so that the vacuum spectrograph can view the plasma perpendicular to the tube axis. This arrangement avoids seeing spectra of lower stages of ionization and impurities; these spectra are most strongly radiated near the ends of the coil and have consequently reduced the value of earlier spectrograms acquired with axial viewing. A microsecond shutter, described in Section III,B, situated in front of the spectrograph entrance slit is opened while the discharge emits the ionization stages of interest. The resultant argon spectrum obtained using a theta pinch shows a complete freedom from impurity lines and no stages of ionization of argon except Ar X to Ar XIV.

A different plasma machine, a plasma focus similar to that described by Mather (1966), has enabled the observation of Ar XVII to Ar XVIII (Peacock et al., 1969) and resonance spectra of Kr XXVII to Kr XXX (Hobby and Peacock, 1973).

B. Laser-Produced Plasmas

The first reports of highly ionized spectra emitted from laser-produced plasmas were by Ehler and Weissler (1966) who observed lines generated by a 10-MW laser and by Fawcett et al. (1966) who classified lines of Fe XV and Fe XVI using a 400-MW ruby laser with 15-nsec pulse duration. There followed a number of publications (Fawcett, 1970a,b; Burgess et al., 1967; Boland et al., 1968; Fawcett et al., 1967a,b, 1972b; Basov et al., 1967; Kononov, 1966, 1969; Boiko et al., 1970; Kononov and Koshelev, 1970; Kononov et al., 1970; Kasyanov et al., 1973; Podobedova et al., 1971) which have demonstrated that laser-produced plasmas provide one of the most powerful means of spectral analysis. The plasmas are generated when the output of a powerful laser is focused onto a solid target situated in a vacuum. Neodymium or carbon dioxide lasers may be preferred to ruby lasers because of the high cost of ruby laser rods. A suitable laser is described by Fawcett (1970a) and has been used extensively for this purpose. It consists of an oscillator and amplifier. The oscillator contained a neodymium-doped glass laser rod of 0.95 cm diameter pumped along its central 11 cm with a helical flash tube.

The oscillator consisted of components aligned in the following order: a fully reflecting dielectric mirror, a quarter-wave-switched KDP pockel cell, the laser rod, and a 50% dielectric output mirror. The output beam of divergence 5.5 arcsec was expanded before it entered the amplifier which comprised a flash-tube-pumped neodymium glass rod 52 cm long and 1.6 cm in diameter. Powers of 3 GW were also used to generate plasmas after adding a second amplifier to the laser. The arrangement of components in this laser prevented pre-lasing which becomes troublesome at high powers when light is reflected from the target into the oscillator. This trouble is severe in the usual neodymium laser designed to output pulses of 1- to 10-nsec duration. In such lasers the oscillator cavity usually has two totally reflecting mirrors, the laser rod, a Glan Thompson polarizer, and a pockel cell. A square-wave pulse applied to the pockel cell switches the laser output beam of the same duration through the Glan Thompson polarizer. Such a system operating at 1.5 GW and flux densities of 10^{12} W/cm^2 has been used by Gallanti and Peacock (1973) to produce good spectra of helium-like lines and their satellites of Mg XI and Al XII.

If light from the laser target causes free lasing it may be necessary to stop it entering the laser with a Faraday rotator or pockel cell combined with polarizers and switched to constitute an optical shutter. This complication has so far impeded work at high powers and the study of the effect of using different high power output pulse lengths. For higher ionization of the plasma, larger laser powers are necessary. Heavy element plasmas radiate more and therefore require greater laser power. Thus for iron X V to X VI 400 MW is necessary, but for C IV, 20 MW will suffice. Below these approximate thresholds the shorter wavelength end of the spectrum consists of a continuum crossed by absorption lines of various stages of ionization. This absorption-type spectrum has so far not been exploited in these sources. To achieve the best results the laser beam must be precisely focused on the target. Defocusing the laser beam on the target is usually the most convenient method of reducing the intensity of the emission from the most highly ionized atoms. The ability to focus the laser beam depends on its beam divergence. The operation of the laser in a single transverse mode and attention paid to the finesse of the laser optical system can give low beam divergence, hence good focusing and hotter plasmas at lower laser powers. This was done at the Spectroscopic Institute in Moscow where such a laser operating at 400 MW produced degrees of ionization which would have otherwise have required 1 GW. As well as defocusing and variation of laser power, the study of the spatial distribution (Irons *et al.*, 1972) of the line emission from laser plasmas provides a method of separating lines from different ionization stages. Spatial resolution in spectrograms is achieved with grazing incidence spectrographs by placing the target about 1 cm from the entrance slit and a second slit 200-μm wide perpendicular to and half way between the entrance slit and the grating. A cross section of the plasma is therefore imaged in each spectral line. With normal incidence spectrographs stigmatic spectra are obtained by placing the target at Sirks focus. Sirks focus lies on the intersection of a line from the grating center through the exit slit and a line which is a tangent to the Rowland circle on the opposite side to the grating center. This separation of lines from different ionization is only one of the reasons why the laser-produced plasma is a valuable source; it will, in addition, radiate spectra almost free from impurity lines of any solid the experimenter chooses as a target. Furthermore, its high electron density of around 10^{19} cm^{-3} gives plasma conditions in which excitation takes place into levels only slowly populated directly from the ground level. Hence, a new range of spectra come into view for investigation. The disadvantages are that the high densities give large stark widths (Burgess *et al.*, 1967; Irons *et al.*, 1972) for sensitive lines and that the mass motion of the plasma, which can be 3×10^7 to 5×10^6 cm/sec, leads to corresponding mass motion doppler widths. These widths limit the accuracy of wavelength determination. The doppler widths are more serious with light element plasmas which expand more rapidly, but they can be minimized by operating the laser at the lowest power level consistent with producing the required ionization stage.

C. The High-Voltage Vacuum Spark

This device (Edlén, 1934) is likely to conserve its place in the spectroscopic laboratory, even though it has been there so long, mainly because of the small linewidths associated with highly ionized emission lines. It can be inferred that the small widths are due to lower ion temperatures and doppler widths than in partially confined plasmas. The circuit consists of a low inductance 60- to 100-kV capacitor of about 0.5 μF connected via a low inductance line to two electrodes situated in a vacuum a few millimeters apart. Increasing the inductance is the most convenient way of obtaining spectra with lower dominant stage of ionization. If a lot of work is to be done a carbon anode is suitable since it does not generate too many spectral lines and such a light element with a high melting point gives a hotter discharge. The cathode can be either purely of the element under consideration or, to maintain a hotter discharge of carbon with the element under investigation, inserted in a small axial hole. A modified version of the spark (Fawcett, 1965) with a 100-kV, 0.5-μF capacitor included a swinging cascade switch and was coaxial throughout with a total inductance of 0.6 μH.

The most highly ionized material in these sparks exists during the first half cycle, and it is successively cooler in each following half cycle partly due to recombination and partly because more cold material enters the discharge. This has been exploited (Fawcett and Gabriel, 1966) to distinguish between ionization stages by using a mechanical shutter described in Section III,B to expose the plates during particular half cycles (see Fig. 1).

Vodar and Astoin (1950) and Bockasten (1955, 1956) have inserted an insulator between the electrodes in such a way that the spark travels along its surface. This so-called Sliding spark can be operated at 30 kV or lower depending on the ionization conditions required.

D. The Low-Inductance High Capacity Spark

Feldman et al. (1967a,b) have made many new classifications and observations of emission lines belonging to the highest stages of ionization produced in the laboratory. Their light source differs from the high voltage spark discussed in the previous section in that it has a higher capacity of 13 μF and lower operating voltage of up to 17 kV. The inductance of the circuit was 160×10^{-9} H and resistance 13×10^{-3} Ω; the low inductance is important in maintaining high spark currents of about 180 kA, so special care had to be taken in the choice of suitable capacitors and low inductance couplings to the spark. The spark gap was 2 mm, but 17 kV will not spark across this gap in a vacuum so some initial ionization had to be produced. Feldman et al. introduced a third electrode placed near one of the other electrodes separated from it by a ceramic surface across which a sliding spark was developed to trigger the main discharge. More recently Lee (1972; Lee and Elton, 1971; Elton and Lee, 1972) triggered the spark by generating a laser-produced plasma near the cathode. The extensive contributions of Feldman, workers at GSFC, and others to the classification of very highly ionized systems will be described in Section V,D. The capability of the source is indicated in a paper by Cohen et al. (1968b) which reports observations of Fe XXV and Fe XXVI and helium-like spectra of Ti XXI to Ni XXVII. The spectra of Fe XXV and its satellites are of special solar significance (see Section IV). A record of Schwob and Fraenkel (1972) is therefore reproduced (see Fig. 10).

E. The Beam Foil Source

An important new development which has great future potential for line classification studies as well as for the measurement of transition probabilities is the beam foil source, which is created in the following way. After a beam of ions is accelerated to energies of up to a few million electron volts, it is incident on a foil where the ions become ionized and excited. They then leave the foil and decay at different distances from the foil. If the foil is located at $x = 0$, the equation for the decay of the excited states at $x > 0$ is

$$dn = -\alpha n(dx/v)$$
$$n = N \exp[-\alpha(x/v)]$$

where n is the number of ions in the excited state at a distance x from the target and $N = n$ at $x = 0$ and $V = $ velocity of ion beam and $\alpha = $ the total radiative transition probability per second.

In a typical experiment (Lindskog et al., 1972) ionized chlorine and oxygen were accelerated in a Van de Graaff machine and observed. The radiation was observed downstream from the target with a grazing incidence spectrograph which recorded spectra of Cl VII–Cl XIII and O V–O V III. In another example Irwin et al. (1972) observed spectra of O I–O IV, F II–F IV, Ne I–Ne VI, Si II–Si IV, and Ar I–Ar VIII between 300 and 1800 Å. There were until now few comprehensive beam foil line classification papers but several spectroscopic papers. Much of the work is reviewed in the Proceedings of the Conference on Beam Foil Spectroscopy at Tucson, 1972 (see Bashkin, 1973).

III. Measuring and Experimental Techniques

A. Measuring Techniques

As the study of spectra moves to more complicated configurations, a combination of speed and accuracy of measurement is needed to obviate excessive demands on the time required to measure spectrograms with many lines. An accuracy of 0.25 μm is desirable for some purposes which can be facilitated with a semiautomatic measuring aid. Fully automatic measurement with 1-μm accuracy can also be achieved and is more than sufficient for most highly ionized spectra which have larger measurement errors dictated by doppler linewidths.

The Abbe-type comparator obtains 0.25-μm accuracy by comparing the spectrogram with a precision scale and by registering readings with a spiral microscope consisting of an archimedean double spiral. It therefore overcomes the 1 μm or more errors usually associated with traveling microscopes. A photoelectric setting device, as added to the Abbe comparator at Lund (Gunnvald, 1962) provides some reduction of labor. With the line thus located it is still necessary to eliminate the need of reading the spiral microscope.

The photoelectric setting device, which is a great advantage for the location of blended or asymmetric lines or faint features, was initiated by Rymer and Halliday (1950) of Reading University, who incorporated some of the principles of the Furth microphotometer (Furth and Oliphant, 1948). Tomkins and Fred (1951) modified the system. Instead of locating the line with microscope crosswires, its profile detected with a specially adapted microphotometer is displayed on an oscilloscope and its fiducial position located by matching it with its mirror image displayed on the same oscilloscope. The special adaption is a rotating prism which not only scans the line profile across the microphotometer exit slit but also provides an ac signal which is matched with the scope scan frequency. The mirror images are obtained by putting the photocell output signal on both scope beams and scanning the two beams in opposite directions. The solution arrived at by Poppe et al. (1964) at the Zeeman Laboratory preserves the photoelectric setting device but registers the position of a special plate with an adapted Michelson interferometer coupled to a fringe counting system and a paper tape puncher. A 5-mW He–Ne laser provides the interferometer with sufficient illumination to deal with the 25-cm plate length. It is essential to avoid air flows, and as with all comparators the temperature of the measuring apparatus must remain constant.

For fully automatic plate measurement the spectral intensity is read with a microphotometer or microdensitometer and at the same time corresponding positions of the plate are located at intervals along the spectrogram. An ideal system could use the Michelson interferometer as above to locate the plate with an on-line computer recording both the fringe count and intensity and subsequently computing the position and peak intensity of spectral lines. Since the computer is unlikely to be able to derive the position of these peaks with an accuracy better than 1 μm, practical applications have utilized Moire fringes with their count fed into an on-line computer. Since Moire fringes are created by two gratings crossing one another, the measurement accuracy will be limited by the mechanical features of the grating ruling machine. Furthermore, a grating of sufficient length and precision is costly. One great advantage of these on-line systems is that the plate can be read in a few minutes—too quickly for temperature changes to affect the measurements. A satisfactory arrangement presently in use at the Zeeman laboratory has a 25-cm long precision scale with rectangular opaque lines and transparent openings (Hoekstra, 1967, 1969) of 5 μm width which move on the same table as the plate. Both scale and plates are read with microphotometers. The scale readings control an encoder so that the microphotometer signals are accepted at equal intervals along the spectrograms and written on a magnetic tape for subsequent processing by a computer. The plate reading speed is 2 mm/min but could be faster if the system were modified to feed into an on-line computer which could also sample at shorter intervals than 5 μm. Less expensive and accurate methods entail the use of digitized lead screws and paper tape. The Joyce Loebel data densitometer has, for instance, a stepping motor which moves the plate at 5-μm intervals when the optical density is recorded on paper or magnetic tape.

B. Experimental Techniques for Recording Time-Resolved Spectrograms

The more ionization stages present on spectrograms the greater is the problem of separating lines belonging to different ions. In Section II it was pointed out that different stages are enhanced relatively at different times during the life history of the plasma. Hence they can be separated with an optical shutter which permits the recording of spectrograms during a short period. Such a shutter reported by Fawcett and Gabriel (1966) has a 100-μm slit mounted on a disc rotating in a vacuum and situated directly in front of the 5-μm entrance slit of a grazing incidence spectrograph. Light can enter the spectrograph during the microsecond or more that the slits are opposite one another. A light beam incident on a mirror attached to the disc axis reflects light into two photomultipliers that provide signals to determine when the slit is open and as a prepulse to fire the light source. The disc is magnetically coupled to an electric motor through a stainless steel window which is part of the exterior wall of the vacuum chamber. Another means of recording time-resolved spectrograms involves a rotating mirror fitted inside a normal incidence spectrograph (Gabriel et al., 1962). The length of the spectral line is limited by placing a second slit perpendicular to the entrance slit and at Sirks focus (see Section II,B). The rotating mirror hence scans the short line across the plateholder and also sends out a pulse to trigger the light source.

C. Separation of Ionization Stages

Methods of separating ionization stages can be conveniently divided into four categories. First, there are those which exploit the time history of ions in either an ionizing or recombining plasma. These have been discussed in conjunction with the theta pinch and spark sources in Section II. Second, the spatial distribution of ions in the light source can be investigated using stigmatic spectra. Spatially resolved solar spectra are useful in this respect since ions emitting in a given temperature region will depend on the ionization equilibrium (Jordan, 1969, 1970). Third, there is the comparison of spectra emitted from the source operated under different conditions. Energy input, applied voltage, pulse duration, series inductance, gas pressure, or geometry of the source can be varied to produce the different conditions in the light source. Fourth, the special properties of certain light sources can be utilized. The section on laser produced plasmas provides such an example.

Lines which decay from high energy levels can also be more intense the higher the temperature of the plasma so care must be taken to distinguish between these and lines belonging to higher ions.

D. SEPARATION OF LINES OF THE SAME ION DUE TO VARIATION OF THEIR INTENSITY WITH SOURCE DENSITY

Spectral lines can be attributed to particular types of transitions owing to their observed intensities in spectra of light sources at different densities because the electron density of the source affects the ratios of intensities of different spectral lines belonging to the same ion. The occurrence of the so-called forbidden lines in the spectra of the solar corona is a well-known example of this. Because of the low electron density of the solar corona, levels with small transition probabilities, which would otherwise be depopulated through electron impact, have time to decay radiatively. Such lines are not observed in laboratory spectra but other lines with moderately low transition probabilities, such as intercombination lines (Jordan, 1971b) between levels of different multiplicity, are similarly enhanced in lower density laboratory light sources relative to those at higher densities.

Another low density effect was noticed in solar spectra by Gabriel *et al.* (1965). Many lines belonging to transitions which do not terminate in the ground level are comparatively weaker in solar corona spectra than in laboratory spectra. The explanation lies in the relative population of the lower energy levels at different densities and in the subsequent excitation from these lower levels to higher levels which then decay to produce the spectral lines. At the low electron density of the solar corona the populations of the excited levels of the ground term can be much less than that in laboratory sources where the density is sufficient for the lines to be thermally populated according to their statistical weights. In the solar corona (Jordan, 1969) the population is mainly in the ground level, and upper levels which are most easily excited from the ground level are preferentially excited. The electric dipole has a dominant effect on both excitation and decay rates so solar coronal lines which have high transition probabilities to the ground level are enhanced. Similarly, density changes in laboratory sources can affect the relative intensities of lines but mainly with differing multiplicities. If the terms with a different multiplicity from the ground term are not low enough to become statistically populated at the density existing in the light source then the intensity of lines due to transitions between levels of this multiplicity can be noticeably decreased.

Many terms which have the same parity as the ground terms become appreciably populated only at higher electron densities. This often occurs because of double excitation. That is, before a level that has been excited from ground terms has had time to decay, it is again excited to a higher level. These transitions from doubly excited levels can be studied in high density plasmas such as laser-produced plasmas (Fawcett, 1970a). Another route by which levels of high principal quantum number may become populated at higher densities is through thermalization with the continuum. Transitions from these levels can therefore also be studied in laser-produced plasma spectra (Fawcett et al., 1966).

IV. Theoretical Calculations

As the analysis of spectra progresses, the simpler systems are described and what remains contains a large proportion of complex spectra. To understand and interpret these it is increasingly necessary to have reliable theoretical calculations of atomic structure available. Fortunately computer programs have been recently developed which employ sophisticated methods. For example, Cowan (1967a, 1968) has developed a Hartree-X (HX) method along with a fully computerized system for the calculation of wavelengths and oscillator strengths of spectral lines. Froese-Fischer (1963) wrote and made available a widely used computer program providing numerical solutions to the Hartree–Fock (HF) equations. C. Froese-Fischer studied how multiconfiguration Hartree–Fock calculations should be done and how to construct an effective computer program for solving them (Froese-Fischer, 1967, 1970, 1971, 1972a,b,c). Goldschmidt and Shadmi have carried out calculations of extremely complex systems (Shadmi, 1965, 1966).

A. The Hartree-X Method

Cowan's program is probably the most systematically constructed for the calculation of atomic spectra, applying Slater–Condon theory (Condon and Shortley, 1935; Slater, 1960). Basis functions for a configuration $(n_1 l_1)^{\omega_2}$ $(n_2 l_2)^{\omega_2} \cdots (n_g l_g)^{\omega_g}$ are constructed by angular-momentum coupling of antisymmetrized products of spin orbitals.

$$\phi_{nlm_l m_s} = [P_{nl}(r)/r] Y_{lm_l(\theta, \phi)} \sigma_{m_s}(s_z)$$

The radial factors $P_{nl}(r)$ may be obtained from Hartree–Fock calculations, but to reduce the required computing time (due partly to convergence difficulties in the HF method) Cowan usually uses his HX method, which makes use of the approximation

$$E(\rho) = -\tfrac{3}{2}(24\rho/\pi)^{1/3}$$

for the average exchange energy of an electron in a uniform free electron gas of density ρ at zero temperature. The HF equations are simplified by replacing the exchange portion of the potential energy operator by

$$V(r) = -K(24\rho/\pi)^{1/3}$$

where $\rho = \rho(r)$ is the local electron density in the atom, and K is a function of ρ and the density ρ_i for an electron in orbital $n_i l_i$ (Cowan, 1967a). From this the method derives its name Hartree plus statistical exchange, abbreviated Hartree-X.

Using the radial functions thus obtained, the program computes numerical values of energy matrix elements between all pairs of basis functions b and b'. These elements are of the form

$$H_{bb'} = E_{av}\sigma_{bb'} + \sum [f_k F^k(l_1 l_j) + g_k G^k(l_1 l_j)] + \sum_i d_i \zeta(l_i)$$

where E_{av} is the total binding energy of the spherically symmetric ion in the configuration of interest, and F^k, G^k, and ζ are the electrostatic and spin-orbit interaction parameters given by

$$F^k(l_i l_j) = \int_0^\infty \int_0^\infty \frac{2r_<^k}{r_>^{k+1}} P_i(r_1)P_j(r_2)P_i(r_1)P_j(r_2)\, dr_1\, dr_2$$

$$G^k(l_i l_j) = \int_0^\infty \int_0^\infty \frac{2r_<^k}{r_>^{k+1}} P_i(r_1)P_j(r_2)P_j(r_1)P_i(r_2)\, dr_1\, dr_2$$

$$\zeta(l_i) = \frac{1}{2}\alpha^2 \int_0^\alpha r^{-1} P_i^2(r) \left(\frac{dV}{dr}\right) dr$$

α being the fine structure constant; the parameter values are in Rydbergs if r is in Bohr units. The parameter coefficients f_k, g_k, and d_i depend on the angular portions of the basis functions and are automatically computed by the program; the summations are over all incompletely filled shells of the configuration.

The energy matrices are then diagonalized to give energy levels and eigenvectors; the latter may be transformed into each of several different pure-coupling representations so as to indicate the existing coupling conditions and the purity of each energy level. Configuration interaction can be included if desired. Finally, from the energy levels, eigenvectors, and radial dipole integral (computed from the P_{nl}), the program derives wavelengths, oscillator strengths, and transition probabilities (Cowan, 1968).

B. The Multiconfiguration Hartree–Fock Method

A configuration interaction (CI) calculation starts with a set of basis functions for the radial functions, say $P_{nl}(r)$. These radial functions are then used to define orbitals

$$\phi_{nlm} = [P_{nl}(r)/r]Y_l^m(\theta, \phi)$$

which, in turn, determine a set of total wave functions $\Phi(\gamma\alpha LS)$ for configurations γ in a particular LS basis state. Here α refers to any other quantum number such as seniority which may be required to uniquely designate the basis state. The wave function for the atomic state, $\psi(LS)$, is then approximated by an expansion of the form

$$\psi(LS) = \sum_{i\gamma} a_i \Phi(\gamma_i \alpha_i LS) \qquad (1)$$

$$a_i^2 = 1 \qquad \langle \Phi(\gamma_i \alpha_i LS) | \Phi(\gamma_j \alpha_i LS) \rangle = \delta_{ij}$$

where the "mixing coefficients" a_i are determined variationally. This leads to the eigenvalue, eigenvector problem

$$\sum_i H_{ij} a_j = E a_j \qquad (2)$$

Here

$$H_{ij} = \langle \Phi(\gamma_i \alpha_i LS) | H | \Phi(\gamma_j \alpha_j LS) \rangle.$$

The multiconfiguration Hartree–Fock (MCHF) method of C. Froese-Fischer differs in that the variational principle is used not only to find the mixing coefficients in Eq. (1) but also the radial functions $P_{nl}(r)$. As a result, the eigenvalue problem of Eq. (2) is augmented by a coupled system of nonlinear integrodifferential equations, one for each radial function. These equations are similar to the Hartree–Fock equations but with additional terms present arising from the interaction. As a result, the radial functions depend on the mixing coefficients. These equations, together with the eigenvalue problem are solved iteratively (Froese-Fischer, 1972a,b).

Generally, one of the coefficients dominates, say a_0. Then the corresponding configuration γ_0 is called the "reference configuration" and orbitals occurring in $\Phi(\gamma_0 \alpha_0 LS)$ are said to be "occupied." Other configurations are obtained from γ_0 by the replacement of electrons and the new orbitals introduced are said to be "unoccupied" or "virtual" orbitals. When a_0 is dominant, the occupied orbitals are similar to Hartree–Fock orbitals and, by Brillouin's theorem, include the interaction with many one-electron substituted configurations, at least to first orders. The virtual orbitals, on the other hand, often are quite different from Hartree–Fock orbitals and tend to have maxima in the same region as the occupied orbitals. In the He $1s^2$ ^1S case (as well as others) the multiconfiguration Hartree–Fock orbitals are the natural orbitals introduced by Lowdin (1955) and shown to represent a transformation of the basis so that the expansion of Eq. (1) has the fastest rate of convergence.

Excited states often enter into the calculations of atomic properties—an example is oscillator strengths. Experience here indicates that a maximum rate of convergence can only be achieved if orbitals for different configurations are allowed to vary. For example, in Mg 3s 3p ^{1}P the single configuration HF results already include the interaction with all 3s np ^{1}P configurations, to first order. A MCHF calculation for

$$\psi = a_0 \Phi(3s\ 3p\ ^1P) + a_1 \Phi(3p\ 3d\ ^1P)$$

would determine functions 3s, 3p, 3d so that the above interactions are still included as well as 3p nd ^{1}P. However, np 3d ^{1}P can only be included if

$$\psi = a_0 \Phi(3s\ 3p\ ^1P) + a_1 \Phi(3p'\ 3d\ ^1P).$$

The introduction of a second 3p radial function greatly improves the accuracy of the two-configuration Hartree–Fock approximation. This multiconfiguration method therefore calculates P_{nl} and gives the energy dependence on *LS* corresponding to the first part of Cowan's program but does not include spin–orbit effects nor does it calculate *J*-dependent wave functions.

When applying theoretical energy level calculations to the interpretation of observed spectra it is first necessary to establish whether the theoretical spectrum is a satisfactory replica of the observed spectrum. If not, important perturbations or interactions may have been overlooked. Alternatively the approximation used could be unjustified. More exact methods may have to be introduced involving much computer time and a high level of theoretical conceptual knowledge.

V. Line Classifications of Highly Ionized Systems

Recent spectral line classifications of highly ionized systems are predominantly in elements lighter than nickel, which is not surprising considering the amount of interest arising from rocket and satellite solar spectra dominated by these spectra of solar-abundant elements. This interest has stimulated intensive laboratory studies. On the other hand, the complexity of heavier element spectra impedes progress on their interpretation. It is convenient to discuss different types of spectra in separate sections which mainly deal with particular sorts of transition arrays.

A. The $2s^2\ 2p^n$–$2s\ 2p^{n+1}$ and $2s\ 2p^n$–$2p^{n+1}$ Transition Arrays

Before turning to the new classifications in these transition arrays it is worth noting the fine quality of the earlier data which have influenced contemporary work. The early data are critically compiled in Atomic Energy Levels (Moore, 1949). It includes the majority of information available regarding elements in the period between lithium and sodium and is derived from observations of conventional sources such as the hollow cathode, electrodeless discharges, arcs and sparks and, most important, the high voltage vacuum spark. High-precision measurements were made of spectra recorded with 5-meter grazing incidence instruments in the extreme ultraviolet, high resolution instruments in the ultraviolet and visible, and the Fabry–Perot interferometer in the visible. The high accuracy was exploited to increase the certainty of the identifications—for instance, by comparing known and predicted wavelength differences. Many wavelengths were derived by Edlén (1963) and Herzberg (1962) by applying the Ritz combination principle. These measurements and identifications provided a reliable basis for isoelectronic extrapolation and therefore aided workers who have recently added classifications in the period between sodium and iron, hence extending the knowledge of these sequences in as many as ten elements along many sequences. The extrapolations are complicated by gradual changes from *LS* to *JJ* coupling along the sequences, so the new identifications generally relied on observations of the spectra of successive elements in isoelectronic sequences. Refined empirical extrapolation techniques aided verification particularly in the case of the terms nearest the ground level. The observations upon which the progress depended were mainly of spectra emitted from laser-produced plasmas or the theta-pinch light sources described in Section II. The work on laser-produced plasmas was carried out in the Astrophysics Research Unit, Culham (Fawcett, 1970a, 1971a,b) and the Lebedev and Spectroscopic Institute, Moscow (Basov *et al.*, 1967; Kononov, 1966, 1969; Kononov *et al.*, 1970; Podobedova *et al.*, 1971; see also Section II,B). It included experiments with all elements in the period between sodium and iron except the gas argon. Using a 5-GW laser Feldman *et al* recently (1973b) identified these spectra in Fe XVIII. The 2s $2p^n$–$2p^{n+1}$ transitions are more intense in laser-produced plasma spectra than in spectra of lower density sources, and they are consequently identified as a result of laser-produced plasma studies in elements lighter than chlorine. The lines belonging to the $2s^2 2p^n$–$2s 2p^{n+1}$ transition arrays can easily be distinguished in laser-produced plasma spectra of scandium, titanium, and vanadium (Fawcett, 1971a). The new spectral identifications referred to in this section have accounted for many lines in solar corona and flare spectra (see Section VI) and furthermore check the identity of many solar forbidden lines. Because of the low electron density of the solar corona, levels such as $2s^2 2p^n$ with small transition probabilities which would otherwise be depopulated through electron impact have time to decay radiatively. The consequent emission lines are the so-called forbidden lines. The values of the $2s^2\ 2p^n$ energy levels hence determine the wavelengths of the forbidden lines which are confirmed since these values can be extracted from the laboratory wavelengths of the $2s^2\ 2p^n$–$2s\ 2p^{n+1}$ lines. For example, in the spectra the intervals between the two sharp ^2P–^2S Ar X lines and the intervals between the ^2P–^2P Ar XIV lines correspond to the wavelengths of the Ar X and Ar XIV forbidden lines which Edlén identified in 1969 using elaborate extrapolation procedures which are of general interest (1969b). To extrapolate the $2s^2\ 2p^5\ ^2P_{1/2}$–$^2P_{1\frac{1}{2}}$ interval he applied the Sommerfeld–Dirac formula with a screened nuclear charge. For each interval there is a corresponding screening value s which can be fitted into an empirical formula relating it to Z. Thus, considering only the first Dirac term the spin–orbit splitting is written as

$$(R\alpha^2/16)[1 + (\alpha/\pi)](Z - s)^4 = A(2p)(Z - s)^4$$

and to allow for high terms

$$s' = s - (5/32)\alpha^2(Z - s')^3 = s - 0.832 \times 10^5(Z - s')^3$$

where $A(2p) = 0.366075$ cm^{-1} and $[1 + (\alpha/\pi)] = 1.002323$. A plot of s against $(Z - s)^{-1}$ is almost linear and therefore provides a suitable means of extrapolation. A similar procedure but with a different constant A in the formula relating the spin–orbit splitting interval to the values plotted enables extrapolation of the $2s^2\ 2p\ ^2P_{3/2}$–$^2P_{1/2}$ interval. Edlén (1972a) also extrapolates the $2s^2\ 2p^4$, $2s^2\ 2p^2$, and $2s^2\ 2p^3$ energy levels by determining the relationship between the Slater energy parameters $F_k\ G_k$ and ζ (see Section IV) and atomic number. The Slater energy parameters can be obtained from theoretical formula (Edlén, 1964) which relate these energy parameters to the splitting between individual levels. For instance for the $2s^2\ 2p^4$ configuration the formulas are

$$2(^1D - ^3P_2) = [(12F_2)^2 + 24F_2\zeta_{2p} + (3\zeta_{2p})^2]^{1/2}$$

$$4(^3P_1 - ^3P_2) = 2(^1D - ^3P_2) - 12F_2 + 3\zeta_{2p}$$

with further formulas involving the $^1S_0$ level. The energy levels derived by Edlén are listed in Table I.

B. The $3s^2\ 3p^n$–$3s\ 3p^{n+1}$ and $3s^2\ 3p^n$–$3s^2\ 3p^{n-1}$ 3d Transition Arrays

It is these transitions in iron which are responsible for the bulk of the intense lines in the far ultraviolet solar spectrum (see Section VI). Almost all the data available on the transitions existing in 1960 were already listed by Moore (1949, 1952). Most of this was in calcium or lighter elements and there were a considerable number of erroneous classifications. The task since

TABLE I

CALCULATED VALUES OF THE LOWER ENERGY LEVELS FOR C I-, N I-, AND O I-LIKE SPECTRA[a,b]

$2s^2 2p^4$	$^3P_2$	$^3P_1$	$^3P_0$	$^1D_2$	$^1S_0$
O I	0	158.3	227.0	15867.9	33792.6
F II	0	340.8	489.3	20872.7	44917.4
Ne III	0	642.5	920.5	25840.8	55748.4
Na IV	0	1106.5	1577.7	30838.0	66500.1
Mg V	0	1784.0	2524.5	35919.2	77288.6
Al VI	0	2735.1	3830.1	41139.4	88203.6
Si VII	0	4029.8	5568.0	46560.3	99335.5
P VIII	0	5749	7812	52254	110788
S IX	0	7987	10636	58305	122689
Cl X	0	10849	14105	64816	135197
Ar XI	0	14456	18274	71907	148513
K XII	0	18944	23181	79716	162886
Ca XIII	0	24460	28835	88400	178621
Sc XIV	0	31170	35223	98136	196086
Ti XV	0	39249	42293	103116	215706
V XVI	0	48887	49971	121548	237965
Cr XVII	0	60285	58157	135651	263387

$2s^2 2p^2$	$^3P_0$	$^3P_1$	$^3P_2$	$^1D_2$	$^1S_0$
C I	0	16.4	43.4	10192.6	21648.0
N II	0	48.7	130.8	15316.3	32688.8
O III	0	113.2	306.2	20273.3	43185.7
F IV	0	227.1	614.5	25236.4	53537.1
Ne V	0	412.4	1110.1	30291.5	63913.6
Na VI	0	697.2	1857.7	35509.0	74423.9
Mg VII	0	1117.3	2932.3	40962.7	85172.4
Al VIII	0	1717.2	4420.1	46738.4	96265.3
Si IX	0	2553	6418	52939	107827
P X	0	3695	9036	59692	120006
S XI	0	5226	12393	67153	132979
Cl XII	0	7248	16619	75512	146957
Ar XIII	0	9882	21852	85003	162185
K XIV	0	13266	28236	95908	178947
Ca XV	0	17558	35920	108561	197567

$2s^2 2p^3$	$^4S_{3/2}$	$^2D_{3/2}$	$^2D_{5/2}$	$^2P_{1/2}$	$^2P_{3/2}$
N I	0	19233.2	19224.5	28838.9	28839.3
O II	0	26830.2	26810.7	40468.6	40467.5
F III	0	34121.5	34088.5	51562.1	51561.2
Ne IV	0	41279.5	41234.6	62434.6	62441.3
Na V	0	48359.3	48313.5	73201.9	73236.4
Mg VI	0	55374.9	55357.0	83920.9	84025.6
Al VII	0	62318.8	62384.9	94626.7	94880.2
Si VIII	0	69166.8	69414.3	105347.4	105882.7
P IX	0	75877	76462	116109	117136
S X	0	82398	83548	126939	128769
Cl XI	0	88659	90693	137867	140936
Ar XII	0	94586	97922	148926	153819
K XIII	0	100099	105261	160149	167612
Ca XIV	0	105121	112736	171571	182524
Sc XV	0	109580	120376	183230	198768

[a] From Edlén (1972a).
[b] Values given in cm^{-1}.

1960 was not only to extend the classifications through the astrophysically important elements calcium, iron, and nickel but also to critically assess old data before relating it to analysis further along the sequences. There was a good reason why, in spite of the industry of the earlier workers, so little progress had been made on these configurations. These workers had reaped such a rich harvest from other configurations using the high voltage vacuum spark and other traditional light sources that they tended not to develop plasma light sources. Unfortunately the traditional spark sources radiated low stages of ionization strongly at later times during the discharge. In the transition elements the $3s^2\ 3p^n - 3s^2\ 3p^{n-1}\ 3d$ lines were obscured by lines of lower ionization stages belonging to the $3p^6\ 3d^n - 3p^5\ 3d^{n+1}$ transition arrays which displayed a numerous profusion of unclassified lines. In iron the former transitions are in Fe IX to Fe XIV and in the latter Fe V to Fe VIII. The relevant spectral region in Fe IV lies beyond the ionization limit. The ionization balance in the solar corona is very different. An equilibrium balance between ionization and recombination is reached, and ionization stages exist which correspond to the electron temperature in the region emitting. Appreciable emission in the extreme ultraviolet requires a minimum electron temperature. Above this temperature the prevalent ions are Fe IX to Fe XVI so the solar spectral region between 170 and 220 Å is dominated by the intense spectra for these ions. This is why the first good quality ultraviolet solar rocket spectra underlined the problem. This effect can be further accentuated by a lower abundance of Fe V to Fe VIII ions compared with other ionization stages in the Sun. Some high temperature plasmas such as Zeta (see Section II) and the theta pinch can operate under conditions in which the emission from Fe IX to Fe XIV is more intense than from lower stages. This is because the plasma is heated rapidly, remains at a high temperature for a longer period during which phase it radiates the high ionization stages, after which the plasma is lost quickly to the walls of the container or elsewhere. Zeta, with its stainless steel liner which injected material into the plasma, therefore fortuitously reproduced the solar spectra of iron and nickel. This in itself did not supply the answer to the problem since isoelectronic extrapolations, from dubious data along sequences where the coupling was varying in some cases rapidly from *LS* to *JJ* coupling, produced unusable predictions of the spectra. The intermediate *LS* to *JJ* coupling diagram for the Si I or Fe XIII levels of the array in Fig. 2 demonstrates this point. The analysis of these spectra was accomplished (Gabriel *et al.*, 1965, 1966; Fawcett and Gabriel, 1966; Fawcett *et al.*, 1967) by plotting the wavenumber of each line in the isoelectronic sequence against atomic number. This depended on the acquisition of spectograms with adequate discrimination between the ionization stages in each of the elements in the period between potassium and nickel. At first this was done by variation of

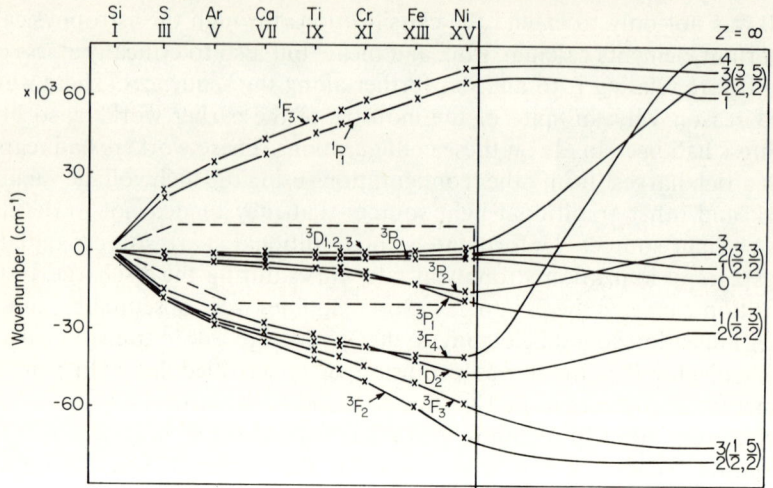

FIG. 2a. The calculated energy levels of the 3p3d configuration of Fe XIII and isoelectronic levels between silicon and nickel. From Fawcett (1971d).

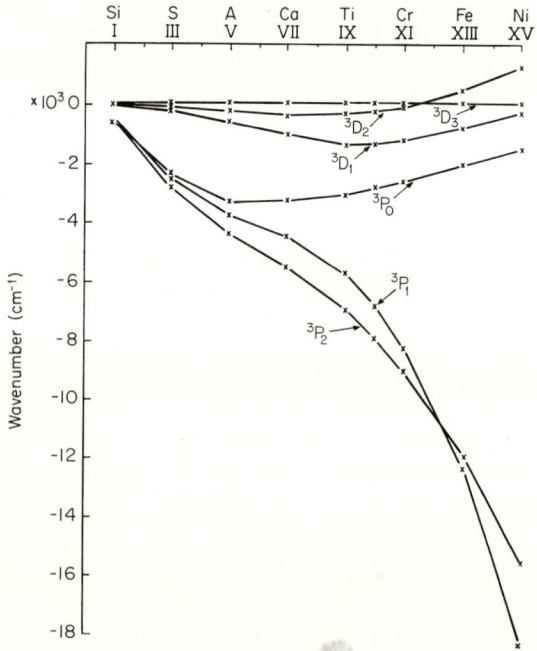

FIG. 2b. An enlarged plot of a section of Fig. 2a showing $3d^3P$ and $3d^3D$ levels. From Fawcett (1971d).

the inductance of the high voltage vacuum spark and the application of the microsecond shutter, described in Section III,B, to this source. The type of configuration interaction which occurs at one point in the isoelectronic sequence where two configurations cross (see Fig. 3) proved helpful in

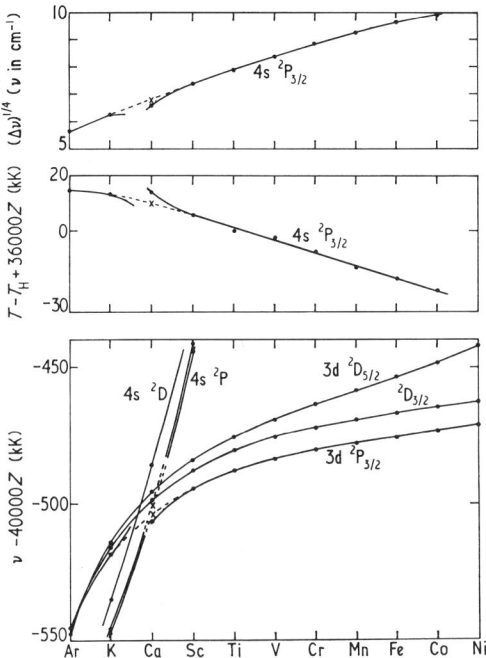

FIG. 3. Configuration interaction. Plotted are chlorine I sequence term values. Terms are given in wave numbers; v relative to the ground state, and T relative to the first ionization limit. From Gabriel et al. (1966).

confirming the terms involved. Theoretical calculations were subsequently applied successfully to these configurations (Cowan and Peacock, 1965, 1966; Fawcett et al., 1968). An important contribution was a critical and comprehensive classification of titanium, scandium, calcium, and potassium spectra (Svensson and Ekberg, 1968; Ekberg and Svensson, 1970; Ekberg, 1971; Svensson, 1971b). A more detailed analysis of the $3s^2\ 3p^n-3s^2\ 3p^{n-1}\ 3d$ transitions and in addition the analysis of the $3s^2\ 3p^n-3s\ 3p^{n+1}$ transitions had to wait for studies of laser-produced plasmas (Fawcett, 1970b) and of the theta-pinch plasma, described in Section II,A, into which small percentages of volatile compounds of the transition elements were introduced (Fawcett, 1971d, e; Fawcett and Hayes, 1972). A 2-meter grazing incidence spectrograph, which viewed the theta pinch from the side and was fitted with the microsecond shutter recorded the spectra during the period when the required ionization stages were emitted most intensely.

The wavelengths of these transitions permit the evaluation of the $3s^2\,3p^n$ energy levels and hence confirm the identity of solar forbidden lines and in the same way as for the analogous $2s^2\,2p^n$ transitions which were described in detail in Section V,A. Conclusive support is hence provided for the classifications of almost all the solar forbidden lines of these elements identified by Edlén (1942) in the visible spectrum and by Gabriel et al. (1971) and Jordan (1971a) in the spectra between 1000 and 2200 Å obtained during the 1970 solar eclipse with a rocket-borne spectrograph.

Edlén based his classic forbidden line classifications on refined extrapolations of the $3s^2\,3p^n$ energy levels. The extrapolations have been recently extended (Svennson, 1971a) into Fe XI, Fe XII, and Fe XIII and through these isoelectronic sequences as far as nickel. Again, as in Section V,A, Slater energy parameters F_k and ζ_{3p} were derived from observational data to form the basis for extrapolation. These results listed in Table II independently

TABLE II

Calculated Values of the $3s^2 3p^2$, $3s^2 3p^3$, and $3s^2 3p^4$ Energy Levels in Transition Elements[a,b] as Calculated by Means of the Parameters F_2 and ζ_p

$3s^2 3p^2$	V X	Cr XI	Mn XII	Fe XIII	Co XIV	Ni XV
$^3P_0$	0	0	0	0	0	0
$^3P_0$	4183.6	5535.7	7222.3	9303.1	11843.4	14913.6
$^3P_2$	9404.7	11974	15010	18561.6	22632	27372.6
$^1D_2$	32495	36987	42136	48065	54960	62871
$^1S_0$	67754	74962	82822	91459	101004	111601
$3s^2 3p^3$	V IX	Cr X	Mn XI	Fe XII	Co XIII	Ni XIV
$^4S_{3/2}$	0	0	0	0	0	0
$^2D_{3/2}$	34963	37092	39376	41560	43678	45781
$^2D_{5/2}$	36309	39443	42686	46083	49686	53558
$^2P_{1/2}$	59030	63935	68943	74101	79458	85083
$^2P_{3/2}$	61224	67165	73555	80512	88158	96631
$3s^2 3p^4$	V VIII	Cr IX	Mn X	Fe XI	Co XII	Ni XIII
$^3P_2$	0	0	0	0	0	0
$^3P_1$	6012.4	7825.2	10025.0	12667.8	15812.5	19522.3
$^3P_0$	7597	9570	11819	14318	17078	19981
$^1D_2$	27076	30287	33822	37744	42118	47015
$^1S_0$	60639	66844	73521	80815	88837	97831

[a] From Svensson (1971a).
[b] Values given in cm^{-1}.

support forbidden line identifications. Attempts to determine these energy levels with *ab initio* calculations inevitably ran into difficulty because of interactions. Nevertheless Burton *et al.* (1967) identified the solar forbidden lines at 1242.15 and 1349.57 Å as $^4S_{1\frac{1}{2}}-^2P_{1\frac{1}{2}}$ and $^4S_{1\frac{1}{2}}-^2P_{1/2}$ in Fe XII based on Cowan's theoretical calculations.

A somewhat different type of $n = 3$ to 3 transition in Mg IV and Al V which are not connected to ground levels have been the subject of recent investigations at Meudon in association with NBS in Washington, D.C. (Artru and Kaufman, 1972; Eidelsberg, 1972).

C. Transitions to Levels of Principal Quantum Number $n = 2$ from Those with Higher Principal Quantum Numbers

Early published work on spectra which result from transitions between levels of principal quantum number 2 and higher principal quantum numbers is referenced by Moore (1949) in the well-known tables which provide the atomic energy levels. Since 1949 there have been several comprehensive additions in the light elements. These are contributed by Bockasten (1956) for C IV; Michels *et al.* (1971) and Hallin (1966a,b) for N IV and N V; Bromander (1969) for O IV; Bockasten and Johansson (1968) for O V, and Palenius (1973) for F IV to F VI. The term diagrams of N IV illustrated in Fig.4 are a fair example of the extent of these studies. There are also several new papers reporting neon classifications (Fawcett *et al.*, 1964; Tondello and

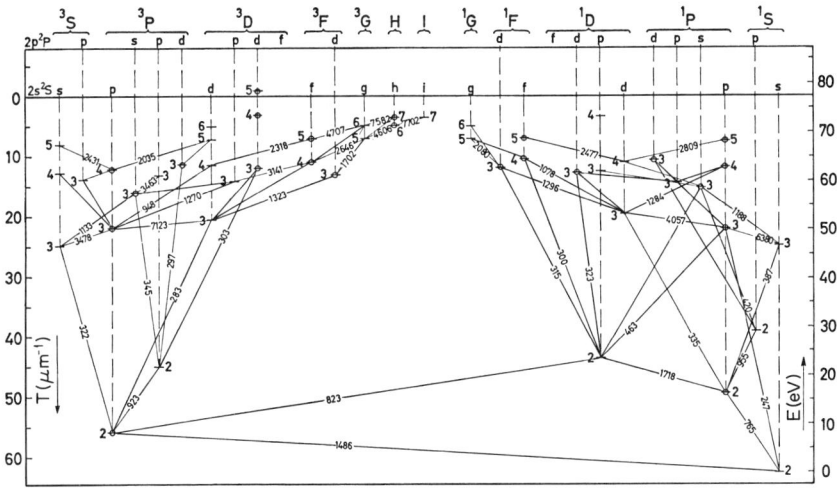

Fig. 4. The term diagram of N IV. From Hallin (1966a).

Paget, 1970; Hermansdorfer, 1972). The light sources applied were theta pinches for the gases and sliding sparks for the solids. The low inductance spark yielded new classifications in beryllium- and boron-like spectra of magnesium, aluminum, silicon (Hoory et al., 1970a, 1971), and phosphorus (Goldsmith et al., 1973) and laser-produced plasmas were used for silicon (Tondello, 1969) and nearby elements (Fawcett et al., 1970). The previously mentioned publications formed the foundations for research aimed at understanding important new solar spectra. One of these was a soft X-ray spectrum of the sun acquired in May 1966 by two rocket-borne slitless Bragg spectrometers (Evans et al., 1967). Some of the lines in the spectrum were classified as Fe XVIII as a result of laboratory observations of laser-produced plasmas (Fawcett et al., 1966). This was significant at the time since an ionization energy of 1270 eV is required to produce Fe XVIII and this was thought to be high for solar conditions. This ignorance of solar conditions was soon to be somewhat enlightened for in March 1966 an impressive spectrum of a solar flare (Fig. 5) taken from the OSO I satellite

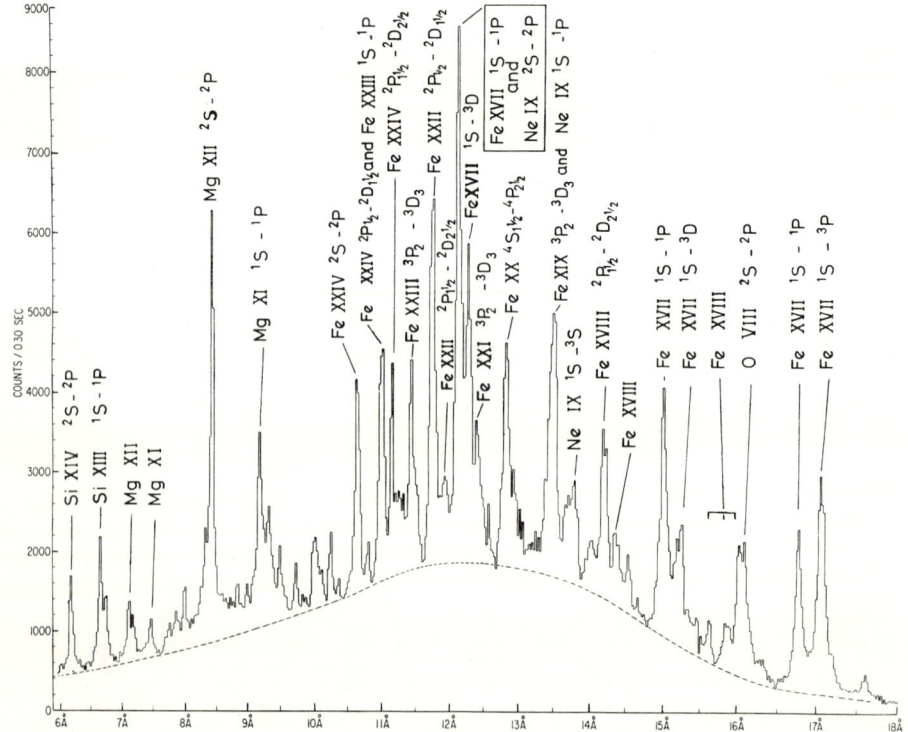

FIG. 5. The spectrum of a solar flare between 7 and 20 Å taken from the OSO 5 satellite by NASA. From Neupert et al. (1973).

(Neupert et al., 1967) included spectra of Fe XVII to XXIV mainly belonging to the $2s^2\ 2p^n$–$2s^2\ 2p^{n-1}$ 3d transitions. New records and suggested identifications have recently resulted from observations with crystal spectrometers from the O VI 10 satellite (Walker and Rugge, 1969), the OSO V satellite (Neupert et al., 1973), and the OSO VI satellite (Doschek et al., 1973b). Not only did these observations have profound implications for the physics of solar flares but they also provided a strong driving force for studying the isoelectronic sequences which account for these lines.

Some of these studies were carried out as a joint project between scientists from Goddard Space Flight Center and Tel Aviv University. They used the low inductance spark (see Section II,D) as a light source and observed spectra of these isoelectronic sequences in the period between scandium and copper, thus contributing some of the most impressive data on extremely highly ionized spectra. They covered most of the sequences which include Fe XVII- to Fe XXIV-like spectra (Feldman and Cohen, 1967a,b, 1968; Feldman et al., 1967a, 1970; Cohen et al., 1968a,b; Goldsmith et al., 1971b, 1972). Most of their classifications depended on isoelectronic extrapolations, but for some they applied the computer program of Froese-Fischer (see Section IV) and theoretical calculations of Garstang (1954) and Tech and Garstang (1965). Their instrument was a 3-meter grazing incidence spectrograph. Although this gave well-resolved spectra, even greater resolving power is needed for further progress with these crowded spectra. More recently, Feldman et al. (1973b) have extended the analysis of the ions K XI through Co XIX and Doscheck et al. (1973a) have made identifications in K XII through Mn XVII.

Data which allow improved isoelectronic extrapolations to the solar flare iron lines are given in a recent paper by Fawcett and Hayes (1973) which includes a comprehensive classification of highly ionized phosphorus, sulfur, chlorine, and argon in the Be I, Bo I, C I, and N I isoelectronic sequences with precise wavelengths. The spectra were emitted from a theta pinch into which a small quantity of hydrogen sulfide, phosphine, chlorine, or argon was added. The same source was used for another contribution in argon (Fawcett et al., 1971).

Theoretical Hartree-X (see Section IV,A) calculations were applied to predict the wavelengths and oscillator strengths of the $2p^n$–$2p^{n-1}$ 3d transitions in Fe XIX to Fe XXIV and isoelectronic spectra and the validity of the calculations checked by comparison with the sulfur data just mentioned. The application of the results in Fe XIX to Fe XXIV (Fawcett et al., 1974a,b) combined with some observational and theoretical considerations affecting solar line intensities leads to a more definitive analysis of solar flare spectra of these ions. Feldman and Cohen (1967b) and Walker and Rugge (1969) also reported $2p^6ml$–$2p^5\ 3sml$ "autoionizing" or "satellite" lines on the longer wavelength side of the Fe XVII 2p–3s lines. These satellites and $2p^6ml$–$2p^5$ $3dml$ satellite lines are more intense in spectra of laser-produced plasmas and have recently been observed in solar spectra (Parkinson, 1973). A satellite line is emitted when an otherwise normal transition takes place in the presence of m outer electrons.

D. HELIUM-LIKE SPECTRA AND THEIR SATELLITES

Satellite lines of helium-like spectra, which are analogous to those discussed at the end of the previous section and result from $1s^2ml$–$1s^1$ $2pml$ transitions, were first reported by Edlén and Tyren (1939) but have just been classified (Gabriel and Jordan, 1969; Gabriel, 1972) as a result of laboratory investigations and theoretical computations. These investigations gained considerable momentum when their results found an important application in the interpretation of new solar X-ray spectra. They also led to the identification of the $1s^2$ $^1S_0$–$1s2s$ $^3S_1$ helium-like forbidden lines which account for dominant lines in the solar X-ray spectrum belonging to all solar abundant elements in the period between carbon and iron. The first solar observation of a line due to this transition was in O VII with crystal spectrometers (Fritz et al., 1967; Rugge and Walker, 1968), but it was only distinguished from satellite lines after accurate wavelengths were determined from grazing incidence solar spectra (Jones et al., 1968) at the ARU, Culham. The lines were at first thought to be satellite lines, but Gabriel and Jordan (1969) pointed out that the wavelengths corresponded to the forbidden lines. Fine solar spectra showing the forbidden line and satellite structure are available in magnesium (Parkinson, 1972; see Fig. 6) and silicon (Walker and Rugge, 1970, 1971; see Fig. 7) and by NRL (see Fig. 8), but the

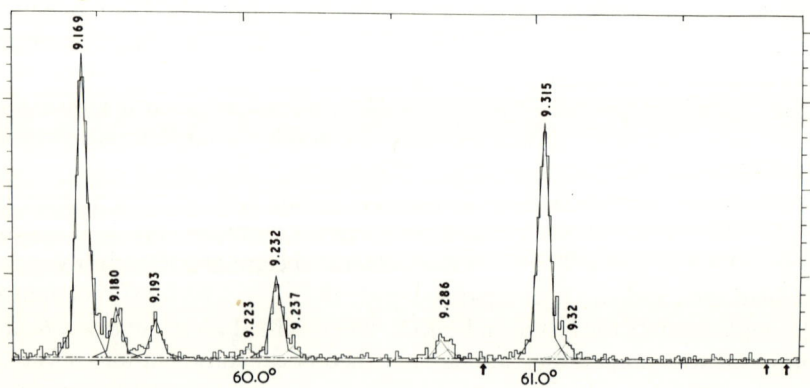

FIG. 6. Mg IX solar lines near 9.0 Å. R is ^1S–^1P, I is ^1S–^3P, F is ^1S–^3S. Others are satellite lines. A Leicester University spectrum taken with a crystal spectrometer. From Parkinson (1972).

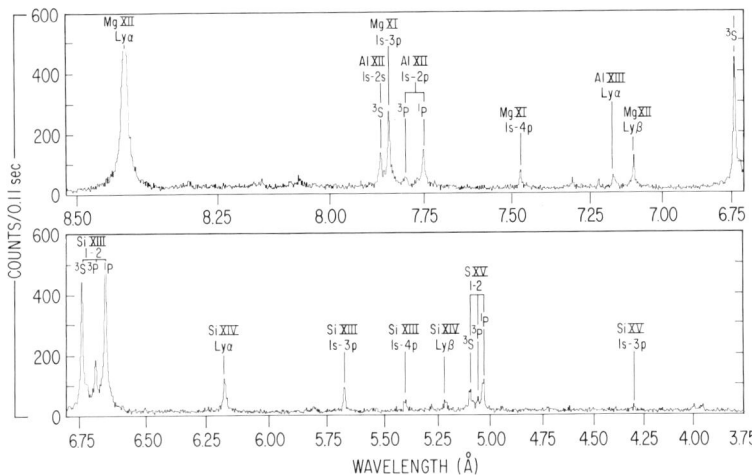

FIG. 7. Spectra between 4.0 and 8.5 Å of an active sun taken from the O VI 17 satellite by Aerospace. From Walker and Rugge (1970).

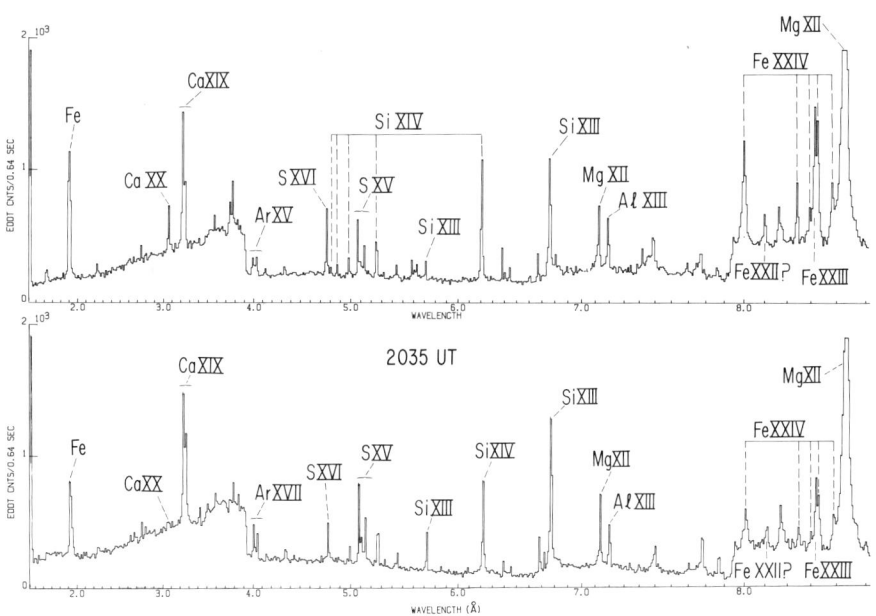

FIG. 8. Solar flare spectra between 1.0 and 6.5 Å taken from the OSO 6 satellite by NRL. From Doschek and Meekins (1971).

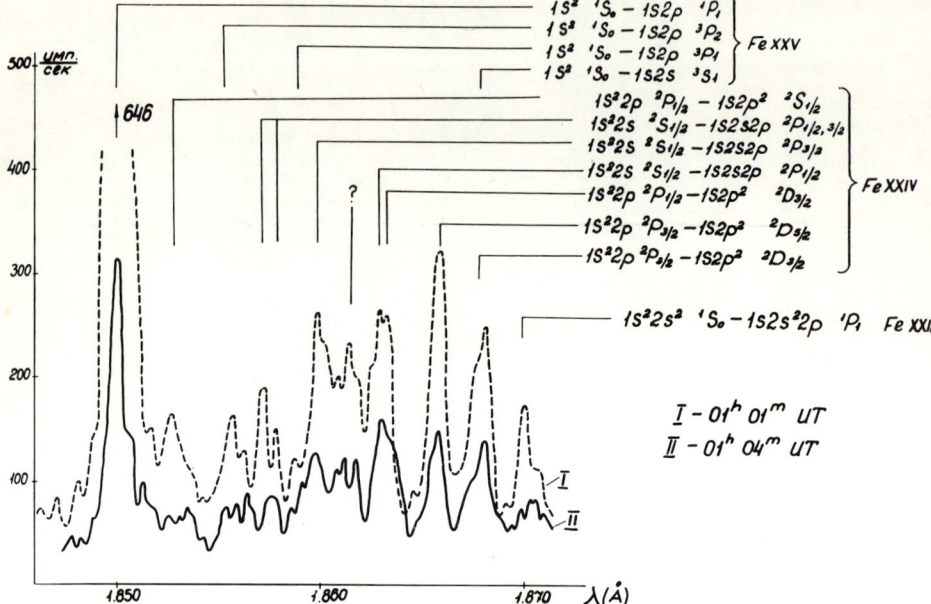

Fig. 9. Solar flare spectra near 1.0 Å of Fe XXV with its satellite lines and intercombination line. Taken from the Intercosmos 4 satellite by the Lebedev Institute. Compare with Fig. 10. From Vasilyev et al. (1972).

most outstanding is in Fe XXV and was acquired with the Intercosmos 4 orbiting vehicle (Grineva et al., 1971, 1973; see Fig. 9). These lines lie near 1Å and are emitted from solar flares along with the resonance line of Fe XXV. They supply surprising information about the temperature and density of solar flares. Laboratory spectra of Fe XXVI and Fe XXV and its satellites have been produced with low inductance sparks (Cohen et al., 1968b; Schwob and Fraenkel, 1972; see Fig. 10). Satellites have also been observed in the laboratory from a number of other elements (Sawyer, 1962; Roth and Elton, 1968; Peacock et al., 1971; Goldsmith, 1969b).

The satellites were classified with the aid of the Hartree–Fock program of Froese-Fischer (see Section IV). To reduce errors in the wavelength calculations, the values computed were the wavelength differences between the satellite lines and the nearby helium-like resonance lines. Configuration interaction and intermediate coupling were included in the calculations (Gabriel, 1972) and electrostatic energies were adjusted where necessary to fit observed wavelengths in carbon, oxygen, and magnesium. Vainstein and Safronova (1971a,b) made calculations independently using a Z-expansion method, which is feasible for two electron spectra and obtained accurate wavelengths which agreed both with the above calculations and others which Cowan obtained with the Hartree-X program. The transition probabilities for decay by autoionization and radiation of Gabriel and Cowan agree and are probably more reliable than those of Vainstein and Safronova which differ. Dielectronic recombination is a predominant process in the formation of these satellites. Other processes such as excitation of three electron ions

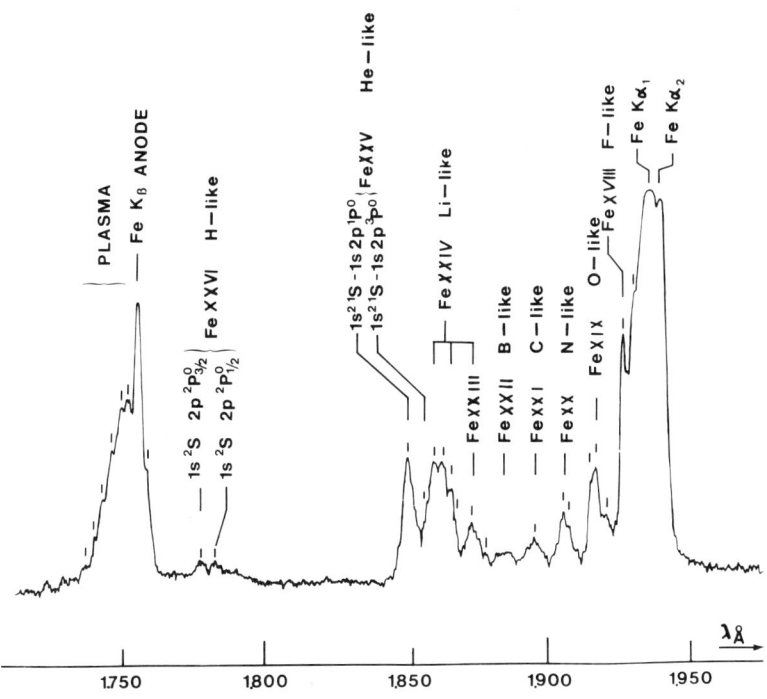

FIG. 10. The Fe XXV helium-like spectra near 1.0 Å showing the satellite lines. The source is a low inductance vacuum spark. From Schwob and Fraenkel (1972).

give contributions which vary according to the atomic number of the ion and the physical conditions under which they exist. The variation of intensity of the satellites is therefore applied to determine the parameters of solar plasmas.

Conventional helium like 2s–np and 2p–nd classifications have recently been extended in C V, N VI, O VII, and Ne IX (Fawcett et al., 1966; Edlén and Lofstrand, 1970).

E. Transitions between Levels of Principal Quantum Number 3 and 4

In spite of the serious deficiencies in the data listed up to 1968 regarding spectra of the $3p^n$–$3p^{n-1}4s$, 4d and $3p^n3d$–$3p^n4p$, 4f transition arrays, there were hardly any relevant papers published during the preceding 30 years. In iron these spectra lie between 40 and 160 Å and are responsible for lines in the solar spectrum but, what is more important, some are blended with other solar lines which would otherwise have been the subject of elaborate experiments requiring the accurate measurement of their intensities. There are two other solar applications for these transition arrays. First, some knowledge of their energy levels is required to allow for cascade processes in the interpretation of solar line intensities. Second, measurement of the $3p^n3d$ energy levels can lead to the identification of solar forbidden lines due to transitions between these levels. In 1968, Cowan computed the theoretical wavelengths and oscillator strengths of these transitions using his Hartree-X program (see Section IV). Time-resolved vacuum spark spectra of titanium and manganese with excellent differentiation between ionization stages were recorded with a 2-meter grazing incidence spectrograph fitted with a microsecond shutter (see Section III,B). In Fig. 11 the observed and computed spectra of titanium are compared, and it can be seen that the theoretical spectra is a replica of the observed one, hence permitting detailed classification (Fawcett et al., 1968; Svensson and Ekberg, 1968). This experimental work was mainly confined to the $3p^n$–$3p^{n-1}4d$ transition arrays in titanium and manganese. Subsequently, comparison of the theoretical spectra with the spectra of the theta pinch containing iron and manganese carbonyl described in Section V,B and with laser-produced plasma spectra of elements in the period between scandium and nickel permitted the classification of the strongest lines of all these transition arrays on each of the ionization stages isoelectronic with Fe IX to Fe XV (Fawcett et al., 1972a,b). A feature of these classifications is that they included the mixed parentage composition of the levels. This is necessary for the adequate description of many complex spectra. Meanwhile Svensson and Ekberg carried out a detailed investigation of potassium, calcium, scandium, and titanium which included 3p–4s and 3p–4d transitions (Svensson and Ekberg, 1968; Ekberg and Svensson, 1970; Ekberg, 1971; Svensson, 1971b). Wagner and House (1971) established classifications of $3p^53d$–$3p^54f$ transitions by comparing observational data with Hartree–Fock calculations. They then estimated the remaining $3p^53d$ energy levels by statistically fitting the measured ones to the computed levels. From the estimated values they attempted to identify solar forbidden lines due to transitions within the $3p^53d$ configuration in iron.

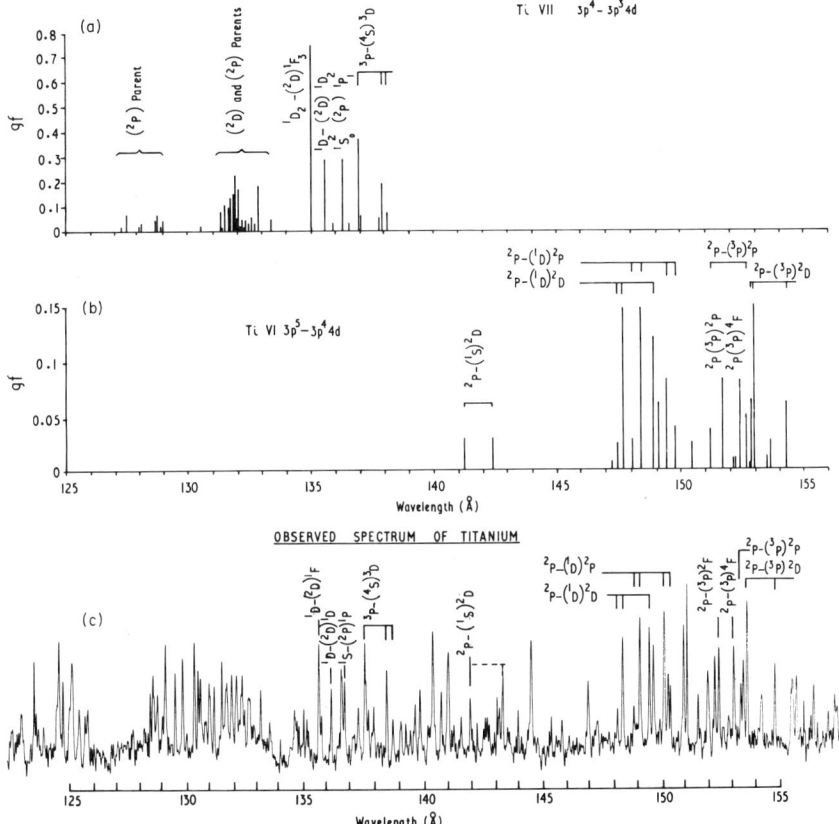

FIG. 11. Titanium spectra calculated using Hartree-X calculations (a, b) compared with observed spectra (c). From Fawcett et al. (1968).

Intensity calibration problems provided the motive for the recent study of 3 to 4 transitions isoelectronic with elements in the first row of the periodic table. A comparison of the spectral intensities of $n = 3$–4 transitions with those of $n = 2$–4 transitions initiating from the same upper level provides a basis for relative intensity calibration of the two spectral regions. Other lines are important to the interpretation of solar intensities (Kunze, 1971). Useful new classifications of this type include the 3^2D–4^2F and $3p^2P$–$4d^2D$ lithium-like lines in the period between fluorine and silicon (Kunze, 1971; Fawcett, 1971b; Druetta et al., 1972).

F. THE FIRST LONG PERIOD

Difficult analytical problems arise because of the degree of complexity of spectra of elements in the first long period between potassium and zinc. $3d^n$

configurations near the ground level complicate the spectra. Edlén (1973) has recently provided a survey of data on $3d^n 4s^m$ configurations and demonstrated relationships between the relative positions of the terms of these configurations with the aid of graphical representations. Even for low stages of ionization the analytical problem is formidable. For example, in the spectrum of Ni II Shenstone (1970) provided an analysis based on 4300 observed lines and supported by the refined theoretical calculations of N. Spector and Y. Shadmi. The Ni II lines were so closely packed on the spectograms that one line occurred in every 2-Å wavelength interval. From this one can understand how severe the task is in the case of highly ionized spectra for which spectra belonging to several different stages of ionization overlap. To tackle the identification of just the strongest lines requires sensitive experimental methods for separating ionization stages, and the use of complex theoretical calculations is usually necessary.

Fortunately the $3p^63d-3p^64f$, $3p^63d-3p^53d^2$, and $3p^63d-3p^64p$ transition arrays in the potassium sequence are relatively simple and were readily analyzed (Alexander *et al.*, 1965; Gabriel *et al.*, 1965, 1966; Even-Zohar and Fraenkel, 1968; Goldsmith and Fraenkel, 1970) in the transition elements up to zinc. A recent thorough analysis of the Cr VI spectrum is reported by Ekberg (1973c). The potassium-like spectra of the $3p^63d-3p^53d$ 4s transition arrays could only be identified after comparing observed spectra with Hartree-X calculations. Cowan (1967b) confirmed their identity in Cr VI, Mn VII, and Fe VIII after adjusting the Slater itegrals so that his theoretical calculations matched the observed spectra. Hoory *et al.* (1970b) extended the analysis as far as Cu XI.

There have been some very impressive detailed papers dealing with the calcium-like spectra of V IV and Cr V. In her analysis of the spectrum of V IV, undertaken at the National Bureau of Standards, Washington, Iglesias (1968) accounted for 340 lines of V IV in the spectral region between 675 and 5940 Å. Isoelectronic spectra of Cr V were subsequently classified at Lund by Ekberg (1973b), who was aided by the unpublished work of B. Edlén and J. W. Swensson in Ti III. The main configurations dealt with were $3d^2$, 3d4s, 3d4p, and 3d5s. $3p^63d^2-3p^63d4f$ transitions belonging to Fe VII, Co VIII; Ni IX, Cu X, and Zn XI are classified (Alexander *et al.*, 1966; Even-Zohar and Fraenkel, 1968; Shadmi, 1966). Shadmi checked and revised the earlier classifications of Cady (1933) and confirmed those of Bowen and Edlén (1939) by applying theoretical calculations. His technique was to derive initial values of the six electrostatic parameters by equating theoretical formulas with observed terms and by acquiring estimates of the spin–orbit interaction parameters for the 3d and 4f electrons from Fe VIII spectra. From these values, matrix elements were derived and after diagonalization, energy levels were calculated. A least squares fit was made between these and observed energy levels. Better estimates of the parameters were hence obtained. Three iterations of this process gave the required convergence.

A great step forward in the interpretation of the complicated scandium-like Cr IV spectra was taken by Ekberg (1973a), who accounted for 275 lines belonging to the $3d^3-3d^24p$ and $3d^24s-3d^24p$ transition arrays in the spectral region between 523 and 2423 Å.

One of the most extensive contributions to the analysis of the spectra discussed in this section is of Fe IV by Edlén (1969a) who located the $3d^4(^5D)4s$, $3d^4(^5D)4p$, and $3d^5$ levels. Edlén predicted the $3d^44s$ term structure from the $3d^4$ terms of Fe V using relationships connecting these configurations (Edlén, 1964, p. 215). From the position of the $3d^44p$ groups he was able to estimate that of the $3d^44s$ groups. He applied Garstang's (1958) calculated values for the $3d^5$ levels to the analysis.

The analysis of the $3d^5$ group allowed Edlén to determine the wavelengths of the forbidden lines in the slowly variable nova-like star RR telescopii. Earlier Edlén had related $3d^2$ ground levels in Fe VII to forbidden lines in Nova Pictoris.

Another contribution was made by Alexander *et al.* (1971) in the Co I, Ni I, and Cu I isoelectronic sequences reported in a paper dealing with classifications of Y IX–Y XIII, Zr X–Zr XIV, Nb XI–Nb XV, and Mo XII–Mo XVI. In the Co I sequence the analysis of the $3d^9-3d^84p$ transitions in Ga V, Ge VI, As VII, and Se VIII was contributed by Kononov (1967) and Zvereva and Kononov (1968).

Future stellar projects based on satellites may benefit from a better knowledge of the spectra discussed in this section especially those of Fe IV, V, VI, and VII which are in urgent need of further investigation. Investigations of lower stages of ionization which are now in progress at several laboratories may help in the analysis of higher stages by permitting a study of isoelectronic sequences. At the Institute of Optics at Madrid such studies are in progress on V II–V III, Cr I–Cr IV, and Mn I–Mn IV.

G. The Spectra of Heavier Elements

The dramatic increase in papers on lighter element spectra since 1960 is not reflected in publications regarding highly ionized heavier element spectra, but there are a few important contributions. For an overall view of heavier element spectra, it helps to bear in mind certain analogies between these and those already discussed in previous sections. Thus elements in the period between Ga and Rb have a similar outer structure to those between Al and K, but with a higher principal quantum number. The ground configuration of the elements Ga to K is $4s^24p^n$ and Al to Ar $3s^23p^n$. The Al I to K I isoelectronic sequences, which include Fe XIV–Fe VIII, have been successfully analyzed, as mentioned earlier in this section. Now, Fe X in the Cl I sequence showed a very good correlation between observed and calculated spectra; and it is not surprising therefore to find that Hartree–Fock calculations were applied successfully to the analogs Br I sequence by Ekberg et al. (1972) who classified Y V, Zr VI, Nb VII, and Mo VIII thereby extending Chagtai's analysis (1970a,b) of highly ionized Zr, Nb, and Mo, and of Reader and Epstein (1972) in Y V. Working independently of Ekberg et al., Even-Zohar and Fraenkel (1972) and Reader et al. (1972) extended Chagtai's analysis along the Br I isoelectronic sequences to Ru IX, Rh X, Pd XI, and Ag XII, and in the same elements identified lines in the Rb I and Kr I sequences which are analogs to Fe VIII and Fe IX. Because of the similarity of structure of ions isoelectronic with those between Ga I and Rb I to those between Al I and K I, further progress can be expected in these sequences in future. With the isoelectronic sequences of next elements in the periodic table, namely, those between Sr I and Cd I, the outlook is not so promising. All the complex problems associated with these spectra are those that arose in connection with their analogous spectra discussed in Section V,F, which dealt with ions with $3d^n$ ground configurations. Even-Zohar and Fraenkel (1972) were able, however, to make contributions to the identification of Sn VI, Sb VII, Te VIII, and I IX in the Rb I sequence and to the identification of some iodine lines in the Pd I, Ag I, and Cd I sequences. Chaghtai et al. have just published several papers on highly ionized yttrium.

Highly ionized spectra isoelectronic with elements heavier than indium have received almost no attention in recent years. Some with simpler outer electronic structure analogous to spectra discussed in Sections V,A–E may be readily interpreted, but there is an increasing probability of inner-shell ionization and other effects complicating the problem. For some complicated systems involving d and f electrons it can be an accomplishment even to distinguish which lines belong to a certain configuration. Confining references to fourth or higher spectra there is an analysis of Lu IV by Sugar and Kaufman (1972) and of Bi IV by Martin et al. (1972). The quantity of lines on heavy element spectograms will make automatic plate-measuring techniques more necessary and require the application of computer techniques to plate analysis, data storage, the theoretical calculation of spectra and the statistical correlation of observed and computed spectra.

H. References to Published Data on Line Classification

The reference list at the end of the article includes references to papers on classifications of the spectra of highly ionized atoms contributed since 1960. These are marked (Cl). Table III provides an index to the classification work on particular ions described in the papers listed in the reference list. This

TABLE III

REFERENCE INDEX FOR CLASSIFICATION PAPERS ON SPECTRA OF HIGHLY IONIZED ATOMS[a]

Ion	Reference number	Ion	Reference number
B IV	62	Si XIII	140
C IV	20, 140, 142	P VI	109
C V	71, 102, 127, 140	P VII	109
N IV	168, 213	P VIII	109
N V	167, 140, 142	P IX	91, 101, 109, 192
N IV	23, 102, 140	P X	91, 101, 109, 192
O IV	29	P XI	91, 101, 109, 192
O V	21	P XII	91, 109, 192
O VI	22, 140, 142	P XIII	91, 109, 140, 192
O VII	84, 102, 140	P XIV	140
F IV	74, 94, 233	S IV	81
F V	74, 94, 233	S V	76
F VI	94, 233	S VIII	66
F VII	22, 94, 140, 142	S IX	50, 110, 195, 239
F VIII	84, 140	S X	50, 101, 109, 110, 114, 195, 239
Ne IV	155, 205, 268	S XI	51, 91, 101, 109, 110, 239
Ne V	155, 169, 205	S XII	77, 91, 101, 109, 110, 239
Ne VI	22, 103, 155, 169, 205	S XIII	91, 101, 109, 239
Ne VII	22, 103, 104, 169, 205, 271	S XIV	91, 101, 109, 140
Ne VIII	22, 58, 103, 104, 140, 142, 169, 202, 271	S XV	140
		Cl VI	92
Ne IX	84, 102, 104, 140, 237	Cl X	50, 56, 110
Na VI	94	Cl XI	51, 101, 110
Na VII	91	Cl XII	51, 91, 101, 110
Na VIII	91	Cl XIII	91, 101, 110
Na IX	91, 94, 140, 142	Cl XIV	91, 101, 110
Na X	140	Cl XV	91, 140
Mg IV	6	Cl XVI	140
Mg VI	91	Ar IV	81, 106, 108
Mg VII	91, 94	Ar V	81, 106
Mg VIII	91, 180	Ar VI	81, 103
Mg IX	91, 178	Ar VII	77
Mg X	91, 94, 122, 140, 142	Ar VIII	58, 103
Mg XI	140	Ar IX	90, 104
Al VII	91, 200	Ar X	66, 90, 104, 110
Al VIII	91, 94, 200	Ar XI	98, 104, 110
Al IX	91, 180, 201	Ar XII	51, 98, 101, 104, 110
Al X	91, 94, 178, 201	Ar XIII	51, 101, 110
Al XI	91, 94, 122, 140, 142, 201	Ar XIV	66, 110
Al XII	140	Ar XV	110
Si IV	272	Ar XVI	140
Si VIII	91, 110, 189	Ar XVIII	42, 140, 237
Si IX	91, 94, 110, 189	K IV	265
Si X	91, 94, 110, 180	K V	81, 106
Si XII	91, 94, 122, 140, 270	K VI	81, 106

(*continued*)

TABLE III—*continued*

Ion	Reference number	Ion	Reference number
K VII	81, 92	Ti XII	78, 117
K VIII	92, 95	Ti XIII	90, 125
K IX	80	Ti XIV	41, 90, 93, 125
K XI	51, 107, 116	Ti XV	56, 93, 114, 115, 157
K XII	51, 56, 107, 115	Ti XVI	93
K XIII	24, 51, 91	Ti XVII	115, 158
K XIV	24	Ti XX	140, 159
K XV	24	Ti XXI	42, 140
K XVII	140, 159	V IV	181
K XVIII	42, 140	V V	78, 92, 146, 181
Ca IV	99, 146, 265	V VI	2, 78, 108, 146, 284
Ca V	99, 146, 265	V VII	99, 108, 112, 146
Ca VI	81, 92, 106, 146	V VIII	92, 99, 112, 146
Ca VII	81, 106	V IX	92, 96, 106, 112, 146
Ca VIII	81	V X	92, 96, 106, 112
Ca IX	77, 92	V XI	92, 111, 146
Ca X	80	V XII	92, 111, 112
Ca XII	10, 107, 125	V XIII	111, 112, 117
Ca XIII	10, 56, 93, 107, 115	V XIV	90, 116
Ca XIV	10, 91, 93, 107	V XV	41, 90, 93, 125
Ca XV	93	V XVI	56, 93, 114, 115, 157
Ca XVIII	140, 159	V XVII	93, 115
Ca XIX	42, 140	V XVIII	115, 158
Sc IV	108, 146, 265	V XIX	115
Sc V	99, 108, 146, 265	V XXI	140, 159
Sc VI	99, 108, 112, 146, 265	V XXII	43, 140
Sc VII	81, 92, 106, 112, 146	Cr IV	78
Sc VIII	81, 92, 106, 112	Cr V	79
Sc IX	81, 92	Cr VI	2, 45, 80, 92, 146
Sc X	81, 92	Cr VII	2, 108, 119, 146, 265, 293
Sc XI	92	Cr VIII	92, 99, 112, 146
Sc XII	90, 116	Cr IX	92, 96, 99, 112, 146
Sc XIII	41, 90, 93, 125	Cr X	92, 96, 106, 112, 146
Sc XIV	93, 114, 115, 157	Cr XI	92, 96, 106, 112
Sc XV	93, 114	Cr XII	92, 96, 106, 111, 146
Sc XVI	93, 158	Cr XIII	92, 106, 111, 112
Sc XIX	140, 159	Cr XIV	92, 106, 112, 117
Sc XX	42, 140	Cr XV	269
Ti IV	72	Cr XVI	41, 93, 125
Ti V	146, 265	Cr XVII	56, 93
Ti VI	99, 112, 146, 265	Cr XXII	140, 159
Ti VII	99, 108, 112, 146, 264, 265	Cr XXIII	42, 140
Ti VIII	92, 94, 106, 112, 146	Mn IV	299
Ti IX	92, 94, 106, 112	Mn VII	45, 146
Ti X	92, 94, 146	Mn VIII	2, 111, 119, 146, 284
Ti XI	77, 92	Mn IX	92, 99, 108, 111, 146

TABLE III—continued

Ion	Reference number	Ion	Reference number
Mn X	92, 96, 99, 108, 111, 146	Fe XXV	42, 140, 254
Mn XI	92, 96, 106, 111, 146	Co VIII	5
Mn XII	92, 96, 106, 111, 112	Co IX	2, 146, 152, 179
Mn XIII	92, 96, 106, 111	Co X	2, 100, 112, 146, 152
Mn XIV	92, 96, 106, 111, 112	Co XI	100, 112, 146, 152
Mn XV	106, 111, 112, 117	Co XII	100, 112, 146, 152
Mn XVII	41, 93, 106, 125	Co XIII	100, 112
Mn XVIII	56, 93, 115	Co XIV	100, 112
Mn XIX	115	Co XV	100, 112
Mn XXIII	140, 159	Co XVI	100, 112, 123
Mn XXIV	41, 140	Co XVII	100, 112, 117, 123, 124
Fe IV	65	Co XVIII	125
Fe VII	261	Co XIX	41, 106, 125
Fe VIII	2, 45, 119, 146	Co XXVI	42
Fe IX	2, 108, 111, 119, 146, 284	Ni IV	149, 240
Fe X	96, 99, 108, 111, 146	Ni VII	26
Fe XI	92, 96, 99, 108, 111, 146, 189	Ni VIII	26
Fe XII	92, 96, 106, 108, 111, 146, 189	Ni IX	3, 88
Fe XIII	92, 96, 106, 108, 111, 189	Ni X	2, 88, 119, 146, 154, 179
Fe XIV	92, 96, 106, 108, 111, 146, 236	Ni XI	2, 88, 100, 108, 112, 119, 146
Fe XV	48, 49, 92, 96, 106, 108, 111, 112, 123, 236	Ni XII	100, 108, 112, 146, 154
		Ni XIII	100, 108, 112, 146, 189
Fe XVI	105, 106, 111, 117, 118, 236	Ni XIV	100, 112, 146, 189
Fe XVIII	24, 41, 106, 124, 125	Ni XV	189
Fe XIX	114, 115	Ni XVI	100, 112
Fe XX	114, 115	Ni XVII	100, 105, 112, 123
Fe XXI	114	Ni XVIII	105, 112, 117, 120, 123
Fe XXII	114	Ni XIX	41, 121, 125
Fe XXIII	114	Ni XXVI	254
Fe XXIV	114, 140, 254	Ni XXVII	42, 254

^a Compiled with the aid of a list by Edlén (1973).

table which only lists fourth and higher stages of ionization was drawn up with the help of a similar table prepared by Edlén (1973), which also lists lower ions.

For listings of classified lines the reader is referred to Kelly (1968) and Kelly and Palumbo (1972) who provide the most extensive tabulation and also to Moore (1965–1971) and Fawcett (1971c,e). Dr. L. Hagan of the National Bureau of Standards is extending the main tabulation of Atomic Energy Level tables by Moore (1949, 1952, 1958). For Bibliographies the reader is referred to Moore (1968–1969) and Hagan et al. (1972). The recent critical review by Edlén (1973) should also be consulted.

VI. Identification of Emission Lines of Highly Ionized Atoms in the Solar Spectrum

A. History

V2 rockets were the first to carry vacuum spectrographs and to acquire solar vacuum ultraviolet spectra (Baum *et al.*, 1946; Johnson *et al.*, 1951). These projects were conducted by NRL, Washington under the leadership of Tousey. The shortest wavelength recorded in these observations was 2100 Å, but most highly ionized spectra are at shorter wavelengths with many important lines below 300 Å. To record spectra in this region requires grazing incidence spectrographs. The University of Colorado (Violett and Rense, 1959) first retrieved solar spectra with a rocket-borne grazing incidence spectrograph, but the shortest wavelength line recorded was helium Lyman alpha at 303.78 Å. Shorter wavelengths were masked by stray light. The first useful grazing incidence solar spectrum was obtained with a photoelectric scanning instrument by AFCRL (Hinteregger, 1960, 1961) and a comparable photographic record was acquired soon afterward (Austin *et al.*, 1962) by NRL. As a result of rockets launched by NRL, AFCRL, and GSFC, and of the OSO I satellite, sufficiently well-resolved solar spectra were available in 1963 for the compilation of the first lists of wavelengths and identifications of coronal emission lines between 10 and 1000 Å (Tousey *et al.*, 1965; Jordan, 1965).

B. Solar Spectra between 20 and 2000 Å

Table IV lists the flights of grazing incidence instruments up to 1973 and Table V those of crystal instruments. It is based on a tabulation by Walker (1972). A selection of some of the best quality data for each spectral region obtained from these flights will now be described and used to form an up-to-date list of solar emission lines below 1000 Å with their identifications (see Table VI). Some of the most accurate wavelengths were recorded with a 3-meter grazing incidence spectrograph which included a 1200 line/mm grating. It was flown by GSFC in conjunction with Tel Aviv University on May 16, 1969 and carried on an Aerobee rocket. The instrument had a much larger grating radius than others previously flown and so was capable of producing better resolved spectra. It photographed the solar spectrum from 60 to 158 Å and from 163 to 385 Å with a resolution of 0.04 Å or better. A section of this spectrum is reproduced in Fig. 12 alongside a spectrum of a theta-pinch plasma containing iron carbonyl (Fawcett, 1971d). The measured wavelengths have an accuracy of 0.008 Å above 100 Å and 0.004 Å

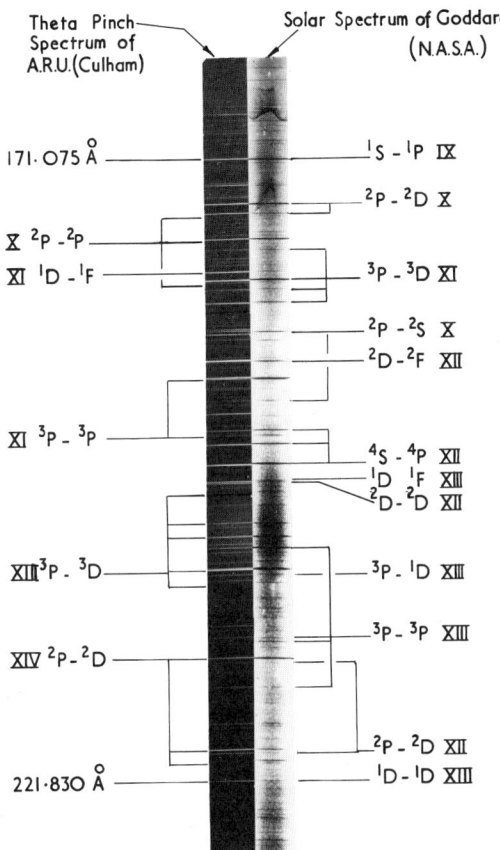

FIG. 12. The solar spectrum between 170 and 222 Å compared with a spectrum of a theta pinch containing iron carbonyl. The solar spectrum was taken by NASA. From Behring et al. (1972).

below 100 Å. The paper describing the experiment (Behring et al., 1972) includes a table of line identifications and measurements.

Three rocket flights of 1-meter grazing incidence instruments by the ARU, Culham, are reported by Freeman and Jones (1970). These instruments utilized 600 line/mm gratings and recorded spectra between 14 and 800 Å, including some lines not listed by Behring et al. (1972). The fourth instrument launched on November 20, 1969 contained a 1200 line/mm grating and gained much of the best solar wavelength data available between 15 and 70 Å, including the helium-like forbidden lines of oxygen, nitrogen, and carbon. In March 1973 the Culham group successfully flew a new grazing incidence experiment in which three spectrographs, each preceded by a grazing incidence confocal paraboloid–hyperboloid mirror recorded spectra

TABLE IV

Observations of the Solar Far Ultraviolet Spectrum with Grating Instruments[a]

Date of flight	Wavelength range (Å)	Laboratory[b]	References	If satellite its name
June 4, 1958	83.9–1216	Univ of Colorado	282	
March 30, 1959	83.9–1216	AFCRL		
Jan 1960	170–400	NRL	171	
June 21, 1961	168–700	NRL	7, 276	
Sept 30, 1961	120–370	GSFC	13, 233	
March 9, 1962–May 1962	170–400	GSFC	13, 223, 229	OSO 1
Aug 22, 1962	170–340	NRL	277	
May 3, 1963	55–310	AFCRL	174	
May 10, 1963	100–500	NRL	8, 278	
Sept 30, 1963	33–180	NRL	278, 289, 291	
Mar 30, 1964	55–310	AFCRL	165	
April 9, 1965	60–300	Culham	32	
Sept 20, 1965	9.5–200	Lebedev Institute	298	
Oct 10, 1965	9.5–200	AFCRL	209, 210	
Nov 3, 1965	30–128	NRL	9	
Feb 1, 1966	33–110			
April 28, 1966 Jun 27, 1966	160–410 170–650	NRL	28, 27	
May 20, 1966	30–60	GSFC	224, 230	
July 27, 1966	33–110 160–410	NRL	293	
Mar 10, 1967 and others	260–1300	AFCRL	164, 173 166	OSO III
Aug 8, 1967	30–128	AFCRL	211	
Oct 1967	300–1300	Harvard	294, 295	OSO IV
July 1967	170–1700	AFCRL	260	OSO IV
March 20, 1968	12–70 140–500	Culham	128, 185	
April 27, 29, 19 Sept 22, 1968 Nov 4, 1969	300–650	NRL	259, 275	
Jan 22, 1969	20–400	GSFC	225	OSO V
April 4, 1969	50–300	AFCRL	261	
May 16, 1969	50–380	GSFC	14	
June 5, 1969	170–1700	Tel Aviv U AFCRL	261	OSO VI
Aug 1969	300–1300	Harvard	59, 232	OSO VI
Nov 20, 1969	7–100	Culham	128, 129	
April 29, 1971	50–210	APW, Frieburg	253	
March 1973	170–1000	Culham	126	

[a] Based on Walker (1972).
[b] Key to abbreviations: AFCRL, Air Force Cambridge Research Lab., USA; NRL, Naval Research Laboratory, USA; GSFC, Goddard Space Flight Center, USA; APW,

TABLE V

OBSERVATIONS OF THE SOLAR SPECTRUM BELOW 25 Å WITH CRYSTAL INSTRUMENTS[a]

Date of flight	Wavelength range (Å)	Laboratory	References	If satellite its name	Date of flight	Wavelength range (Å)	Laboratory	References	If satellite its name
July 25, 1963	13–25	NRL	16–18		Aug 8, 1967	3–15	U Leicester	11	
May 27, 1964	18–34	LASL	85		Oct 1967	1–8	NRL	215	OSO IV
July 23, 1964					Jan 9, 1969	1–25	GSFC	226, 228, 231	OSO V
Aug 5, 1964	12–25	UC London	86, 87		March 1969	1–25	Aerospace	288, 289	OVI 17
May 5, 1966		U Leicester			Aug 1969	0.6–15	NRL	53, 54, 57	OSO VI
					Oct 1970	1.8	Lebedev	161, 281	Intercosmos 4
Oct 4, 1966	1.5–25	NRL	15, 130		Dec 6, 1970	4–22.5	U Leicester	234	
Nov 12, 1966	18–34	LASL	5		April 29, 1971	13.2–13.8	Lockheed	1	
Dec 1966–April 1966	8–25	Aerospace	249, 287	OVI 10		21.4–22.3			
March 1967	1–4	GSFC	230	OSO III	Nov 30, 1971	8–23	U Leicester	235	
	9–25				1972		GSFC	231	OSO VII

[a] Based on Walker (1972).

TABLE VI

Solar Spectrum from 1 to 2000 Å

Wavelength of solar line	Intensity	Ion	Term	J–J	Configuration	Laboratory classification otherwise theoretical (Ref.)	Listed solar wavelength (Ref.)
1.59	3^a	Ni XXVII	$^1S–^1P$	0–1	$1s^2–1s2p$	42	225
1.78	1^a	Fe XXVI	$^2S–^2P$		$1s–2p$	42	225
1.850	9^a	Fe XXV	$^1S–^1P$	0–1	$1s^2–1s2p$	42	161, 161a, 281
1.856	1^a	Fe XXV	$^1S–^3P$	0–1	$1s^2–1s2p$	140, 254	161, 161a, 281
1.860	3^a	Fe XXIV	$^2S–^2P$	$\frac{1}{2}–1\frac{1}{2}$	$1s^22s–1s2p2s$	140, 254	161, 161a, 281
1.863	4^a	Fe XXIV	$^2P–^2D$	$\frac{1}{2}–1\frac{1}{2}$	$1s^22p–1s2p^2$	140, 254	161, 161a, 281
1.866	4^a	Fe XXIV	$^2P–^2D$	$1\frac{1}{2}–2\frac{1}{2}$	$1s^22p–1s2p^2$	140, 254	161, 161a, 281
1.868	3^a	Fe XXV	$^1S–^3S$	0–1	$1s^2–1s2p$	140, 254	161, 161a, 281
1.870	2^a	Fe XXIV	$^2P–^4P$	$\frac{1}{2}–1\frac{1}{2}$	$1s^22p–1s2p^2$	140, 254	161, 161a, 281
2.01	1^a	Mn XXIV	$^1S–^1P$	0–1	$1s^2–1s2p$	42	57, 216, 225
2.20	1^a	Cr XXIII	$^1S–^1P$	0–1	$1s^2–1s2p$	42	57, 216, 225
2.70	1^a	Ca XIX	$^1S–^1P$	0–1	$1s^2–1s3p$	b	57, 216, 225
3.02	3^a	Ca XX	$^2S–^2P$		$1s–2p$	b	57, 216, 225
3.17	7^a	Ca XIX	$^1S–^1P$	0–1	$1s^2–1s2p$	42, 140	57, 216, 225
3.19	4^a	Ca XIX	$^1S–^3P$	0–1	$1s^2–1s2p$	140	57, 216
3.21	5^a	Ca XIX	$^1S–^3S$	0–1	$1s^2–1s2s$	140	57, 216, 225
3.37	1^a	Ar XVII	$^1S–^1P$	0–1	$1s^2–1s3p$	233	57, 216, 225
3.53	3^a	K XVIII	$^1S–^1P$	0–1	$1s^2–1s2p$	140	57, 216, 225
3.55	1^a	K XVIII	$^1S–^3P$	0–1	$1s^2–1s2p$	140	57, 216, 225
3.57	2^a	K XVIII	$^1S–^3S$	0–1	$1s^2–1s2s$	140	57, 216, 225
3.73	5^a	Ar XVIII	$^2S–^2P$		$1s–2p$	140, 237	57, 216, 225
3.95	6^a	Ar XVII	$^1S–^1P$	0–1	$1s^2–1s2p$	140, 237	57, 216, 225
3.97	2^a	Ar XVII	$^1S–^3P$	0–1	$1s^2–1s2p$	140, 237	225
3.99	5^a	Ar XVII	$^1S–^3S$	0–1	$1s^2–1s2s$	140	57, 216, 225
4.30	1^a	S XV	$^1S–^1P$	0–1	$1s^2–1s3p$	b	287, 288
4.733	6^a	S XVI	$^2S–^2P$		$1s–2p$	193	287, 288
5.039	3	S XV	$^1S–^1P$	0–1	$1s^2–1s2p$	140	287, 288
5.065	1	S XV	$^1S–^3P$	0–1	$1s^2–1s2p$	140	287, 288
5.100	2	S XV	$^1S–^3S$	0–1	$1s^2–1s2s$	140	287, 288
5.219	2	Si XIV	$^2S–^2P$		$1s–3p$	193	287, 288
5.401	1	Si XIII	$^1S–^1P$	0–1	$1s^2–1s4p$	b	287, 288
5.68	6	Si XIII	$^1S–^1P$	0–1	$1s^2–1s^13p$	b	287, 288
6.184	5	Si XIV	$^2S–^2P$		$1s–2p$	193	287, 288
6.649	8	Si XIII	$^1S–^1P$	0–1	$1s^2–1s2p$	140	287, 288
6.684	3	Si XIII	$^1S–^3P$	0–1	$1s^2–1s2p$	140	287, 288
6.717	2	Si XII	$^2S–^2P$		$1s^22s–1s2s2p$	140	287, 288
6.739	5	Si XIII	$^1S–^3S$	0–1	$1s^2–1s2s$	140	287, 288
7.111	3	Mg XII	$^2S–^2P$		$1s–3p$	193	287, 288
7.179	1	Al XIII	$^2S–^2P$		$1s–2p$	193	287, 288
7.316	1	Mg XI	$^1S–^1P$	0–1	$1s^2–1s5p$	193	287, 288
7.474	2	Mg XI	$^1S–^1P$	0–1	$1s^2–1s4p$	193	287, 288
7.759	3	Al XII	$^1S–^1P$	0–1	$1s^2–1s2p$	193	287, 288
7.805	1	Al XII	$^1S–^3P$	0–1	$1s^2–1s2p$	140	287, 288

TABLE VI—continued

Wavelength of solar line	Intensity	Ion	Term	J–J	Configuration	Laboratory classification otherwise theoretical (Ref.)	Listed solar wavelength (Ref.)
7.850	4	Mg XI	^1S–^1P	0–1	$1s^2$–$1s3p$	193	287, 288
7.873	2	Al XII	^1S–^3S	0–1	$1s^2$–$1s2s$	140	287, 288
7.99	5^a	Fe XXIV	^2S–^2P		$2s$–$4p$	b	57, 216
8.09	1^a						57, 216
8.17	1^a						57, 216
8.23	2^a	Fe XXIV	^2P–^2D	$\frac{1}{2}$–$\frac{3}{2}$	$2p$–$4d$	b	57, 216
8.30	4^a	Fe XXIV	^2P–^2D	$\frac{3}{2}$–$\frac{5}{2}$	$2p$–$4d$	b	57, 216
8.421	9	Mg XII	^2S–^2P		$1s$–$2p$	193	287, 288
9.169	9	Mg XI	^1S–^1P	0–1	$1s^2$–$1s2p$	193	234
9.180	2						234
9.193	2						234
9.232	3	Mg XI	^1S–^3P	0–1	$1s^2$–$1s2p$	140	234
9.28	1						234
9.315	7	Mg XI	^1S–^3S	0–1	$1s^2$–$1s2s$	140	234
10.25	1	Ne X	^2S–^2P		$1s$–$3p$		249, 286
10.61	3^a	Fe XXIV	^2S–^2P	$\frac{1}{2}$–$1\frac{1}{2}$	$2s$–$3p$	114	231
10.65	3^a	Fe XXIV	^2S–^2P	$\frac{1}{2}$–$\frac{1}{2}$	$2s$–$3p$	114	231
11.01	5^a	Fe XXIV	^2P–^2D	$\frac{1}{2}$–$\frac{3}{2}$	$2p$–$3d$	114	231
		Fe XXIII	^1S–^1P	0–1	$2s^2$–$2s3p$	114	231
11.17	4^a	Fe XXIV	^2P–^2D	$\frac{3}{2}$–$\frac{5}{2}$	$2p$–$3d$	114	231
11.45	4^a	Fe XXIII	^3P–^3D	2–3	$2s2p$–$2s3d$	114	231
11.57	2	Ne IX	^1S–^1P		$1s^2$–$1s3p$	237	249, 286
11.76	9^a	Fe XXII	^2P–^2D	$\frac{1}{2}$–$\frac{3}{2}$	$2p$–$3d$	114	231
11.93	1^a	Fe XXII	^2P–^2D	$\frac{3}{2}$–$\frac{5}{2}$	$2p$–$3d$	114	231
12.13	9	Ne X	^2S–^2P		$1s$–$2p$	237	86
12.26	7	Fe XVII	^1S–^3D	0–1	$3p^6$–$3p^54d$	118	86
12.38	5	Fe XXI	^3P–^3D	2–3	$2p^2$–$2p3d$	114	231
12.45	4^a	Ni XIX	^1S–^1P	0–1	$3p^6$–$3p^53d$	121	249, 286
12.64	4^a	Ni XIX	^1S–^3D	0–1	$3p^6$–$3p^53d$	121	86, 249, 286
12.82	7^a	Fe XX	^4S–^4P	$\frac{3}{2}$–$\frac{5}{2}$	$2p^3$–$2p^23d$	114	231
13.45	6	Ne IX	^1S–^1P	0–1	$1s^2$–$1s2p$	104	86
13.49	6^a	Fe XIX	^3P–^3D	2–3	$2p^4$–$2p^33d$	114	231
13.55	2	Ne IX	^1S–^3P	0–1	$1s^2$–$1s2p$	104	86
13.71	4	Ne IX	^1S–^3S	0–1	$1s^2$–$1s2s$	140	234
13.77	3^a	Ni XIX	^1S–^1P	0–1	$2p^6$–$3p$	121	86
13.824	3	Fe XVII	^1S–^1P	0–1	$2p^6$–$3p$	194	234
13.888	2	Fe XVII	^1S–^3P	0–1	$2p^6$–$3p$	194	234
14.02	3^a	Ni XIX	^1S–^3P		$2p^6$–$3p$	121	249, 286
14.25	2	Fe XVIII	^2P–$^2D(^3P)$	$\frac{3}{2}$–$\frac{5}{2}$	$2p^5$–$2p^43d$	106	86
14.40	1	Fe XVIII	^2P–$^2D(^1D)$	$\frac{3}{2}$–$\frac{5}{2}$	$2p^5$–$2p^43d$	106	86
15.01	42	Fe XVII	^1S–^1P	0–1	$3p^6$–$3p^53d$	194	128
15.17	1	O VIII	^2S–^2P		$1s$–$3p$	193	86
15.26	38	Fe XVII	^1S–^3D	0–1	$3p^6$–$3p^53d$	194	128
15.45	22	Fe XVII	^1S–^3P	0–1	$3p^6$–$3p^53d$	194	128
15.61	1	Fe XVIII	^2P–^2D	$1\frac{1}{2}$–$2\frac{1}{2}$	$3p^5$–$3p^43s$	106	128

(continued)

TABLE VI—continued

Wavelength of solar line	Intensity	Ion	Term	J–J	Configuration	Laboratory classification otherwise theoretical (Ref.)	Listed solar wavelength (Ref.)
15.88	1	Fe XVIII	^2P–^2P	$1\frac{1}{2}$–$1\frac{1}{2}$	$3p^5$–$3p^43s$	106	249, 286
16.01	20	O VIII	^2S–^2P		1s–2p	193	128, 249, 286
		Fe XVIII	^2P–^4P	$1\frac{1}{2}$–$1\frac{1}{2}$	$3p^5$–$3p^43s$	106	
16.77	46	Fe XVII	^1S–^1P	0–1	$3p^6$–$3p^53s$	194	128
17.05	48	Fe XVIII	^1S–^3P	0–1	$3p^6$–$3p^53s$	194	128
17.11	46	Fe XVI					128
17.42	1	Fe XVI	^2S–^2P		$3p^63s$–$3p^53s^2$	117	249, 286
17.64	1				$2p^6nl$–$2p^5nlml$	118	249, 286
17.78	10	O VII	^1S–^1P	0–1	$1s^2$–$1s4p$	193	128
18.63	20	O VII	^1S–^1P	0–1	$1s^2$–$1s3p$	193	128
18.97	84	O VIII	^2S–^2P		1s–2p	193	128
21.60	42	O VII	^1S–^1P	0–1	$1s^2$–$1s2p$	193	128
21.80	20	O VII	^1S–^3P	0–1	$1s^2$–$1s2p$	193	128
22.09	38	O VII	^1S–^3S	0–1	$1s^2$–$1s2p$	140	128
24.78	5	N VII	^2S–^2P		$1s$–$1s2s$	193	128
28.47	11	C VI	^2S–^2P		1s–3p	193	128
28.79	21	N VI	^1S–^1P	0–1	$1s^2$–$1s2p$	193	128
29.52	19	N VI	^1S–^3S	0–1	$1s^2$–$1s2s$	140	128
30.02	16						128
30.43	14	S XIV	^2S–^2P	$\frac{1}{2}$–$\frac{3}{2}$	2s–3p	101	128
33.74	70	C VI	^2S–^2P		1s–2p	193	128
35.22	10	S XII	^2P–^2P	$\frac{1}{2}$–$\frac{1}{2}$	$2s^22p$–$2s^23p$	101	128
35.58	13						128
35.70	18						128
36.39	12	S XII	^2P–^2D	$\frac{1}{2}$–$\frac{3}{2}$	$2s^22p$–$2s^23p$	101	128
39.26	17	S XI	^3P–^3D		$2p^2$–$2p3d$	101	128
40.27	31	C V	^1S–^1P	0–1	$1s^2$–$1s2p$	193	128
40.73	18	C V	^1S–^3P	0–1	$1s^2$–$1s2p$	193	128
40.91	23	Si XIII	^2S–^2P		2s–3p	193	128
41.47	18	C V	^1S–^3S	0–1	$1s^2$–$1s2s$	140	128
42.53	15	S X	^4S–^4P		$2p^2$–$2p3d$	101	128
43.74	32	Si XI	^1S–^1P		$2s^2$–$2s2p$	193	128
44.02	20	Si XII	^2P–^2D	$\frac{1}{2}$–$\frac{3}{2}$	2p–3d	193	128
44.17		Si XII	^2P–^2D	$\frac{3}{2}$–$\frac{5}{2}$	2p–3d	193	148
45.52	14	Si XII	^2P–^2S	$\frac{1}{2}$–$\frac{1}{2}$	2p–3s	193	128
45.68	18	Si XII	^2P–^2S	$\frac{1}{2}$–$\frac{3}{2}$	2p–3s	193	128
46.41	17	Si XI	^3P–^3D	2–3	$2s2p$–$2s3d$	193	128
49.22	46	Si XI	^3P, ^1P–^1D, ^3S	1–2	$2s2p$–$2s3d$	193	128
50.36	23	Fe XVI	^2S–^2P	$\frac{1}{2}$–$1\frac{1}{2}$	3s–4p	194	128
50.53	35	Si X	^2P–^2D	$\frac{1}{2}$–$\frac{3}{2}$	2p–3d	193	128
50.69	37	Si X	^2P–^2D	$\frac{3}{2}$–$\frac{5}{2}$	2p–3d	193	128
52.30	28	Si XI	^1P–^1S	1–0	$2s2p$–$2s3s$	193	128
52.87	12	Fe XV	^1S–^1P	0–1	$3s^2$–$3s4p$	194	128
54.15	16	S IX	^3P–^3D	2–3	$2p^4$–$2p^33s$	193	128
		S VIII	^2P–^2D	$\frac{3}{2}$–$\frac{5}{2}$	$2p^5$–$3p^43d$	193	

HIGHLY IONIZED ATOM SPECTRA CLASSIFICATION

TABLE VI—continued

Wavelength of solar line	Intensity	Ion	Term	J–J	Configuration	Laboratory classification otherwise theoretical (Ref.)	Listed solar wavelength (Ref.)
		Fe XVI	^{2}P–^{2}D	$\frac{1}{2}$–$\frac{3}{2}$	3p–4d	194	128
54.70	24	Fe XVI	^{2}P–^{2}D	$\frac{3}{2}$–$\frac{5}{2}$	3p–4d	194	128
55.06	15						128
55.30	37	Si IX	^{3}P–^{3}P, ^{3}D		2p^2–2p3d	193	128
56.12	13	S IX	^{3}P–^{3}S	2–1	2p^4–2p^{3}3d	193	128
56.92	14						128
57.21	10						128
57.37	12						128
57.56	12						128
57.88	30	Mg X	^{2}S–^{2}P	$\frac{1}{2}$–$\frac{3}{2}$	2s–3p	193	128
58.97	14	Fe XIV	^{2}P–^{2}D	$\frac{1}{2}$–$\frac{3}{2}$	3p–4d	194	128
		Mg X	^{2}S–^{2}P	$\frac{1}{2}$–$\frac{1}{2}$	2s–3p	193	
59.42	12	Fe XV	^{1}P–^{1}D	1–2	3s3p–3s4d	111	128
59.62	14	Fe XIV	^{2}P–^{2}D	$\frac{3}{2}$–$\frac{5}{2}$	3p–4d	193	128
61.009	15	Si VIII	^{4}S–^{4}P	$\frac{3}{2}$–$\frac{3}{2}$	2p^3–2p^{2}3d	193	14
61.081	15	Si VIII	^{4}S–^{4}P	$\frac{3}{2}$–$\frac{5}{2}$	2p^3–2p^{2}3d	193	14
61.66	10	S VIII	^{2}P–^{2}D	$\frac{3}{2}$–$\frac{5}{2}$	2p^5–2p^{4}3d	193	128
61.912	10	S VIII	^{2}P–^{2}D	$\frac{1}{2}$–$\frac{3}{2}$	2p^5–2p^{4}3d	193	14
62.35	10						128
61.755	10	Mg IX	^{1}S–^{1}P	0–1	2s^2–2s3p	193	14
62.876	20	Fe XVI	^{2}P–^{2}S	$\frac{1}{2}$–$\frac{1}{2}$	3p–4s	194	14
63.153	20	Mg X	^{2}P–^{2}D	$\frac{1}{2}$–$\frac{3}{2}$	2p–3d	193	14
63.294	30	Mg X	^{2}P–^{2}D	$\frac{3}{2}$–$\frac{5}{2}$	2p–3d	193	14
		S VIII	^{2}P–^{2}P	$\frac{3}{2}$–$\frac{3}{2}$	2p^5–2p^{4}3s	193	
63.714	30	Fe XVI	^{2}P–^{2}S	$\frac{3}{2}$–$\frac{1}{2}$	3p–4s	194	14
		S VIII	^{2}P–^{2}P	$\frac{1}{2}$–$\frac{3}{2}$	2p^5–2p^{4}3s	193	14
64.141	10					193	14
65.60	10	Mg X	^{2}P–^{2}S	$\frac{1}{2}$–$\frac{1}{2}$	2p–3s	193	208
65.84^d	25	Mg X	^{2}P–^{2}S	$\frac{3}{2}$–$\frac{1}{2}$	2p–3s	193	14
66.251	20	Fe XVI	^{2}D–^{2}F	$\frac{3}{2}$–$\frac{5}{2}$	3d–4f	194	14
66.356	30	Fe XVI	^{2}D–^{2}F	$\frac{5}{2}$–$\frac{7}{2}$	3d–4f	194	14
67.132	20	Mg IX	^{3}P–^{3}D	1–2	2s2p–2s3d	193	14
67.35	10	Ne VIII	^{2}S–^{2}P		2s–4p	104	128
69.658	45	Si VIII	^{4}S–^{4}P	$\frac{3}{2}$–$\frac{5}{2}$	2p^3–2p^{2}3s	193	14
		Fe XIV	^{4}P–^{4}P	$\frac{5}{2}$–$\frac{5}{2}$	3p^2–3p4s	111	
69.825^d	15	Si VIII	^{4}S–^{4}P	$\frac{3}{2}$–$\frac{3}{2}$	2p^3–2p^{2}3s	193	14
69.939	15	Si VIII	^{4}S–^{4}P	$\frac{3}{2}$–$\frac{1}{2}$	2p^3–2p^{2}3s	193	14
70.051	15	Fe XV	^{3}D–^{3}F	3–4	3s3d–3s4f	194	14
70.54	10						128
71.919	35	Mg X	^{3}P–^{3}S	1–1	2s2p–2s3s	193	14
72.030	20	Mg IX	^{3}P–^{3}S	2–1	2s2p–2s3s	193	14
		S VII	^{1}S–^{1}P	0–1	2p^6–2p^5(^{2}P)3s	193	
72.311	30	Mg IX	^{1}P–^{1}D	1–2	2s2p–2s3d	193	14
72.63	10	Fe XI	^{3}P–^{3}D	2–3	2p^4–2p^{3}4d	111	128
72.898	10						14

(continued)

TABLE VI—continued

Wavelength of solar line	Intensity	Ion	Term	J–J	Configuration	Laboratory classification otherwise theoretical (Ref.)	Listed solar wavelength (Ref.)
73.15	10						128
73.471^d	30	Fe XV	^1D–^1F	2–3	3s3d–3s4f	111	14
74.854^d	25	Fe XIII	^3P–^3P	2–2	$3p^2$–3p4s	111	14
		Mg VIII	^2P–^2D	$\frac{1}{2}$–$\frac{3}{2}$	$2s^22p$–$2s^23d$	193	
75.034	30	Mg VIII	^2P–^2D	$\frac{3}{2}$–$\frac{5}{2}$	$2s^22p$–$2s^23d$	193	14
76.023	35	Fe XIV	^2D–^2F	$\frac{5}{2}$–$\frac{7}{2}$	3s3d–3s4f	111	14
		Fe X	^2P–^2P	$\frac{3}{2}$–$\frac{3}{2}$	$3p^5$–$3p^44d$	111	14
76.113	20	Fe XIII	^1D–^1P	2–1	$3p^2$–3p4s	111	14
		Fe XIV	^2D–^2F	$\frac{5}{2}$–$\frac{7}{2}$	3s3d–3s4f	111	
76.507	30	Fe XVI	^2D–^2P	$\frac{5}{2}$–$\frac{3}{2}$	3d–4p	111	14
		Fe X	^2P–^2D	$\frac{1}{2}$–$\frac{3}{2}$	$3p^5$–$3p^44d$	111	
76.867	30	Fe XVI	^2D–^2P	$\frac{3}{2}$–$\frac{1}{2}$	3d–4p	111	14
		Fe X	^2P–^2P	$\frac{1}{2}$–$\frac{1}{2}$	$3p^5$–$3p^44d$	111	
77.741	35	Mg IX	^1P–^1S	1–0	2s2p–2s3s	193	14
77.854	20	Fe X	^2P–^2D	$\frac{3}{2}$–$\frac{5}{2}$	$3p^5$–$3p^44d$	111	14
79.494	30	Fe XII	^4S–^4P	$\frac{3}{2}$–$\frac{5}{2}$	$3p^3$–$3p^24s$	111	14
80.020^d	15	Fe XII	^4S–^4P	$\frac{3}{2}$–$\frac{3}{2}$	$3p^3$–$3p^24s$	111	14
80.510	35	Fe XII	^4S–^4P	$\frac{3}{2}$–$\frac{1}{2}$	$3p^3$–$3p^24s$	111	14
81.94	10						128
82.425	20	Fe IX	^1S–^1P	0–1	$3p^6$–$3p^54d$	2	14
82.672	32						14
83.336	20						14
83.631	20						14
84.024	20	Mg VII	^3P–^3D	2–3	$2s^22p^2$–$2s^22p3d$	193	14
84.491^d	25	Fe XII	^4F–^4G		$3p^23d$–$3p^24f$	111	14
85.470	30	Fe XII	^2G–^2H	$4\frac{1}{2}$–$5\frac{1}{2}$	$3p^23d$–$3p^24f$	111	14
86.765	30	Fe XI	^3P–^3D	2–3	$3p^4$–$3p^34s$	194	14
87.018	25	Fe XI	^3P–^3D	2–2	$3p^4$–$3p^34s$	194	14
88.082^d	25	Ne VIII	^2S–^2P		2s–3p	104	14
88.933	35						14
89.087	20	Fe XI	^1D–^1D	2–2	$3p^4$–$3p^34s$	194	14
89.178	25	Fe XI	^3P–^3S	2–1	$3p^4$–$3p^34s$	194	14
91.808	25	Ni X	^2D–^2F	$\frac{5}{2}$–$\frac{7}{2}$	$3p^63d$–$3p^64f$	2	14
92.178	25						14
93.618	30						14
93.933	20	Fe XVIII	^2P–^2S	$1\frac{1}{2}$–$\frac{1}{2}$	$2s^22p^5$–$2s2p^6$	24	14
94.016	35	Fe X	^2P–^2D	$\frac{3}{2}$–$\frac{5}{2}$	$3p^5$–$3p^44s$	194	14
95.370^d	15	Fe X	^2P–^2D	$\frac{1}{2}$–$\frac{3}{2}$	$3p^5$–$3p^44s$	194	14
96.007	30						14
96.119	30	Fe X	^2P–^2P	$\frac{3}{2}$–$\frac{3}{2}$	$3p^5$–$3p^44s$	194	14
97.122	25	Fe X	^2P–^4P	$\frac{3}{2}$–$\frac{3}{2}$	$3p^5$–$3p^44s$	194	14
97.58	10	Ne VII	^1S–^1P	0–1	$2s^2$–2s3p		208
97.839	20	Fe X	^2P–^4P	$\frac{3}{2}$–$\frac{5}{2}$	$3p^5$–$3p^44s$	194	14
98.126	25	Ne VIII	^2P–^2D	$\frac{1}{2}$–$\frac{3}{2}$	2p–3d	193	14
98.263	30	Ne VIII	^2P–^2D	$\frac{3}{2}$–$\frac{5}{2}$	2p–3d	193	14

TABLE VI—continued

Wavelength of solar line	Intensity	Ion	Term	J–J	Configuration	Laboratory classification otherwise theoretical (Ref.)	Listed solar wavelength (Ref.)
98.517	15						14
100.575	35						14
101.559	25						14
102.88	10	Ne VIII	$^2P-^2S$	$\frac{1}{2}-\frac{1}{2}$	2p–3s	104	208
103.085	20	Ne VIII	$^2P-^2S$	$\frac{3}{2}-\frac{1}{2}$	2p–3s	104	
103.564	35	Fe IX	$^1S-^1P$	0–1	$3p^6-3p^54s$	194	14
103.928^d	30	Fe XVIII	$^2P-^2S$	$\frac{1}{2}-\frac{1}{2}$	$2s^22p^5-2s2p^6$	24	14
105.209	30	Fe IX	$^1S-^3P$	0–1	$2p^6-3p^54s$	194	14
105.820	10						14
111.261	40						14
111.557^d	25						
111.724	10						
113.8	20						211
115.8	6						211
116.75	9						211
122.72	9	Ne VI	$^2P-^2D$	$\frac{3}{2}-\frac{5}{2}$	2p–3d	104	211
123.50	8						211
							211
127.70	7						211
129.87	4	O VI	$^2P-^2D$	$\frac{3}{2}-\frac{5}{2}$	2p–4d	193	14
130.713	10						14
130.943	20	Fe VIII	$^2D-^2F$	$\frac{3}{2}-\frac{5}{2}$	3d–4f	194	14
131.247	25	Fe VIII	$^2D-^2F$	$\frac{5}{2}-\frac{7}{2}$	3d–4f	194	14
133.923	20						14
138.202	20						14
141.032	30	Ca XII	$^2P-^2S$	$\frac{3}{2}-\frac{1}{2}$	$2s^22p^5-2s2p^6$	107	14
142.019	20						14
142.750^d	15						14
144.212	30	Ni X	$^2D-^2D$	$\frac{3}{2}-\frac{3}{2}$	$3p^63d-3p^53d^2$	141	14
144.986	35	Ni X	$^2D-^2D$	$\frac{5}{2}-\frac{5}{2}$	$3p^63d-3p^63d^2$	141	14
145.656	10						14
145.734	25						14
146.083	20						14
146.937	30						14
147.274	25	Ca XII	$^2P-^2S$	$\frac{1}{2}-\frac{1}{2}$	$2s^22p^5-2s2p^6$	107	14
147.653	15						14
148.374	60	Ni XI	$^1S-^1P$	0–1	$3p^6-3p^53d$	141	14
150.089^d	40	O VI	$^2S-^2P$		2s–3p	193	14
151.568	15						14
151.979	20						14
152.154	50	Ni XII	$^2P-^2D$	$\frac{3}{2}-\frac{5}{2}$	$3p^5-3p^43d$	141	14
152.703	15						14
152.95		Ni XII	$^2P-^2D$	$\frac{1}{2}-\frac{3}{2}$	$3p^5-3p^43d$	141	14
154.179	45	Ni XII	$^2P-^2P$	$\frac{3}{2}-\frac{3}{2}$	$3p^5-3p^43d$	141	14
157.730	35	Ni XIII	$^3P-^3D$	2–3	$3p^4-3p^33d$	100	14

(continued)

TABLE VI—continued

Wavelength of solar line	Intensity	Ion	Term	J–J	Configuration	Laboratory classification otherwise theoretical (Ref.)	Listed solar wavelength (Ref.)
158.38	10	Ni X	$^2D-^2F$	$\frac{5}{2}-\frac{7}{2}$	$3p^63d-3p^53d^2$	141	208
159.94	10	Ni X	$^2D-^2F$	$\frac{3}{2}-\frac{5}{2}$	$3p^63d-3p^53d^2$	141	208
164.146	35	Ni XIII	$^3P-^3P$	2–2	$3p^4-3p^33d$	100	14
		Ni XIV	$^2D-^2F$	$\frac{5}{2}-\frac{3}{2}$	$3p^3-3p^23d$	100	14
164.194	30						14
165.509							14
167.495	40	Fe VIII	$^2D-^2D$	$\frac{3}{2}-\frac{3}{2}$	$3p^63d-3p^53d^2$	145	14
168.176	40	Fe VIII	$^2D-^2D$	$\frac{5}{2}-\frac{5}{2}$	$3p^63d-3p^53d^2$	145	14
168.548	30	Fe VIII	$^2D-^2P$	$\frac{5}{2}-\frac{3}{2}$	$3p^63d-3p^53d^2$		14
168.933^d	30	Fe VIII	$^2D-^2P$	$\frac{3}{2}-\frac{1}{2}$	$3p^63d-3p^53d^2$		14
169.616	20	Fe				c	14
169.680	30	Ni XIV	$^4S-^4P$	$1\frac{1}{2}-1\frac{1}{2}$	$3p^3-3p^23d$	100	14
169.915	30	Fe				c	14
171.075	90	Fe IX	$^1S-^1P$	0–1	$3p^6-3p^53d$	145	14
171.359	40	Ni XIV	$^4S-^4P$	$1\frac{1}{2}-2\frac{1}{2}$	$3p^3-3p^23d$	100	14
171.533	30	Fe				c	14
172.174	20	O V	$^1S-^1P$	0–1	$2s^2-2s3p$	193	14
172.936	30	O VI	$^2P-^2D$	$\frac{1}{2}-\frac{3}{2}$	2p–3d	193	14
173.081	40	O VI	$^2P-^2D$	$\frac{3}{2}-\frac{5}{2}$	2p–3d	193	14
174.534	90	Fe X	$^2P-^2D$	$\frac{3}{2}-\frac{5}{2}$	$3p^5-3p^43d$	145	14
175.266	50	Fe X	$^2P-^2D$	$\frac{1}{2}-\frac{3}{2}$	$3p^5-3p^43d$	145	14
175.474	30	Fe X	$^2P-^2P$	$\frac{3}{2}-\frac{1}{2}$	$3p^5-3p^43d$	99	14
176.694^d	40	Ni XV	$^3P-^3P$	0–1	$3p^2-3p3d$	100	14
176.982	40	Fe				c	14
177.243	80	Fe X	$^2P-^2P$	$\frac{3}{2}-\frac{3}{2}$	$3p^5-3p^43d$	145	14
177.597^d	30	Fe				c	14
177.732^d	10						14
178.060	40	Fe XI	$^3P-^3D$	2–2	$3p^4-3p^33d$	145	14
178.725	20	Fe				c	14
179.270	20	Ni XV	$^3P-^3D$	2–3	$3p^2-3p3d$	100	14
179.762	40	Fe XI	$^1D-^1F$	2–3	$3p^4-3p^33d$	145	14
180.407	90	Fe XI	$^3P-^3D$	2–3	$3p^4-3p^33d$	145	14
180.600	30	Fe XI	$^3P-^3P$	1–1	$3p^4-3p^33d$	96	14
181.140	40	Fe XI	$^3P-^3D$	0–1	$3p^4-3p^33d$	145	14
182.173	60	Fe XI	$^3P-^3D$	1–2	$3p^4-3p^33d$	145	14
182.310	30	Fe X	$^2P-^2P$	$\frac{1}{2}-\frac{3}{2}$	$3p^5-3p^43d$	99	14
183.951	20	O VI	$^2P-^2S$	$\frac{1}{2}-\frac{1}{2}$	$2p-3s^2S_{\frac{1}{2}}$	193	14
184.125	30	O VI	$^2P-^2S$	$\frac{3}{2}-\frac{1}{2}$	$2p-3s^2S_{\frac{1}{2}}$	193	14
184.542	60	Fe X	$^2P-^2S$	$\frac{3}{2}-\frac{1}{2}$	$3p^5-3p^43d$	99	14
184.800	30	Fe XI	$^1D-^1D$	2–2	$3p^4-3p^33d$	99	14
185.225	50	Fe VIII	$^2D-^2F$	$\frac{5}{2}-\frac{7}{2}$	$3p^63d-3p^53d^2$	145	14
185.24	20	Ca XIV	$^4S-^4P$	$\frac{3}{2}-\frac{3}{2}$	$2s^22p^3-2s2p^4$	93	14
186.609	40	Fe VIII	$^2D-^2F$	$\frac{5}{2}-\frac{7}{2}$	$3p^63d-3p^53d^2$	145	14
186.86^d	30	S XI	$^3P-^3S$	0–1	$2s^22p^2-2s2p^3$	51	14
186.885	60	Fe XII	$^2D-^2F$	$\frac{5}{2}-\frac{7}{2}$	$3p^3-3p^23d$	145	14

HIGHLY IONIZED ATOM SPECTRA CLASSIFICATION 273

TABLE VI—continued

Wavelength of solar line	Intensity	Ion	Term	J–J	Configuration	Laboratory classification otherwise theoretical (Ref.)	Listed solar wavelength (Ref.)
186.983	20	Fe					14
187.231	30	Fe VIII	$^2D-^2F$	$\frac{5}{2}-\frac{7}{2}$	$3p^63d-3p^53d^2$		14
188.219	70	Fe XI	$^3P-^3P$	2–2	$3p^4-3p^33d$	99	14
188.305	70						14
188.498	40	Fe				c	14
188.673^d	30	S XI	$^3P-^3S$	1–1	$2s^22p^2-2s2p^3$	51	14
189.003	30	Fe				c	14
189.128	30	Fe				c	14
189.739	30	Fe				c	14
189.945	30	Fe				c	14
190.044	50	Fe X	$^2P-^2S$	$\frac{1}{2}-\frac{1}{2}$	$3p^5-3p^43d$	99	14
190.378^d	30						14
191.051	20	Fe XII	$^2P-^2D$	$1\frac{1}{2}-2\frac{1}{2}$	$3p^3-3p^23d$	96	14
191.261^d	40	S XI	$^3P-^3S$	2–1	$2s^22p^2-2s2p^3$	51	14
		Fe XIII	$^1D-^1P$	2–1	$3p^2-3p3d$	96	
192.00		Fe				c	14
192.402	70	Fe XII	$^4S-^4P$	$\frac{3}{2}-\frac{1}{2}$	$3p^3-3p^23d$	106	14
192.637	20	Fe				c	
192.819	50	Fe XI	$^3P-^3P$	1–2	$3p^4-3p^33d$	99	14
192.9	20^a	Ca XVII	$^1S-^1P$	0–1	$2s^2-2s2p$	243	243
193.517	80	Fe XII	$^4S-^4P$	$\frac{3}{2}-\frac{3}{2}$	$3p^3-3p^23d$	106	14
193.880	20	Ca XIV	$^4S-^4P$	$\frac{3}{2}-\frac{5}{2}$	$2s^22p^3-2s2p^4$	93	
195.127	90	Fe XII	$^4S-^4P$	$\frac{3}{2}-\frac{5}{2}$	$3p^3-3p^23d$	106	14
196.531	50	Fe XIII	$^1D-^1F$	2–3	$3p^2-3p3d$	106	14
196.649	50	Fe XII	$^2D-^2D$	$\frac{5}{2}-\frac{5}{2}$	$3p^3-3p^23d$	96	14
197.038	30	Fe				c	14
197.443	30	Fe XIII	$^3P-^3D$	0–1	$3p^2-3p3d$	293	14
197.857^d	30						14
198.561	40	S VIII	$^2P-^2S$	$\frac{3}{2}-\frac{1}{2}$	$2s^22p^5-2s2p^6$	193	14
		Fe XII	$^2P-^2P$	$\frac{1}{2}-\frac{1}{2}$	$3p^3-3p^23d$	96	14
200.033	60	Fe XIII	$^3P-^3D$	1–2	$3p^2-3p3d$	96	14
201.134^d	70	Fe XIII	$^3P-^3D$	1–1	$3p^2-3p3d$	96	14
		Fe XII	$^2P-^2P$	$\frac{3}{2}-\frac{3}{2}$	$3p^3-3p^23d$	96	
201.53	30	Fe				c	14
201.58^d	30	Fe				c	14
201.745	30	Fe XII	$^2P-^2P$	$\frac{1}{2}-\frac{1}{2}$	$3p^3-3p^23d$	96	14
202.056	80	Fe XIII	$^3P-^3P$	0–1	$3p^2-3p3d$	96	14
202.432	40	Fe					14
202.620	30	S VIII	$^2P-^2S$	$\frac{1}{2}-\frac{1}{2}$	$2s^22p^5-2s2p^6$	193	14
202.723	30	Fe XI	$^1D-^3P$	2–1	$3s^23p^4-3s3p^5$	96	14
203.118^d	40						14
203.739^d	50	Fe				c	14
203.835	70	Fe XIII	$^3P-^3D$	2–3	$3p^2-3p3d$	106	14
204.274	40	Fe XIII	$^3P-^1D$	1–2	$3p^2-3p3d$	96	14
204.951	50	Fe XIII	$^3P-^3D$	2–1	$3p^2-3p3d$	293	14

(continued)

TABLE VI—continued

Wavelength of solar line	Intensity	Ion	Term	J–J	Configuration	Laboratory classification otherwise theoretical (Ref.)	Listed solar wavelength (Ref.)
206.180	30	Fe				c	14
							14
206.264	30	Fe				c	14
206.381	30	Fe				c	14
207.124	40	Fe				c	14
207.457^d	40	Ni XVII	^3P–^3D	2–3	3s3p–3s3d	112	14
207.935^d	30	Fe				c	14
208.333	20	Fe					14
208.690^d	20	Fe XIII	^1S–^1P	0–1	$3p^2$–3p3d	96	14
208.69	20^a	Ca XV	^3P–^3D	1–2	$2s^22p^2$–$2s2p^3$	95	243
209.634	40	Fe XIII	^3P–^3P	1–2	$3p^2$–3p3d	96	14
209.776	30	Fe				c	14
209.927	50	Fe XIII	^3P–^3P	2–1	$3p^2$–3p3d	96	14
211.328	80	Fe XIV	^2P–^2D	$\frac{1}{2}$–$\frac{1}{2}$	3p–3d	146	14
211.440	20						14
211.749^d	30	Fe XII	^2D–^2P	$\frac{3}{2}$–$\frac{1}{2}$	$3p^3$–$3p^23d$	96	14
212.127^d	20	S XII	^2P–^2P	$\frac{1}{2}$–$\frac{3}{2}$	$2s^22p$–$2s2p^2$	110	14
213.781^d	40	Fe XIII	^3P–^3P	2–2	$3p^2$–3p3d	96	14
214.415	20						14
214.75	10	Si VIII	^2D–^2P	$\frac{3}{2}$–$\frac{1}{2}$	$2s^22p^3$–$2s2p^4$	193	208
215.165^d	30	S XII	^2P–^2P	$\frac{1}{2}$–$\frac{1}{2}$	$2s^22p$–$2s2p^2$	110	14
215.48	10^a	Ca XV	^3P–^3D	2–3	$2s^22p^2$–$2s2p^3$	95	243
215.757	30						14
216.86^d	35	Ni XVII	^1P–^1D	1–2	3s3p–3s3d	96	14
216.92^d	40	Si VIII	^2D–^2P	$\frac{5}{2}$–$\frac{3}{2}$	$2s^22p^3$–$2s2p^4$	193	14
217.108	40	Fe IX	^1S–^3D	0–1	$3p^6$–$3p^53d$	265	14
217.283	30	Fe XII	^2D–^2P	$\frac{3}{2}$–$\frac{3}{2}$	$3p^3$–$3p^23d$	96	14
218.193^d	30	S XII	^2P–^2P	$\frac{3}{2}$–$\frac{3}{2}$	$2s^22p$–$2s2p^2$	110	14
218.957	30						14
219.135	60	Fe XIV	^2P–^2D	$\frac{3}{2}$–$\frac{5}{2}$	3p–3d	146	14
219.453	40	Fe XII	^2D–^2P	$\frac{5}{2}$–$\frac{3}{2}$	$3p^3$–$3p^23d$	96	14
220.095	60	Fe XIV	^2P–^2D	$\frac{3}{2}$–$\frac{3}{2}$	3p–3d	236	14
220.262	30						14
220.882	30	Fe					14
221.16	10^a	Ar XV	^1S–^1P	0–1	$2s^2$–2s2p	112	243
221.26	10	S IX	3p–3p	2–1	$2s^22p^4$–$2s2p^5$	95	208
221.429^d	30	S XII	^2P–^2P	$\frac{3}{2}$–$\frac{1}{2}$	$3s^23p$–$2s2p^2$	95	14
221.830	40	Fe XIII	^1D–^1D	2–2	$3p^2$–3p3d	96	14
223.004^d	30	Ni XVI	^2P–^2P	$\frac{1}{2}$–$\frac{1}{2}$	$3s^23p$–$3s3p^2$	100	14
223.212	30	S IX	^3P–^3P	1–0	$2s^22p^4$–$2s2p^5$	95	14
223.758^d	30	Si IX	^3P–^3S	0–1	$2s^22p^2$–$2s2p^3$	95	14
224.353	40						14
224.745	40	Fe XV	^3P–^3D	0–1	3s3p–3s3d	236	14
		S IX	^3P–^3P	2–2	$2s^22p^4$–$2s2p^5$	95	208
225.031	50	Si IX	^3P–^3S	1–1	$2s^22p^2$–$2s2p^3$	104	14

TABLE VI—continued

Wavelength of solar line	Intensity	Ion	Term	J–J	Configuration	Laboratory classification otherwise theoretical (Ref.)	Listed solar wavelength (Ref.)
225.170	40						14
225.867	30	Fe				c	14
226.330	30	Fe				c	14
227.006	50	Si IX	^3P–^3S	2–1	$2s^22p^2$–$2s2p^3$	104	14
227.22^d	20	Fe XV	^3P–^3D	1–2	$3s3p$–$3s3d$	236	14
227.482	30	S XII	^2P–^2S	$\frac{1}{2}$–$\frac{1}{2}$	$2s^22p$–$2s2p^2$	95	14
228.057	30	Fe				c	14
228.167	40	S X	^2D–^2D	$\frac{3}{2}$–$\frac{3}{2}$	$2s^22p$–$2s2p^4$		14
228.866^d	20	S IX	^3P–^3P	1–2	$2s^22p^4$–$2s2p^5$	95	14
229.755	20						14
230.129	40	Fe				c	14
231.449	30	He II	^2S–^2P		$1s$–$8p$	193	14
232.571^d	30	He II	^2S–^2P		$1s$–$7p$	193	14
233.240^d	30	Fe				c	14
233.444	20						14
233.647	20						14
233.865	30	Fe XV	^3P–^3D	2–3	$3s3p$–$3s3d$	236	14
234.360	50	He II	^2S–^2P		$1s$–$6p$	193	14
236.486	30						14
237.345^d	50	He II	^2S–^2P		$1s$–$5p$	193	14
238.571^d	20	O IV	^2P–^2D	$\frac{3}{2}$–$\frac{5}{2}$	$2p$–$3d$	193	14
239.032	50						14
240.400	30						14
240.713	50	Fe XIII	^3P–^3S	0–1	$3s^23p^2$–$3s3p^3$	96	14
241.739	60	Fe IX	^1S–^3P	0–2	$3p^6$–$3p^53d$	126	14
242.208^d	30						14
242.853	20						14
243.029^d	50	He II	^2S–^2P		$1s$–$4p$	193	14
243.418^d	10						14
243.783	50	Fe XV	^1P–^1D	1–2	$3s3p$–$3s3d$	96	14
244.152	10						14
244.912	40	Fe IX	^1S–^3P	0–1	$3p^6$–$3p^53d$	126	14
246.210	50	Fe XIII	^3P–^3S	1–1	$3s^23p^2$–$3s3p^3$	96	14
248.47	10	O V	^1P–^1S	1–0	$2s2p$–$2s3s$	193	14
249.180	40	Ni XVII	^1S–^1P	0–1	$3s^2$–$3s3p$	100	14
249.389	40	Fe				c	14
251.063	40	Fe XVI	^2P–^2D	$\frac{1}{2}$–$\frac{3}{2}$	$3p$–$3d$	96	14
251.949	50	Fe XIII	^3P–^3S	2–1	$3s^23p^2$–$3s3p^3$	96	
252.190	40	Fe XIV	^2P–^2P	$\frac{1}{2}$–$\frac{3}{2}$	$3s^23p$–$3s3p^2$	96	14
253.791	40	Si X	^2P–^2P	$\frac{1}{2}$–$\frac{3}{2}$	$2s^22p$–$2s2p^2$	96	14
254.591	20						14
256.30	70	He II	^2S–^2P		$1s$–$3p$	193	14
	50	Si X	^2P–^2P	$\frac{1}{2}$–$\frac{1}{2}$	$2s^22p$–$2s2p^2$	91	14
256.37		Fe XIII	^1D–^1P	2–1	$3s^23p^2$–$3s3p^3$	96	
256.676	50	S XIII	^1S–^1P	0–1	$2s^2$–$2s2p$	109	14

(continued)

TABLE VI—continued

Wavelength of solar line	Intensity	Ion	Term	J–J	Configuration	Laboratory classification otherwise theoretical (Ref.)	Listed solar wavelength (Ref.)
256.919	40	Fe				c	14
257.133	10	S X	^4S–^4P	$\frac{3}{2}$–$\frac{1}{2}$	$2s^22p^3$–$2s2p^4$	51	14
257.255	50						14
257.385	50	Fe XIV	^2P–^2P	$\frac{1}{2}$–$\frac{1}{2}$	$3s^23p$–$3s3p^2$	96	14
257.534	40						14
257.762	30						14
258.073^d	20	Si IX	^1D–^1D	2–2	$2s^22p^2$–$2s2p^3$	91	14
258.365	60	Si X	^2P–^2P	$\frac{3}{2}$–$\frac{3}{2}$	$2s^22p$–$2s2p^2$	91	14
259.482	40	S X	^4S–^4P	$\frac{3}{2}$–$\frac{3}{2}$	$2s^22p^3$–$2s2p^4$	51	14
259.961	20						14
260.289	20						14
261.045	50	Si X	^2P–^2P	$\frac{3}{2}$–$\frac{1}{2}$	$2s^22p$–$2s2p^2$	91	14
262.973	40	Fe XVI	^2P–^2D	$\frac{3}{2}$–$\frac{5}{2}$	$3p$–$3d$	236	14
264.215	40	S X	^4S–^4P	$\frac{3}{2}$–$\frac{5}{2}$	$2s^22p^3$–$2s2p^4$	51	14
264.779	60	Fe XIV	^2P–^2P	$\frac{3}{2}$–$\frac{3}{2}$	$3s^23p$–$3s3p^2$	96	14
270.512	50	Fe XIV	^2P–^2P	$\frac{3}{2}$–$\frac{1}{2}$	$3s^23p$–$3s3p^2$	91	14
271.979	50	Si X	^2P–^2S	$\frac{1}{2}$–$\frac{1}{2}$	$2s^22p$–$2s2p^2$	95	14
272.20	10	Si VII	^3P–^3P	2–1	$2s^22p^4$–$2s2p^5$	95	208
274.191	60	Fe XIV	^2P–^2S	$\frac{1}{2}$–$\frac{1}{2}$	$3s^23p$–$3s3p^2$	96	14
275.364	30	Si VII	^3P–^3P	2–2	$2s^22p^4$–$2s2p^5$	193	14
277.00	30	Mg VII	^3P–^3S	1–1	$2s^22p^2$–$2s2p^3$	95	128
		Si VIII	^2D–^2D	$\frac{5}{2}$–$\frac{5}{2}$	$2s^22p^3$–$2s2p^4$	95	
277.261	40	Si X	^2P–^2S	$\frac{3}{2}$–$\frac{1}{2}$	$2s^22p$–$2s2p^2$	91	14
278.40	20	Mg VII	^3P–^3S	2–1	$2s^22p^2$–$2s2p^3$	95	208
281.409^d	20	S XI	^3P–^3D	0–1	$2s^22p^2$–$2s2p^3$	95	14
284.147	80	Fe XV	^1S–^1P	0–1	$3s^2$–$3s3p$	242	14
285.57	10	S XI	^3P–^3D	1–1	$2s^22p^2$–$2s2p^3$	95	208
285.85	10	S XI	^3P–^3D	1–2	$2s^22p^2$–$2s2p^3$	95	208
288.378	20	S XII	^2P–^2D	$\frac{1}{2}$–$\frac{3}{2}$	$2s^22p$–$2s2p^2$	95	14
289.17	10	Fe XIV	^2P–^2S	$\frac{3}{2}$–$\frac{1}{2}$	$3s^23p$–$3s3p^2$	96	208
290.72	20	Si IX	^3P–^3P	0–1	$2s^22p^2$–$2s2p^3$	91	208
290.997^d	20	Fe XII	^2D–^2P	$2\frac{1}{2}$–$1\frac{1}{2}$	$2s^22p^3$–$2s2p^4$	96	14
291.63	10	S XI	^3P–^3D	2–3	$2s^22p^2$–$2s2p^3$	95	208
292.10	10	Ni XVIII	^2S–^2P	$\frac{1}{2}$–$\frac{3}{2}$	$3s$–$3p$	112	208
292.778^d	10	Si IX	^3P–^3P	1–1	$2s^22p^2$–$2s2p^3$	91	14
296.111^d	30	Si IX	^3P–^3P	2–2	$2s^22p^2$–$2s2p^3$	91	14
296.226	20						14
303.318	60	Si XI	^1S–^1P	0–1	$2s^2$–$2s2p$	270	14
303.771^d	90	He II	^2S–^2P		$1s$–$2p$	193	14
312.166	20						14
312.55	5	Fe XV	^3P–^1D	1–2	$3s3p$–$3p^2$	48, 49	48, 49
313.743	30	Mg VII	^2P–^2P	$\frac{1}{2}$–$\frac{1}{2}$	$2s^22p$–$2s2p^2$	193	14
314.351	30	Si VIII	^4S–^4P	$\frac{3}{2}$–$\frac{1}{2}$	$2s^22p^3$–$2s2p^4$	193	14
315.025	40	Mg VIII	^2P–^2P	$\frac{3}{2}$–$\frac{3}{2}$	$2s^22p$–$2s2p^2$	193	14
316.231	30	Si VIII	^4S–^4P	$\frac{3}{2}$–$\frac{3}{2}$	$2s^22p^3$–$2s2p^4$	193	14

TABLE VI—continued

Wavelength of solar line	Intensity	Ion	Term	J–J	Configuration	Laboratory classification otherwise theoretical (Ref.)	Listed solar wavelength (Ref.)
317.02	10	Mg VIII	$^2P_{1\frac{1}{2}}-^2P_{\frac{1}{2}}$		$2s^22p-2s2p^2$	193	126
319.845^d	40	Si VIII	$^4S-^4P$	$\frac{3}{2}-\frac{5}{2}$	$2s^22p^3-2s2p^4$	193	14
320.814	20						14
332.783	20	Al X	$^1S-^1P$	0–1	$2s^2-2s2p$	126	14
334.178	50	Fe XIV	$^2P-^2D$	$\frac{1}{2}-\frac{3}{2}$	$3s^23p-3s3p^2$	96	14
335.410	60	Fe XVI	$^2S-^2P$	$\frac{1}{2}-\frac{1}{2}$	$3s-3p$	236	14
339.03		Mg VIII	$^2P_{1\frac{1}{2}}-^2S_{\frac{1}{2}}$		$2s^22p-2s^22p^2$	193	126
341.115	10	Fe XI	$^3P-^3P$	2–1	$3s^23p^4-3s3p^5$	96	14
345.127	20	Si IX	$^3P-^3P$	1–2	$2s^22p^2-2s2p^3$	91	14
345.75^d	10	Fe X	$^2P-^2S$	$1\frac{1}{2}-\frac{1}{2}$	$3s^23p^5-3s3p^6$	96	14
346.859^d	20	Fe XII	$^4S-^4P$	$\frac{3}{2}-\frac{1}{2}$	$3s^23p^3-3s3p^4$	96	14
347.417	40	Si X	$^2P-^2D$	$\frac{1}{2}-\frac{3}{2}$	$2s^22p-2s2p^2$	91	14
348.199	40						14
349.892	20	Si X	$^3P-^3D$	2–3	$2s^22p^2-2s2p^3$	91	14
352.115	30	Fe XII	$^4S-^4P$	$\frac{3}{2}-\frac{3}{2}$	$3s^23p^3-3s3p^4$	96	14
352.680	30	Fe XI	$^3P-^3P$	2–2	$3s^23p^4-3s3p^5$	96	14
353.838	30	Fe XIV	$^2P-^2D$	$\frac{3}{2}-\frac{5}{2}$	$3s^23p-3s3p^2$	96	14
		Ar XVI	$^2S-^2P$	$\frac{1}{2}-\frac{3}{2}$	$2s-2p$	59	
356.051^d	40	Si X	$^2P-^2D$	$\frac{3}{2}-\frac{5}{2}$	$2s^22p-2s2p^2$	91	14
356.59	5	Fe XIV	$^2P-^2D$	$\frac{3}{2}-\frac{3}{2}$	$3s^23p-3s3p^2$	96	14
359.73^d	10	Fe XIII	$^3P-^3D$	1–2	$3s^23p^2-3s3p^3$	96	14
360.796	60	Fe XVI	$^2S-^2P$	$\frac{1}{2}-\frac{1}{2}$	$3s-3p$	236	14
364.477	40	Fe XII	$^4S-^4P$	$\frac{3}{2}-\frac{5}{2}$	$3s^23p^3-3s3p^4$	96	14
365.24	5	Mg VII	$^3P-^3P$	1–1	$2s^22p^2-2s2p^3$	193	14
367.64^d	10	Mg VII	$^3P-^3P$	2–2	$2s^23p^2-2s2p^3$	193	14
368.071	50	Mg IX	$^1S-^1P$	0–1	$2s^2-2s2p$	193	14
		Fe XIII	$^3P-^3D$	2–3	$3s^23p^2-3s3p^3$	96	
384.6	5	Mn XV	$^2S-^2P$	$\frac{1}{2}-\frac{1}{2}$	$3s-3p$	96	59
389.1	6	Ar XVI	$^2S-^2P$	$\frac{1}{2}-\frac{1}{2}$	$2s-2p$	59	59
		Cr XIV	$^2S-^2P$	$\frac{1}{2}-\frac{3}{2}$	$3s-3p$	96	59
401.7	9	Ne VI	$^2P-^2P$	$\frac{3}{2}-\frac{3}{2}$	$2s^22p-2s2p^2$	193	59
411.6	5	Na VIII	$^1S-^1P$	0–1	$2s^2-2s2p$	193	59
		Cr XIV	$^2S-^2P$	$\frac{1}{2}-\frac{1}{2}$	$3s-3p$	96	
417.24	14	Fe XV	$^1S-^3P$	0–1	$3s^2-3s3p$	48, 49	48, 49
417.71	14	S XIV	$^2S-^2P$		$2s-2p$	91	48, 49
430.5	4	Mg VIII	$^2P-^2D$	$\frac{1}{2}-\frac{3}{2}$	$2s^22p-2s2p^2$	193	59
434.94		Mg VII	$^3P-^3D$	2–3	$2s^22p^2-2s2p^3$		
436.1	6	Ne VI	$^2P-^2P$		$2s^22p-2s2p^2$	193	
436.73		Mg VIII	$^2P-^2D$	$\frac{3}{2}-\frac{5}{2}$	$2s^22p-2s2p^2$	193	
445.7	5	S XIV	$^2S-^2P$	$\frac{1}{2}-\frac{1}{2}$	$2s-2p$	91	59
459.5	1	C III	$^3P-^3D$		$2s2p-2s3d$	193	59
465.22	12	Ne VII	$^1S-^1P$	0–1	$2s^2-2s2p$	103	128
482.1	3	Fe XV	$^1P-^1D$	1–2	$3s3p-3p^2$	96	
		Ne V	$^3P-^3D$		$2s^22p^3-2s2p^3$	193	59
499.35	33	Si XII	$^2S-^2P$	$\frac{1}{2}-\frac{3}{2}$	$2s-2p$	91	128

(*continued*)

TABLE VI—continued

Wavelength of solar line	Intensity	Ion	Term	J–J	Configuration	Laboratory classification otherwise theoretical (Ref.)	Listed solar wavelength (Ref.)
515.6	1	He I	^1S–^1P	0–1	$1s^2$–$1s5p$	193	59
520.62	17	Si XII	^2S–^2P	$\frac{1}{2}$–$\frac{1}{2}$	$2s$–$2p$	91	128
522.2	2	He I	^1S–^1P	0–1	$1s^2$–$1s4p$	193	
525.8	1	O III	^1D–^1P	2–1	$2s^22p^2$–$2s2p^3$	193	59
537.00	5	He I	^1S–^1P	0–1	$1s^2$–$1s3p$	193	128
550.0	2	Al XI	^2S–^2P	$\frac{1}{2}$–$\frac{3}{2}$	$2s$–$2p$	91	59
554.04	6	O IV	^2P–^2P	$\frac{1}{2}$–$\frac{1}{2}$	$2s^22p$–$2s2p^2$	193	128
554.54	6	O IV	^2P–^2P	$\frac{3}{2}$–$\frac{3}{2}$	$2s^22p$–$2s2p^2$	193	128
562.8	2	Ne VI	^2P–^2D	$\frac{3}{2}$–$\frac{5}{2}$	$2s^22p$–$2s2p^2$	193	59
568.5	1	Al XI	^2S–^2P	$\frac{1}{2}$–$\frac{1}{2}$	$2s$–$2p$	91	59
572.3	1	Ne V	^3P–^3D		$2s^22p^2$–$2s2p^3$	193	59
580.85	1	Si XI	^1S–^3P	0–1	$2s^2$–$2s2p$	31	31
584.33	43	He I	^1S–^1P	0–1	$1s^2$–$1s2p$	193	31
599.69	2	O III					
609.82	23	Mg X	^2S–^2P	$\frac{1}{2}$–$\frac{3}{2}$	$2s$–$2p$	91	31
616.6	3	O II					
624.91	12	Mg X	^2S–^2P	$\frac{1}{2}$–$\frac{1}{2}$	$2s$–$2p$	91	31
629.74	26	O V	^1S–^1P	0–1	$2s^2$–$2s2p$	193	31
639.2	1	Ca VII	^3P–^3D	2–3	$3s^23p^2$–$3s3p^3$	194	59
657.3	1	S IV	^2P–^2D	$\frac{1}{2}$–$\frac{3}{2}$	$3s^23p$–$3s^23d$	193	59
661.4	1	S IV	^2P–^2D	$\frac{3}{2}$–$\frac{5}{2}$	$3s^23p$–$3s^23d$	193	59
681.7	1	Na IX	^2S–^2P	$\frac{1}{2}$–$\frac{3}{2}$	$2s$–$2p$	194	59
685.7	1	N III	^2P–^2P		$2s^22p$–$2s2p^3$	193	59
694.2	1	Na IX	^2S–^2P	$\frac{1}{2}$–$\frac{1}{2}$	$2s$–$2p$	91	59
703.04	4	O III	^3P–^3P		$2s^22p^2$–$2s2p^4$	193	59
706.06	1	Mg IX	^1S–^3P	0–1	$2s^2$–$2s2p$	247	247
718.48	1	O II	^2D–^2D	$\frac{5}{2}$–$\frac{5}{2}$	$2s^22p^3$–$2s2p^4$	193	247
760.49	3	O V	^3P–^3P	2–2	$2s2p$–$2p^2$	193	247
765.17	4	N IV	^1S–^1P	0–1	$2s^2$–$2s2p$	193	31
770.42	10	Ne VIII	^2S–^2P	$\frac{1}{2}$–$\frac{3}{2}$	$2s$–$2p$	103	31
780.28	5	Ne VIII	^2S–^2P	$\frac{1}{2}$–$\frac{1}{2}$	$2s$–$2p$	103	31
786.46	1	S V	^1S–^1P	0–1	$3s^2$–$3s3p$	193	31
787.74	4	O IV	^2P–^2D	$\frac{1}{2}$–$\frac{3}{2}$	$2s^22p$–$2s2p^2$	193	31
790.19	6	O IV	^2P–^2D	$\frac{3}{2}$–$\frac{5}{2}$	$2s^22p$–$2s2p^2$	193	31

From 800 Å listing is restricted to fourth or higher ionization stages (for forbidden lines see Table II)

895.12	1	Ne VII	^1S–^3P	0–1	$2s2p$–$2p^2$	247	247
933.40	5	S VI	^2S–^2P	$\frac{1}{2}$–$\frac{3}{2}$	$3s$–$3p$	193	31
944.44	4	S VI	^2S–^2P	$\frac{1}{2}$–$\frac{1}{2}$	$3s$–$3p$	193	31
1031.92	50	O VI	^2S–^2P	$\frac{1}{2}$–$\frac{3}{2}$	$2s$–$2p$	193	31
1037.64	40	O VI	^2S–^2P	$\frac{1}{2}$–$\frac{1}{2}$	$2s$–$2p$	193	31
1062.54	4	S IV	^2P–^2D	$\frac{1}{2}$–$\frac{3}{2}$	$3s^23p$–$3s3p^2$	193	31
1072.98	10	S IV	^2P–^2D	$\frac{3}{2}$–$\frac{5}{2}$	$3s^23p$–$3s3p^2$	193	31

TABLE VI—continued

Wavelength of solar line	Intensity	Ion	Term	J–J	Configuration	Laboratory classification otherwise theoretical (Ref.)	Listed solar wavelength (Ref.)
1122.50	10	Si IV	^2P–^2D	$\frac{1}{2}$–$\frac{3}{2}$	3p–3d	193	31
1128.52	7	Si IV	^2P–^2D	$\frac{3}{2}$–$\frac{5}{2}$	3p–3d	193	31
1145.61	1	Ne V	^3P–^5S	2–2	$2s^22p^2$–$2s2p^3$	247	247
1204.30	4	S V	^1S–^3P	0–1	3s–3p	31	31
1218.31	30	O V	^1S–^3P	0–1	$2s^2$–$2s2p$	31	31
1238.81	40	N V	^2S–^2P	$\frac{1}{2}$–$\frac{3}{2}$	2s–2p	193	31
1242.02	20	N V	^2S–^2P	$\frac{1}{2}$–$\frac{1}{2}$	2s–2p	193	31
1371.26	15	O V	^1P–^1D	1–2	$2s2p$–$2p^2$	193	31
1393.77	150	Si IV	^2S–^2P	$\frac{1}{2}$–$\frac{3}{2}$	3s–3p	193	31
1397.19	5	O IV	^2P–^4P	$\frac{1}{2}$–$\frac{3}{2}$	$2s^22p$–$2s2p^2$	193	31
1399.77	20	O IV	^2P–^4P	$\frac{1}{2}$–$\frac{1}{2}$	$2s^22p$–$2s2p^2$	193	31
1401.13	100	O IV	^2P–^4P	$\frac{3}{2}$–$\frac{5}{2}$	$2s^22p$–$2s2p^2$	193	31
1402.75	150	Si IV	^2S–^2P	$\frac{1}{2}$–$\frac{1}{2}$	3s–3p	193	31
1404.77	70	O IV	^2P–^4P	$\frac{3}{2}$–$\frac{3}{2}$	$2s^22p$–$2s2p^2$	193	31
1406.00	30	S IV	^2P–^4P	$\frac{5}{2}$–$\frac{5}{2}$	$3s3p$–$3p^2$	193	31
1407.40	20	O IV	^2P–^4P	$\frac{3}{2}$–$\frac{1}{2}$	$2s^22p$–$2s2p^2$	193	31
1416.94	10	S IV	^2P–^4P	$\frac{3}{2}$–$\frac{5}{2}$	$3s3p$–$3p^2$	193	31
1486.52	70	N IV	^1S–^3P	0–1	$2s^2$–$2s2p$	31	31
1548.20	400	C IV	^2S–^2P	$\frac{1}{2}$–$\frac{3}{2}$	2s–2p	193	31
1550.77	400	C IV	^2S–^2P	$\frac{1}{2}$–$\frac{1}{2}$	2s–2p	193	31

[a] Reported in solar flare spectra.
[b] Based on calculated wavelength.
[c] Identified in this publication.
[d] Accuracy uncertain.

from a discrete area (10 arcmin × 20 arcsec) of the solar disc and also from a layer approximately 20 arcsec deep centered 20 arcsec above the solar limb. The instruments were calibrated photometrically and covered the wavelength range from 150 to 870 Å (Firth et al., 1974). Improved data are needed in the 60–100 Å wavelength region. Data are available from the NRL flights of February 1, 1966 when the Sun was quiet and of September 20, 1963 when the Sun was active and of AFCRL (Manson, 1972). In the latter case, the recording was photoelectric. An important new photometric contribution between 50 and 300 Å is by Malinovsky and Héroux (1973). Their photoelectric grazing incidence solar spectrum, of excellent quality, carries an intensity calibration with an accuracy of 20% and wavelength accuracy of 0.06 Å. Sections of their spectrum between 55 and 100 Å and 220 and 300 Å are shown in Figs. 13 and 14. However, most of the best wavelengths are

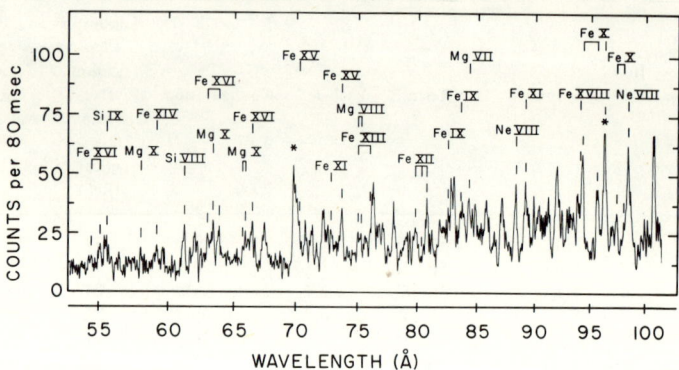

Fig. 13. Photometric spectrum of the Sun from 50 to 100 Å. From Malinovsky and Heroux (1973).

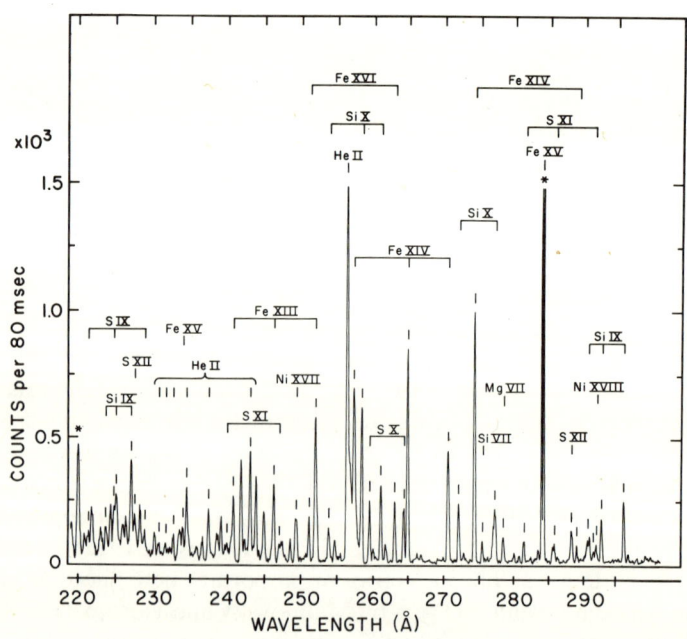

Fig. 14. Photometric spectrum of the Sun from 220 to 300 Å. From Malinovsky and Heroux (1973).

given by Behring et al. (1972). The prominent feature in the solar spectra between 50 and 300 Å is the group of intense Fe IX to Fe XIV lines between 170 and 220 Å which dominate the entire spectrum (see Fig. 12). They belong to the $3p^n$–$3p^{n-1}$ 3d transition array and were classified at the ARU, Culham as a result of laboratory experiments (see Section V,B).

Records of outstanding value in the 300–400 Å region were taken with slitless spectroheliographs on September 22, 1968 and November 4, 1969 by NRL (Widing et al., 1971) which produce spectrally dispersed stigmatic images of the Sun. These instruments contained platinum-coated concave gratings in Wadsworth configurations. In this mounting the instruments are not sensitive to the second order of the intense iron lines between 170 and 200 Å which would otherwise obscure spectra in the 340 to 400-Å-range. On these spectra the relative intensity of the line emission from solar limb, disc, active region, or flare provided information regarding the ionization stage of the emitting ion. Lines from the $3s^2 3p^n$–$3s\ 3p^{n+1}$ transitions in Fe X to Fe XIV were identified in the spectroheliograms; these lines were classified (Fawcett, 1971d) in the laboratory, see Section V,B. Purcell and Widing (1972) launched another spectroheliograph during a solar flare and recorded spectra of Ar XII–Ar XVI, Ca XIV–Ca XVIII, and Fe XXIV due to the $2s^2 2p^n$–$2s 2p^{n+1}$ transitions discussed in Section V,A. Lines from these transitions emitted from a solar active region were recorded by the Harvard scanning spectrometer on OSO 6 (Dupree et al., 1973). These data provide the wavelengths between 380 and 700 Å listed in Table VI to an accuracy of one decimal place.

Wavelengths measured to an accuracy of ± 0.05 Å are listed by Burton and Ridgeley (1970) for wavelengths greater than 500 Å. They were recorded with a normal incidence spectrograph containing a 2400 line/mm grating flown on a Skylark rocket by the ARU, Culham. Servo-controlled alignment systems stabilized an image of the Sun with an accuracy of a few seconds of arc relative to the spectrograph slit which could be set on the limb or disc. The slit width corresponded to 3000 km on the solar surface. Other normal incidence records obtained by Culham and NRL will not be reviewed here.

C. Eclipse Solar Ultraviolet Spectra

Perhaps the most spectacular solar spectra acquired recently were taken during the 1970 solar eclipse (Gabriel et al., 1971) in a collaborative experiment carried out by the ARU, Culham; Harvard College Observatory, Imperial College, London, and York University, Canada. A slitless Wadsworth grating spectrograph mounted on an Aerobee rocket recorded a series of 35 exposures from before second contact to after mid-totality. These exposures were of dispersed stigmatic spectra of excellent quality covering wavelengths from 1052 to 2162 Å. A 1-meter 1000 line/mm grating was employed. This experiment may rate as the most elaborate and successful solar rocket experiment to date. It was necessary to fly the rocket on a predetermined trajectory for optimum viewing of the eclipse, operate the camera at the correct times for a sequence of exposures, engineer and operate a complex sea recovery system, and maintain the pointing control of the rocket using a solar position error detector linked to an inertial reference frame so that a pointing accuracy was achieved consistent with a spatial resolution that permitted wavelength measurements accurate to ± 0.2 Å. The stigmatic spectra retrieved contained a wealth of information. An intensity calibration of the whole instrument allowed the spatial emission of spectral lines to be compared in active and nonactive areas. The spatial study was greatly strengthened because of the movement of the Moon relative to the Sun between different exposures. The most important feature on the spectra as far as classification of highly ionized ions is concerned is the appearance of coronal lines with their characteristic spatial distribution. Several of the unidentified ones turned out to be coronal forbidden lines (see Sections V,A and B) like those identified by Edlén (1942) in the visible spectrum. Burton et al. (1967) had identified two lines at 1242.2 and 1349.6 Å as Fe XII forbidden lines. The remaining forbidden lines were classified (see Table VII) by

TABLE VII

Coronal Lines in the EUV Spectrum[a]

λ obs[b] (Å)	Ion	Transition	T_e (10^6 °K)	λ obs (Å)	Ion	Transition	T_e (10^6 °K)
1190.2*	Mg VII	$^3P_1$–$^1S_0$	0.69	1715.3	S IX	$^3P_2$–$^1D_2$	1.2
1197.1				1826.0	S XI	$^3P_2$–$^1D_2$	2.0
1213.0	Fe XIII	$^3P_1$–$^1S_0$	1.85	1847.1			$\approx (2.0)$
1242.2	Fe XII	$^4S_{3/2}$–$^2P_{3/2}$	1.66	1866.9	Ni XIV	$^4S_{3/2}$–$^2D_{5/2}$	2.24
1349.6	Fe XII	$^4S_{3/2}$–$^2P_{1/2}$	1.66	1918.3	Fe X?		$\approx (1.2)$
1409.1*			(1.6)	1985.0	Si IX	$^3P_1$–$^1D_2$	1.20
1428.7			$\geq (1.2)$	2000.3[c]			$\approx (2.0)$
1446.0	Si VIII	$^4S_{3/2}$–$^2D_{3/2}$	0.93	2042.2[c]	Fe IX?	$^3P_2$–$^1F_3$	$\approx (0.9)$
1463.6			$\approx (0.9)$	2085.7[c]	Ni XV	$^3P_1$–$^1D_2$	2.5
1467.0	Fe XI	$^3P_1$–$^1S_0$	1.45	2126.0*[c]	Ni XIII	$^3P_2$–$^1D_2$	2.0
1582.0			$\approx (1.2)$	2147.4*[c]	Si VII	$^3P_2$–$^1D_2$	0.69
1603.2	Al VII?	$^4S_{3/2}$–$^2D_{3/2}$	$\approx (0.7)$	2149.5*[c]	Si IX	$^3P_2$–$^1D_2$	1.2
1614.6	S XI	$^3P_1$–$^1D_2$	2.0	2169.7*[c]	Fe XII	$^4S_{3/2}$–$^2D_{5/2}$	1.66
1624.0	O VII	$^3S_1$–$^3P_2$	(2.0)	2185.1*[c]	Ni XIV	$^4S_{3/2}$–$^2D_{3/2}$	2.24

[a] From Jordan (1971a); see also Gabriel et al. (1971).
[b] Wavelength accuracy is 0.2 Å, except those marked * when it is 1 Å.
[c] Air wavelengths above 2000 Å.

Gabriel et al. (1971) and Jordan (1971a), based on their predicted wavelengths and a study of their line intensities in quiet and active solar regions. The classification is also independently confirmed by laboratory observations (Fawcett, 1971d) (see Sections V,A and B) and by the isoelectronic extrapolations of Svensson (1971a).

D. Solar Spectra between 1 and 25 Å

Crystal scanning spectrometers have recorded the solar spectrum between 1 and 25 Å. Each crystal covers a limited spectral range so the available data are contained on different records. For instance, a LiF crystal is satisfactory for 0.6–4 Å, EDDT for 1–8.5 Å, KAP for 6–25 Å, and ADP for 3–8 Å. Spectra of solar flares have other emission lines than spectra of active regions, so a complete line list must include both types of spectra. The instruments have been flown in both satellites and rockets. The main groups responsible for the satellite experiments are the Goddard Space Flight Center for OSO 3 and OSO 5, Naval Research Laboratory for OSO 4 and OSO 6, Aerospace for OV1 10 and OV1 17 and the Lebedev Institute for Intercosmos 4. Rocket experiments were conducted by Lockheed Corporation and Leicester University. Table V, which is based on a tabulation by Walker (1972), lists the flights. Observed wavelengths and classifications are included in Table VI. For this spectral region a number of tentatively identified lines are omitted and no attempt has been made to include the wavelengths of the satellite lines to the long wavelength side of the helium-like lines. Most of the wavelengths are extracted from observations during the following flights. For the wavelength region between 9 and 15 Å the flights organized by Leicester University on November 30, 1971 and of OV1 10 on February 23, 1967 gained active region spectra and OSO III on March 22, 1967 recorded flare spectra. For wavelengths between 5.0 and 8.0 Å OV1 17 on March 20, 1969 recorded active region spectra. Below 6.5 Å spectra were obtained by OSO 6 on November 16, 1970 and below 2 Å by Intercosmos 4. Some of these records are reproduced in Figs. 5–9. It is apparent that the active region spectra show helium-like lines of solar abundant elements lighter than silicon and also lines of Fe XVII and Fe XVIII, whereas the flare spectra have helium-like lines of elements as heavy as nickel and show strong spectra of Fe XXVI to Fe XVII. Of special note are the strong flare lines of Fe XXIV to Fe XVII in Fig. 5 obtained by Neupert et al. (1973) from the OSO V satellite of NASA; their classification is discussed in Section 5,C. The satellite lines which appear on the long wavelength side of the helium-like lines are discussed in Section 5,D. The highest stages of ionization seen on these spectra imply temperatures in solar flares of over 20×10^6 °K.

VII. Discussion

There are projects for mounting in orbiting satellites spectrographs capable of recording well-resolved stellar spectra. Although in most stars the ultraviolet spectrum is dominated by absorption lines, there are some objects such as novae, planetary nebulae, and a few of the O-type stars which emit spectra that show dominant lines of atoms or ions not abundant in the Sun and whose highly ionized spectra are still unclassified. Attempts to interpret such spectra may provide a motive for the investigation of more complex configurations. This will necessitate the gathering of well-resolved laboratory spectra. Large 10-meter radius grazing and normal incidence spectrographs exist with this capability. The methods for distinguishing between ionization stages can be applied such as those dealt with in Section III,B. Some of the new light sources reported in Section II, such as the thetatron, laser-produced plasmas, and the beam foil, have not yet been fully exploited. Complex spectra whose spectrograms contain vast numbers of lines may require the application of the automatic measurement techniques outlined in Section III,A. Theoretical spectra computed with the sophisticated techniques mentioned in Section IV will become more essential for future investigations. Computer techniques can also be applied but with great caution to correlating observed and theoretical spectra using statistical methods in cases where the time involved would otherwise be prohibitive. The computer also provides the answer to data storage. It is easier, however, to make meaningful deductions from the line intensities of simpler spectra about the parameters of the plasmas which emit them. They are therefore of more use to astrophysics. It would therefore seem more sensible to concentrate on the remaining problems, especially the fine detail, surrounding these simpler spectra and to place priority on configurations involving forbidden lines which also have astrophysical applications.

Higher resolution solar spectra are required between 40 and 160 Å and 300 and 800 Å. This requires spectrographs with large radius, high dispersion gratings, and long exposures. Intensity calibrated instruments are also desirable. Space shuttle experiments would permit both the long exposures and film recovery. Although the present program does not include such experiments, two of the experiments which operated on the NASA Apollo Telescope Mount SKYLAB (Reeves *et al.*, 1972) were designed to study highly ionized solar spectra. The first is the Harvard Ultraviolet Spectrometer–Spectroheliometer. It obtains solar photoelectric observations in the spectral region 280–1350 Å with a spectral resolution of 1.6 Å and a spatial resolution of 5 arcsec which corresponds approximately to 3500 km^2 on the solar surface. Six other channels with spectral resolutions between 4 and 8.3 Å can be set to monitor spectral lines. Movement of an off-axis parabolic mirror which images the Sun on the entrance slit of the half-meter concave grating spectrometer permits the acquisition of a two-dimensional raster of 2×10^5 km^2 on the solar surface every 5 min. It can be controled manually or remotely. The second manned experiment is the NRL XUV spectroheliograph covering the wavelength region between 150 and 630 Å. The slitless 200-cm concave grating spectrograph produces stigmatic spectrally dispersed images on film of the Sun with a spectral resolution of up to 0.13 Å and a spatial resolution of 2 to 10 arcsec depending on the wavelength.

A significant point to make in concluding this paper is that the information it reports is required to interpret the observations of the multimillion pound SKYLAB experiment and that much of this information is acquired from laboratory spectral classification projects which can operate on modest budgets.

ACKNOWLEDGMENTS

The author thanks Professor B. Edlén and Dr. A. H. Gabriel for reading and checking the text, Dr. R. D. Cowan and Professor C. Froese-Fischer for so generously contributing to the section dealing with theoretical calculations, and Dr. E. Meinders for useful comments on measurement techniques.

REFERENCES[1]

1. Acton, L. W., Catura, R. C., Meyerott, H. J., and Culhane, J. L. (1971). *Nature (London)* **223**, 75.
2. Alexander, E., Feldman, U., and Fraenkel, B. S. (1965). *J. Opt. Soc. Amer.* **55**, 650 (Cl).
3. Alexander, E., Feldman, U., Fraenkel, B. S., and Hoory, S. (1966). *J. Opt. Soc. Amer.* **56**, 651 (Cl).
4. Alexander, E., Even-Zohar, M., Fraenkel, B. S., and Goldsmith, S. (1971). *J. Opt. Soc. Amer.* **61**, 508 (Cl).
5. Argo, H. V., Bergey, J. A., Evans, W. D., and Singer, S. (1968). *Sol. Phys.* **5**, 551.
6. Artru, M. C., and Kaufman, V. (1972). *J. Opt. Soc. Amer.* **62**, 949 (Cl).
7. Austin, W. E., Purcell, J. D., and Tousey, R. (1962). *Astron. J.* **67**, 110.
8. Austin, W. E., Purcell, J. D., Tousey, R., and Widing, K. G. (1966). *Astrophys. J.* **145**, 373.
9. Austin, W. E., Purcell, J. D., Snider, C. B., Tousey, R., and Widing, K. G. (1967). *Space Res.* **7**, 1252.
9a. Bashkin, S., ed. (1973). Proc. Int. Conf. Beam Foil Spectrosc., Tucson, 1972, *Nucl. Instrum. Methods* **110**.
10. Basov, N. G., Boiko, V. A., Voinov, Y. P., Kononov, E. Y., Mandelshtam, S. L., and Sklizkov, G. V. (1967). *JETP Lett.* **5**, 141 (Cl).
11. Batstone, R. M., Evans, K., Parkinson, J. H., and Pounds, K. A. (1970). *Sol. Phys.* **13**, 389.
12. Baum, W. A., Johnson, F. S., Oberly, J. J., Rockwood, C. W., Strain, C. V., and Tousey, R. (1946). *Phys. Rev.* **70**, 781.

[1] The numbers on the left-hand side refer to Tables III–IV (Cl) after a reference indicates a paper on classification of the spectra of highly ionized atoms contributed since 1960.

13. Behring, W. E., Neupert, W. M., and Lindsey, J. C. (1963). *Space Res.* **3**, 814.
14. Behring, W. E., Cohen, L., and Feldman, U. (1972). *Astrophys. J.* **175**, 493.
15. Blake, R. L., and House, L. L. (1971). *Astrophys. J.* **166**, 423.
16. Blake, R. L., Chubb, R. A., Friedman, H., and Unzicker, A. E. (1964). *Science* **146**, 1037.
17. Blake, R. L., Chubb, T. A., Friedman, H., and Unzicker, A. E. (1965a). *Ann. Astrophys.* **28**, 583.
18. Blake, R. L., Chubb, T. A., Friedman, H., and Unzicker, A. E. (1965b). *Astrophys. J.* **142**, 1.
19. Bockasten, K. (1955). *Ark. Fys.* **9**, 457 (Cl).
20. Bockasten, K. (1956). *Ark. Fys.* **10**, 567 (Cl).
21. Bockasten, K., and Johansson, K. B. (1968). *Ark. Fys.* **38**, 563 (Cl).
22. Bockasten, K., Hallin, R., and Hughes, T. P. (1963). *Proc. Phys. Soc., London* **81**, 522 (Cl).
23. Bockasten, K., Hallin, R., Johansson, K. B., and Tsui, P. (1964). *Phys. Lett.* **8**, 181 (Cl).
24. Boiko, V. A., Voinov, Yu P., Gribkov, V. A., and Sklizkov, G. V. (1970). *Opt. Spectrosc. (USSR)* **29**, 545 (Cl).
25. Boland, B. C., Irons, F. E., and McWhirter, R. W. P. (1968). *Proc. Phys. Soc., London (At. Mol. Phys.)* [2] **1**, 1180.
26. Bowen, I. S. (1960). *Astrophys. J.* **132**, 1 (Cl).
27. Bowen, I. S., and Edlén, B. (1939). *Nature (London)* **143**, 374 (Cl).
28. Broadman, W. J., and Billings, D. E. (1969). *Astrophys. J.* **156**, 731 (Cl).
29. Bromander, J. (1969). *Ark. Fys.* **40**, 257 (Cl).
30. Burgess, D. D., Fawcett, B. C., and Peacock, N. J. (1967). *Proc. Phys. Soc., London* **92**, 805.
31. Burton, W. M., and Ridgeley, A. (1970). *Sol. Phys.* **14**, 3.
32. Burton, W. M., and Wilson, R. (1965). *Nature (London)* **267**, 61.
33. Burton, W. M., Ridgeley, A., and Wilson, R. (1967). *Mon. Notic. Roy. Astron. Soc.* **135**, 207.
34. Cady, W. M. (1933). *Phys. Rev.* **43**, 324 (Cl).
35. Chaghtai, M. S. Z. (1970a). *Phys. Scripta* **1**, 31 (Cl).
36. Chaghtai, M. S. Z. (1970b). *Phys. Scripta* **1**, 104 (Cl).
37. Chaghtai, M. S. Z., and Ali, Z. (1971). *J. Opt. Soc. Amer.* **61**, 1264. (Cl).
38. Chaghtai, M. S. Z., Ali, Z., and Yadava, H. L. (1971). *Indian J. Phys.* **45**, 28 (Cl).
39. Chaghtai, M. S. Z., Ali, Z., and Khatoon, S. (1972). *Proc. Phys. Soc., London (At. Mol. Phys.)* [2] **5**, 1 (Cl).
40. Chaghtai, M. S. Z., Ali, Z., and Singh, S. P. (1974). *Z. Phys.* (in press) (Cl).
41. Cohen, L., Feldman, U., and Kastner, S. O. (1968a). *J. Opt. Soc. Amer.* **58**, 331 (Cl).
42. Cohen, L., Feldman, U., Swartz, M., and Underwood, J. H. (1968b). *J. Opt. Soc. Amer.* **58**, 843.
43. Condon, E. U., and Shortley, G. H. (1935). "The Theory of Atomic Spectra." Cambridge Univ. Press, London and New York.
44. Cowan, R. D. (1967a). *Phys. Rev.* **163**, 54.
45. Cowan, R. D. (1967b). *Astrophys. J.* **147**, 377.
46. Cowan, R. D. (1968). *J. Opt. Soc. Amer.* **58**, 808.
47. Cowan, R. D., and Peacock, N. J. (1965). *Astrophys. J.* **142**, 390.
47. Cowan, R. D., Peacock, N. J. (1966). *Astrophys. J.* **143**, 283.
48. Cowan, R. D., and Widing, K. G. (1972). *Bull. Amer. Astron. Soc.* **4**, 3 (Cl).
49. Cowan, R. D., and Widing, K. G. (1973). *Astrophys. J.* **180**, 285.
50. Deutschman, W. A., and House, L. L. (1966). *Astrophys. J.* **144**, 435 (Cl).
51. Deutschman, W. A., and House, L. L. (1967). *Astrophys. J.* **149**, 451 (Cl).
52. Doschek, G. A. (1972). *Space Sci. Rev.* **13**, 765.
53. Doschek, G. A., and Meekins, J. F. (1971). *Solar Phys.* **13**, 220.
54. Doschek, G. A., Meekins, J. F., Kreplin, R. W., Chubb, T. A., and Friedman, H. (1971). *Astrophys. J.* **164**, 165; **170**, 573.
55. Doschek, G. A., Meekins, J. F., and Cowan, R. D. (197). *Astrophys. J.* **177**, 261–269.

56. Doschek, G. A., Feldman, U., and Cohen, L. (1973a). *J. Opt. Soc. Amer.* **63**, 1463.
57. Doscheck, G. A., Meekins, J. F., and Cowan, R. D. (1973b). *Sol. Phys.* **29**, 125.
58. Druetta, M., Datla, R. U., and Kunze, H. J. (1972). *Astrophys. J.* **174**, 215.
59. Dupree, A. K., Huber, M. C. E., Noyes, R. W., Parkinson, W. H., Reeves, E. M., and Withbroe, G. L. (1973). *Astrophys. J.* **182**, 321.
60. Edlén, B. (1934). *Nova Acta Regiae Soc. Sci. Upsal.* [4] **9**, No. 6 (Cl).
61. Edlén, B. (1942). *Z. Astrophys.* **20**, 30 (Cl).
62. Edlén, B. (1952). *Ark. Fys.* **4**, 441 (Cl).
63. Edlén, B. (1963). *Rep. Progr. Phys.* **26**, 182.
64. Edlén, B. (1964). In "Handbuch der Physik" (S. Flügge, ed.), Vol. 27, p. 80. Springer-Verlag, Berlin and New York.
65. Edlén, B. (1969a). *Mon. Notic. Roy. Astron. Soc.* **144**, 391 (Cl).
66. Edlén, B. (1969b). *Sol. Phys.* **9**, 439.
67. Edlén, B. (1972a). *Sol. Phys.* **24**, 356.
68. Edlén, B. (1972b). *Opt. Pura Apl.* **5**, 101.
69. Edlén, B. (1973). *Phys. Scripta* **7**, 93.
70. Edlén, B., and Boden, E. (1974). *Phys. Scripta* (to be published).
71. Edlén, B., and Lofstrand, B. (1970). *Proc. Phys. Soc., London.* (*At. Mol. Phys.*) [2] **3**, 1380 (Cl).
72. Edlén, B., and Swensson, J. W. (1974). *Phys. Scripta* (to be published).
73. Edlén, B., and Tyren, F. (1939). *Nature (London)* **143**, 940 (Cl).
74. Edlén, B., Palenius, H., Bockasten, K., Hallin, R., and Bromander, J. (1969). *Sol. Phys.* **9**, 432 (Cl).
75. Ehler, A. W., and Weissler, G. L. (1966). *Appl. Phys. Lett.* **8**, 89.
76. Eidelsberg, M. (1972). *Phys. Proc. Soc., London* (*At. Mol. Phys.*) [2] **5**, 1031 (Cl).
77. Ekberg, J. O. (1971). *Phys. Scripta* **4**, 101 (Cl).
78. Ekberg, J. O. (1973a). *Phys. Scripta* **7**, 55 (Cl).
79. Ekberg, J. O. (1973b). *Phys. Scripta* **7**, 59 (Cl).
80. Ekberg, J. O. (1973c). *Phys. Scripta* **8**, 35 (Cl).
81. Ekberg, J. O., and Svensson, L. A. (1970). *Phys. Scripta* **2**, 283 (Cl).
82. Ekberg, J. O., Hanson, J. E., and Reader, J. (1972). *J. Opt. Soc. Amer.* **62**, 1134, 1139, and 1143 (Cl).
83. Elton, R. C., and Lee, T. N. (1972). *Space Sci. Rev.* **13**, 747.
84. Engelhardt, W., and Sommer, J. (1971). *Astrophys. J.* **167**, 201.
85. Evans, D. W., Argo, H. V., Bergey, J. A., Henke, B. L., and Montgomery, M. D. (1966). *Abstr. Conf. U.V. X-Ray Spec.*, 1966 p. 38.
86. Evans, K., and Pounds, K. A. (1968). *Astrophys. J.* **152**, 319.
87. Evans, K., and Pounds, K. A., and Culhane, J. L. (1967). *Nature (London)* **214**, 41.
88. Even-Zohar, M., and Fraenkel, B. S. (1968). *J. Opt. Soc. Amer.* **58**, 1420 (Cl).
89. Even-Zohar, M., and Fraenkel, B. S. (1972). *Proc. Phys. Soc., London* (*At. Mol. Phys.*) [2] **5**, 1596 (Cl).
90. Fawcett, B. C. (1965). *Proc. Phys. Soc., London* **86**, 1087 (Cl).
91. Fawcett, B. C. (1970a). *Proc. Phys. Soc., London* (*At. Mol. Phys.*) [2] **3**, 1152 (Cl).
92. Fawcett, B. C. (1970b). *Proc. Phys. Soc., London* (*At. Mol. Phys.*) [2] **3**, 1732 (Cl).
93. Fawcett, B. C. (1971a). *Proc. Phys. Soc., London* (*At. Mol. Phys.*) [2] **4**, 981 (Cl).
94. Fawcett, B. C. (1971b). *Proc. Phys. Soc., London* (*At. Mol. Phys.*) [2] **4**, 1115 (Cl).
95. Fawcett, B. C. (1971c). Astrophy. Res. Unit, Culham, Rep. HM Stationery Office, London (Cl).
96. Fawcett, B. C. (1971d). *Proc. Phys. Soc., London* (*At. Mol. Phys.*) [2] **4**, 1577 (Cl).
97. Fawcett, B. C. (1971e). Astrophys. Res. Unit, Culham, Rep. HM Stationery Office, London (Cl).

98. Fawcett, B. C., and Gabriel, A. H. (1964). *Proc. Phys. Soc., London* **84**, 1038.
99. Fawcett, B. C., and Gabriel, A. H. (1966). *Proc. Phys. Soc., London* **88**, 262 (Cl).
100. Fawcett, B. C., and Hayes, R. W. (1972). *Proc. Phys. Soc., London (At. Mol. Phys.)* [2] **5**, 366 (Cl).
101. Fawcett, B. C., and Hayes, R. W. (1973). *Phys. Scripta* **8**, 244.
102. Fawcett, B. C., and Irons, F. E. (1966). *Proc. Phys. Soc., London* **89**, 1063 (Cl).
103. Fawcett, B. C., Jones, B. B., and Wilson, R. (1961). *Proc. Phys. Soc., London* **78**, 1223 (Cl).
104. Fawcett, B. C., Gabriel, A. H., Jones, B. B., and Peacock, N. J. (1964). *Proc. Phys. Soc., London* **84**, 257 (Cl).
105. Fawcett, B. C., Gabriel, A. H., Irons, F. E., Peacock, N. J., and Saunders, P. A. H. (1966). *Proc. Phys. Soc., London* **88**, 1051 (Cl).
106. Fawcett, B. C., Gabriel, A. H., and Saunders, P. A. H. (1967a). *Proc. Phys. Soc., London* **90**, 863 (Cl).
107. Fawcett, B. C., Burgess, D. D., and Peacock, N. J. (1967b). *Proc. Phys. Soc., London* **91**, 970 (Cl).
108. Fawcett, B. C., Peacock, N. J., and Cowan, R. D. (1968). *Proc. Phys. Soc., London (At. Mol. Phys.)* [2] **1**, 295–306 (Cl).
109. Fawcett, B. C., Hardcastle, R. A., and Tondello, G. (1970). *Proc. Phys. Soc., London (At. Mol. Phys.)* [2] **3**, 564 (Cl).
110. Fawcett, B. C., Gabriel, A. H., and Paget, T. M. (1971). *Proc. Phys. Soc., London (At. Mol. Phys.)* [2] **4**, 986 (Cl).
111. Fawcett, B. C., Cowan, R. D., Kononov, E. Y., and Hayes, R. W. (1972a). *Proc. Phys. Soc., London (At. Mol. Phys.)* [2] **5**, 1255 (Cl).
112. Fawcett, B. C., Cowan, R. D., and Hayes, R. W. (1972b). *Proc. Phys. Soc., London (At. Mol. Phys.)* [2] **5**, 2143 (Cl).
113. Fawcett, B. C., Cowan, R. D., and Hayes, R. W. (1973). *Sol. Phys.* **31**, 339.
114. Fawcett, B. C., Cowan, R. D., and Hayes, R. W. (1974a). *Astrophys. J.* **187**, p. 377.
115. Fawcett, B. C., Galanti, M., and Peacock, N. J. (1974b). *Proc. Phys. Soc., London (At. Mol. Phys.)* [2] **7**, 106 and 1149.
116. Feldman, U., and Cohen, L. (1967a). *Astrophys. J.* **149**, 265 (Cl).
117. Feldman, U., and Cohen, L. (1967b). *J. Opt. Soc. Amer.* **57**, 1128 (Cl).
118. Feldman, U., and Cohen, L. (1968). *Astrophys. J.* **151**, L55 (Cl).
119. Feldman, U., Fraenkel, B. S., and Hoory, S. (1965). *Astrophys. J.* **142**, 719 (Cl).
120. Feldman, U., Cohen, L., and Swartz, M. (1967a). *J. Opt. Soc. Amer.* **57**, 535 (Cl).
121. Feldman, U., Cohen, L., and Swartz, M. (1967b). *Astrophys. J.* **148**, 585 (Cl).
122. Feldman, U., Cohen, L., and Behring, W. (1970). *J. Opt. Soc. Amer.* **60**, 891 (Cl).
123. Feldman, U., Katz, L., Behring, W., and Cohen, L. (1971). *J. Opt. Soc. Amer.* **61**, 91 (Cl).
124. Feldman, U., Doschek, G. A., Nagel, D. J., Behring, W. E., and Cohen, L. (1973a). *Astrophys. J.* **183**, L43.
125. Feldman, U., Doschek, G. A., Cowan, R. D., and Cohen, L. (1973b). *J. Opt. Soc. Amer.* **63**, 1145.
126. Firth, R. S., Freeman, F. F., Gabriel, A. H., Jones, B. B., Jordan, C., Negus, C. R., Shenton, D. B., and Turner, R. F. (1974). *Mon. Notic. Roy. Astron. Soc.* **166**, 543.
127. Francisco, P. J., Goorvitch, V., and Goorvitch, J. (1972). *Astrophys. J.* **178**, 271.
128. Freeman, F. F., and Jones, B. B. (1970). *Sol. Phys.* **15**, 288.
129. Freeman, F. F., Gabriel, A. H., Jones, B. B., and Jordan, C. (1971). *Phil. Trans. Roy. Soc. London, Ser. A* **270**, 127.
130. Fritz, G., Kreplin, R. W., Meekins, J. F., Unziker, A. E., and Friedman, H. (1967). *Astrophys. J.* **148**, L133.
131. Froese-Fischer, C. (1963). *Can. J. Phys.* **41**, 1895.

132. Froese-Fischer, C. (1967). *Bull. Astron. Inst. Neth.* **19**, 86.
133. Froese-Fischer, C. (1970). *Proc. Phys. Soc., London (At. Mol. Phys.)* [2] **3**, 779.
134. Froese-Fischer, C. (1971). *Can. J. Phys.* **49**, 1205.
135. Froese-Fischer, C. (1972a). *Proc. Phys. Soc., London (At. Mol. Phys.)* [2] **5**, 1302.
136. Froese-Fischer, C. (1972b). *Comput. Phys.* **10**, 211.
137. Froese-Fischer, C. (1972c). *Comput. Phys. Commun.* (in press).
138. Furth, R., and Oliphant, W. D. (1948). *J. Sci. Instrum.* **25**, 289.
139. Gabriel, A. H. (1970). *Nucl. Instrum. & Methods* **4**, 157.
140. Gabriel, A. H. (1972). *Mon. Notic. Roy. Astron. Soc.* **160**, 90.
141. Gabriel, A. H., and Fawcett, B. C. (1965). *Nature (London)* **206**, 808 (Cl).
142. Gabriel, A. H., and Jordan, C. (1969). *Nature (London)* **221**, 947 (Cl).
143. Gabriel, A. H., and Paget, T. M. (1972). *Proc. Phys. Soc., London (At. Mol. Phys.)* [2] **5**, 673.
144. Gabriel, A. H., Niblett, G. B. F., and Peacock, N. J. (1962). *J. Quant. Spectrosc. & Radiat. Transfer* **2**, 491.
145. Gabriel, A. H., Fawcett, B. C., and Jordan, C. (1965). *Nature (London)* **206**, 390 (Cl).
146. Gabriel, A. H., Fawcett, B. C., and Jordan, C. (1966). *Proc. Phys. Soc., London* **87**, 825 (Cl).
147. Gabriel, A. H., Garton, W. R. S., Goldberg, L., Jones, T. J. L., Jordan, C., Morgan, F. J., Nicholls, R. W., Parkinson, W. H., Paxton, H. J. B., Reeves, E. M., Shenton, D. B., Speer, R. J., and Wilson, R. (1971). *Astrophys. J.* **169**, 595 (Cl).
148. Galanti, M., and Peacock, N. J. (1973). *Proc. Conf. Controlled Fusion Plasma Phys., 6th, 1973* Vol. 1, p. 427.
149. Garcia-Riquelme, O. (1968). *Physica* **40**, 27 (Cl).
150. Garstang, R. H. (1954). *Mon. Notic. Roy. Astron. Soc.* **114**, 118.
151. Garstang, R. H. (1958). *Mon. Notic. Roy. Astron. Soc.* **118**, 572.
152. Goldsmith, S. (1969a). *J. Opt. Soc. Amer.* **59**, 1678 (Cl).
153. Goldsmith, S. (1969b). *Proc. Phys. Soc., London (At. Mol. Phys.)* [2] **2**, 1075 (Cl).
154. Goldsmith, S., and Fraenkel, B. S. (1970). *Astrophys. J.* **161**, 317 (Cl).
155. Goldsmith, S., and Kaufman, A. S. (1963). *Proc. Phys. Soc., London* **81**, 544 (Cl).
156. Goldsmith, S., and Starkland, J. (1970) *Proc. Phys. Soc., London (At. Mol. Phys.)* [2] **3**, L141 (Cl).
157. Goldsmith, S., Feldman, U., and Cohen, L. (1971a). *J. Opt. Soc. Amer.* **61**, 615 (Cl).
158. Goldsmith, S., Feldman, U., Crooker, A., and Cohen, L. (1971b). *J. Opt. Soc. Amer.* **62**, 260 (Cl).
159. Goldsmith, S., Feldman, U., Oren, L., and Cohen, L. (1972). *Astrophys. J.* **174**, 209 (Cl).
160. Goldsmith, S., Oten, L., and Cohen, L. (1973). *J. Opt. Soc. Amer.* **63**, 352.
161. Grineva, Yu. I., Karev, V. I., Korneev, V. V., Krutov, V. V., Mandelshtam, S. L., Vainshtein, L. A., Vasiljev, B. N., and Zitnik, I. A. (1971). *COSPAR Conf. Proc., 14th, 1972*, p. 1553.
161a. Grineva, Y. I., Karev, V. I., Korneev, V. V., Krutov, V. V., Mandelshtam, S. L., Vainshtein, L. A., Vasiljev, B. N., and Zitnik, I. A. (1973). *Sol. Phys.* **29**, 441.
162. Gunnvald, P. (1962). *Ark. Fys.* **22**, 333.
163. Hagan, L., and Martin, W. C. (1972). *Nat. Bur. Stand.* (U.S.), *Spec. Publ.* **363**.
164. Hall, L. A., and Hinteregger, H. E. (1970). *J. Geophys. Res.* **75**, 6959.
165. Hall, L. A., Schweitzer, W., Heroux, L., and Hinteregger, H. E. (1965). *Astrophys. J.* **142**, 13.
166. Hall, L. A., Higgins, J. E., Chagon, C. W., and Hinteregger, H. E. (1969). *J. Geophys. Res.* **74**, L181.
167. Hallin, R. (1966a). *Ark. Fys.* **31**, 511 (Cl).
168. Hallin, R. (1966b). *Ark. Fys.* **32**, 201 (Cl).
169. Hermansdorfer, H. (1972). *J. Opt. Soc. Amer.* **62**, 1149 (Cl).
170. Herzberg, G. (1962). *Trans. Int. Astron. Union* **11A**, 97.

171. Hinteregger, H. E. (1960). *Astrophys. J.* **132**, 801.
172. Hinteregger, H. E. (1961). *J. Geophys. Res.* **66**, 2367.
173. Hinteregger, H. E., and Hall, L. A. (1969). *Sol. Phys.* **6**, 175.
174. Hinteregger, H. E., Hall, L. A., and Schweitzer, W. (1964). *Astrophys. J.* **140**, 319.
175. Hobby, M. G., and Peacock, N. J. (1973). Submitted for publication.
176. Hoekstra, R. (1967). *Appl. Opt.* **6**, 807.
177. Joekstra, R. (1969). Thesis, Amsterdam, University.
178. Hoory, S., Feldman, U., Goldsmith, S., Behring, W., and Cohen, L. (1970a). *J. Opt. Soc. Amer.* **60**, 1449 (Cl).
179. Hoory, S., Goldsmith, S., Fraenkel, B. S., and Feldman, U. (1970b). *Astrophys. J.* **160**, 781 (Cl).
180. Hoory, S., Feldman, U., Goldsmith, S., Behring, W., and Cohen, L. (1971). *J. Opt. Soc. Amer.* **61**, 504 (Cl).
181. Iglesias, L. (1968). *J. Res. Nat. Bur Stand., Sect. A* **72**, 295 (Cl).
182. Irons, F. E., McWhirter, R. W. P., and Peacock, N. J. (1972). *Proc. Phys. Soc., London (At. Mol. Phys.)* [2] **5**, 1975.
183. Irwin, D. J. G., Livingston, A. E., and Kernahan, J. A. (1972). *Proc. Int. Conf. Beam Foil Spectrosc., 3rd, 1973, Nucl. Instrum. Methods* **110**, 105.
184. Johnson, F. S., Purcell, J. D., and Tousey, R. (1951). *J. Geophys. Res.* **56**, 583.
185. Jones, B. B., Freeman, F. F., and Wilson, R. (1968). *Nature (London)* **219**, 252.
186. Jordan, C. (1965). *Commun. Univ. London Observ.* No. 68.
187. Jordan, C. (1969). *Mon. Notic. Roy. Astron. Soc.* **142**, 499.
188. Jordan, C. (1970). *Mon. Notic. Roy. Astron. Soc.* **148**, 17.
189. Jordan, C. (1971a). *Sol. Phys.* **21**, 381 (Cl).
190. Jordan, C. (1971b). *Highlights Astron.* p. 519.
191. Jordan, C. (1972). *Space Sci. Rev.* **13**, 595.
192. Kasyanov, Y. S., Kononov, E. Y., Korobkin, V. V., Koshelev, K. N., and Serov, R. V. (1972). *Opt. Spectrosc. (USSR)* **6**, 586.
193. Kelly, R. L. (1968). NRL Rep. No. 6648. US Govt. Printing Office, Washington, D.C.
194. Kelly, R. L., Palumbo, L. J. (1973). NRL Rep. No. 7599. U.S. Govt. Printing Office, Washington, DC.
195. Kononov, E. Y. (1966). *Opt. Spectrosc. (USSR)* **20**, 303 (Cl).
196. Kononov, E. Y. (1967). *Opt. Spectrosc. (USSR)* **23**, 170.
197. Kononov, E. Y. (1968). *Opt. Spectrosc. (USSR)* **24**, 827.
198. Kononov, E. Y. (1969). *Astron. Zh.* **46**, 340 (Cl).
199. Kononov, E. Y., and Basov, N. G. (1970). *Opt. Spectrosc. (USSR)* **29**, 216 (Cl).
200. Kononov, E. Y., and Koshelev, K. N. (1970). *Opt. Spectrosc. (USSR)* **29**, 115 (Cl).
201. Kononov, E. Y., Koshelev, K. N., and Ryabtsev, A. N. (1970). *Opt. Spectrosc. (USSR)* **30**, 534 (Cl).
202. Kunze, H. J. (1971). *Phys. Rev.* **4**, 111 (Cl).
203. Lee T. N. (1972). *Proc. APS Conf. High Plasmas, 2nd, 1972.* N.R.L. Rep. 2502, Washington, D.C.
204. Lee, T. N., and Elton, R. C. (1971). *Phys. Rev.* **3**, 865.
205. Lindeberg, S. (1972). UUIP-758, 759, and 760. Uppsala Univ. Inst. Phys.
206. Lindskog, J., Hallin, R., Marelius, A., Bromander, J. P., and Sjodin, R. (1972). *Proc. Int. Conf. Beam Foil Spectrosc., 3rd, 1973. Nucl. Instrum. Methods* **110**, 517.
207. Lowdin, P. O. (1955). *Phys. Rev.* **97**, 1474 and 1490.
208. Malinovsky, M., and Heroux, L. (1973). *Astrophys. J.* **181**, 1009.
209. Manson, J. (1967). *Astrophys. J.* **147**, 703.
210. Manson, J. (1968). *Astrophys. J.* **153**, L191.
211. Manson, J. (1972). *Sol. Phys.* **27**, 107.

212. Martin, W. C., Sugar, J., and Tech, J. L. (1972). *Phys. Rev.* **6**, 2022 (Cl).
213. Mather, J. W. (1966). *Plasma Phys. Contr. Nucl. Fusion Res., Proc. Conf., 2nd, 1965* Vol. 2, p. 389.
214. Meekins, J. F., and Doschek, G. A. (1970). *Sol. Phys.* **13**, 213.
215. Meekins, J. F., Kreplin, R. W., Chubb, T. A., and Friedman, H. (1968). *Science* **162**, 891.
216. Meekins, J. F., Doschek, G. A., Friedman, H., Chubb, T. A., and Kreplin, R. W. (1970). *Sol. Phys.* **13**, 198.
217. Michels, D. J., Tilford, S. G., and Quinn, J. W. (1971). *J. Opt. Soc. Amer.* **61**, 625 (Cl).
218. Moore, C. E. (1949). "Atomic Energy Levels," Nat. Bur. Stand. US Govt. Printing Office, Washington, D.C.
219. Moore, C. E. (1952). "Atomic Energy Levels," Nat. Bur. Stand. US Govt. Printing Office, Washington, D.C.
220. Moore, C. E. (1958). "Atomic Energy Levels," Nat. Bur. Stand. US Govt. Printing Office, Washington, D.C.
221. Moore, C. E. (1965–1971). "Selected Tables of Atomic Spectra NSRDS—NBS," Vol. 3, pp. 1–4. US Govt. Printing Office, Washington, D.C.
222. Moore, C. E. (1968–1969). *Nat. Bur. Stand. (U.S.), Spec. Publ.* **308**, 1–4.
223. Neupert, W. M. (1967). *Sol. Phys.* **2**, 294.
224. Neupert, W. M. (1969). *Annu. Rev. Astron. Astrophys.* **1**, 121.
225. Neupert, W. M. (1970). *Proc. NATO Conf. Sol. Corona, 1971* p. 237.
226. Neupert, W. M. (1971). *Sol. Phys.* **18**, 474.
227. Neupert, W. M., and Behring, W. E. (1962). *J. Quant. Spectrosc. Radiat. & Transfer* **2**, 527.
228. Neupert, W. M., and Swartz, M. (1970). *Astrophys. J.* **160**, L189.
229. Neupert, W. M., Behring, W. E., and Lindsay, J. C. (1964). *Space Res.* **4**, 719.
230. Neupert, W. M., Gates, W., Swartz, M., and Young, R. (1967). *Astrophys. J.* **149**, 79.
231. Neupert, W. M., Swartz, M., and Kastner, S. O. (1973). *Sol. Phys.* **31**, 171.
232. Noyes, R. W. (1971). *Annu. Rev. Astron. Astrophys.* **9**, 209.
233. Palenius, H. (1974). Physics Dept. University of Lund (to be published).
234. Parkinson, J. H. (1972). *Nature (London), Phys. Sci.* **236**, 68; **233**, 38.
235. Parkinson, J. H. (1973). *Astron. Astrophys.* **24**, 215.
236. Peacock, N. J., Cowan, R. D., and Sawyer, G. A. (1965). *Proc. Int. Conf. Ioniz. Gases, 7th, 1966,* p. 599.
237. Peacock, N. J., Speer, R. J., and Hobby, M. G. (1969). *Proc. Phys. Soc., London (At. Mol. Phys.)* [2] **2**, 798 (Cl).
238. Peacock, N. J., Hobby, M. G., and Morgan, P. D. (1971). *Plasma Phys. Contr. Fusion* **1**, 537.
239. Podobedova, L. I., Kononov, E. Y., and Koshelev, K. N. (1971). *Opt. Spectrosc. (USSR)* **30**, 217.
240. Poppe, R. (1968). *Physica* **40**, 17.
241. Poppe, R., Hoekstra, R., and Klinkenberg, P. F. A. (1964). *Appl. Sci. Res., B* **11**, 293.
242. Proceedings of The Conference on Beam Foil Spectroscopy. (1973). *Nucl. Instrum. & Methods* **110**, 1–522.
243. Purcell, J. C., and Widing, K. G. (1972). *Astrophys. J.* **176**, 239.
244. Reader, J., and Epstein, G. L. (1972). *J. Opt. Soc. Amer.* **62**, 619.
245. Reader, J., Epstein, G. L., and Ekberg, J. O. (1972). *J. Opt. Soc. Amer.* **62**, 273.
246. Reeves, E. M., Noyes, R. W., and Withroe, G. L. (1972). *Sol. Phys.* **27**, 251.
247. Ridgeley, A., and Burton, W. M. (1972). *Sol. Phys.* **27**, 280.
248. Roth, N. V., and Elton, R. C. (1968). NRL Rep. No. 6638. US Govt. Printing Office, Washington, D.C.
249. Rugge, H. R., and Walker, A. B. C. (1968). *Space Res.* **8**, 439.
250. Rymer, T. B., and Halliday, J. S. (1950). *J. Sci. Instrum.* **50**, 50.

251. Sawyer, G. A. (1962). *J. Quant. Spectrosc. Radiat. & Transfer* **2**, 467.
252. Sawyer, G. A., Jahoda, F. C., Ribe, F. L., and Stratton, T. F. (1962). *J. Quant. Spectrosc. Radiat. Transfer* **2**, 491.
253. Schweizer, W., and Schmidke, G. (1971). *Astrophys. J.* **169**, L27.
254. Schwob, J. L., and Fraenkel, B. S. (1972). *Phys. Lett. A* **40**, 83.
255. Shadmi, Y. (1965). *J. Opt. Soc. Amer.* **56**, 647.
256. Shadmi, Y. (1966). *J. Res. Nat. Bur. Stand., Sect. A* **70**, 435.
257. Shenstone, A. G. (1970). *Nat. Bur. Stand., Sect. A* **74**, 801 (Cl).
258. Slater, J. C. (1960). "Quantum Theory of Atomic Structures," Vols. I and II. McGraw-Hill, New York.
259. Space Science Board. (1969). "U.S. Space Science Program Report to COSPAR," Vol. XII, p. 29. Nat. Acad. Sci.—Nat. Res. Counc., Washington, D.C.
260. Space Science Board. (1970). "U.S. Space Science Program Report to COSPAR," Vol. XIII, p. 28. Nat. Acad. Sci.—Nat. Res. Counc., Washington, D.C.
261. Space Science Board. (1971). "U.S. Space Science Program Report to COSPAR," Vol. XIV, p. 75. Nat. Acad. Sci.—Nat. Res. Counc., Washington, D.C.
262. Sugar, J., and Kaufman, V. (1972). *J. Opt. Soc. Amer.* **4**, 562 (Cl).
263. Svensson, L. A. (1971a). *Sol. Phys.* **18**, 232 (Cl).
264. Svensson, L. A. (1971b). *Phys. Scripta* **4**, 111 (Cl).
265. Svensson, L. A., and Ekberg, J. O. (1968). *Ark. Fys.* **37**, 65 (Cl).
266. Swartz, M., Kastner, S., Rothe, E., and Neupert, W. (1971). *Proc. Phys. Soc., London (At. Mol. Phys.)* [2] **4**, 1747.
267. Tech, J. L., and Garstang, R. H. (1965). *J. Res. Nat. Bur. Stand., Sect. A* **69**, 401.
268. Tilford, S. G., and Giddings, L. E. (1965). *Astrophys. J.* **141**, 1222.
269. Tomkins, F. S., and Fred, M. (1951). *J. Opt. Soc. Amer.* **41**, 641.
270. Tondello, G. (1969). *Proc. Phys. Soc., London (At. Mol. Phys.)* [2] **2**, 727 (Cl).
271. Tondello, G., and Paget, T. M. (1970). *Proc. Phys. Soc., London (At. Mol. Phys.)* [2] **3**, 1757 (Cl).
272. Toresson, Y. G. (1960). *Ark. Fys.* **17**, 179 (Cl).
273. Tousey, R. (1967a). *Appl. Opt.* **6**, 2044.
274. Tousey, R. (1967b). *Astrophys. J.* **149**, 239.
275. Tousey, R. (1971). *Phil. Trans. Roy. Soc. London* **270**, 59.
276. Tousey, R., Austin, W. E., Purcell, J. D., and Widing, K. G. (1963). *Space Res.* **3**, 773.
277. Tousey, R., Purcell, J. D., Austin, W. E., Garrett, D. L., and Widing, K. G. (1964). *Space Res.* **4**, 703.
278. Tousey, R., Austin, W. E., Purcell, J. D., and Widing, K. G. (1965). *Ann. Astrophys.* **4**, 755.
279. Vainstein, L. A., and Safronova, U. I. (1971a). *Sov. Astron.—AJ* **15**, 175.
280. Vainstein, L. A., and Safronova, U. I. (1971b). "Short Communications in Physics." Lebedev Institute, Academy of Sciences, USSR.
281. Vasilyev, B. N., Grineva, Yu. I., Žhitnik, I. A., Karev, V. I., Koneev, V. V., Krutov, V. V., and Mandelshtam, S. L. (1972). "Short Communications in Physics," Vol. 3, p. 29. Lebedev Institute, Academy of Sciences, USSR.
282. Violett, T., and Rense, W. A. (1959). *Astrophys. J.* **130**, 954.
283. Vodar, B., and Astoin, N. (1950). *Nature (London)* **166**, 1029.
284. Wagner, W. J., and House, L. L. (1971). *Astrophys. J.* **166**, 683 (C1).
285. Walker, A. B. C. (1972). *Space Sci. Rev.* **13**, 672.
286. Walker, A. B. C., and Rugge, H. R. (1969). "Solar Flares and Space Research," p. 102. North-Holland Publ., Amsterdam.
287. Walker, A. B. C., and Rugge, H. R. (1970). *Astron. Astrophys.* **5**, 4.
288. Walker, A. B. C., and Rugge, H. R. (1971). *Astrophys. J.* **164**, 181.
289. Widing, K. G. (1966). *Astrophys. J.* **145**, 380.

290. Widing, K. G. (1972). *Proc. Int. Conf. Beam-foil Spectrosc. 3rd, 1973 Nucl. Instrum. Methods* **110**, 361.
291. Widing, K. G., and Porter, J. R. (1965). *Ann. Astrophys.* **28**, 779.
292. Widing, K. G., and Sandlin, G. D. (1968). *Astrophys. J.* **152**, 545.
293. Widing, K. G., Sandlin, G. D., and Cowan, R. D. (1971). *Astrophys. J.* **169**, 405 (Cl).
294. Withbroe, G. L. (1970a). *Sol. Phys.* **11**, 42.
295. Withbroe, G. L. (1970b). *Sol. Phys.* **11**, 208.
296. Yarosewick, S. J., and Moore, F. L. (1967). *J. Opt. Soc. Amer.* **57**, 1381.
297. Zhitnik, I. A., Krutov, U. U., Majaukin, L. P., Mandelstam, S. L., and Cheremukin, G. S. (1965). *Kosm. Issled.* **5**, 276.
298. Zhitnik, I. A., Krutov, U. U., Majaukin, L. P., Mandelstam, S. L., and Cheremukin, G. S. (1967). *Space Res.* **7**, 1263.
299. Zvereva, L. I., and Kononov, E. Y. (1968). *Opt. Spectrosc.* (*USSR*) 24, 445.

A REVIEW OF JOVIAN IONOSPHERIC CHEMISTRY

WESLEY T. HUNTRESS, JR.

Jet Propulsion Laboratory
California Institute of Technology
Pasadena, California

I. Introduction	295
II. Production of Ions by Photoionization	297
III. Ion–Neutral Reactions	300
A. H_2^+	300
B. He^+	304
C. H^+	305
D. HeH^+	307
IV. Terminal-Ion Loss Processes	309
A. Ion–Electron Recombination	309
B. Radiative Association	311
C. Three-Body Processes	314
D. Reactions with Methane	319
E. Reactions with Excited Neutral Species	321
V. Ion Loss Rates	322
A. H^+ and He^+	322
B. H_3^+	325
C. H_2^+, He_2^+, He_2H^+, HeH^+, and HeH_2^+	326
VI. Ionization Processes of Photoelectrons	328
A. Production of Metastable Excited Species and Secondary Ionization	328
B. Electron Impact Excitation and Dissociation of Ions	332
C. Production of Negative Ions	332
VII. Additional Photon Impact Processes	334
A. Formation and Ionization of Metastables by Photon Impact	334
B. Photodissociation of Ions	334
C. Photoionization of Methane	335
VIII. Concluding Remarks	335
References	338

I. Introduction

The first exploratory missions by unmanned spacecraft to Jupiter and the outer planets will take place in this decade. The initial flyby missions will probe only the local environment and upper atmospheres of these bodies to pave the way for the orbiters and atmospheric probes that will follow. It is due in part to the intriguing chemistry of the outer planets that this exploration is being undertaken. This review is written on the chemistry of the ionosphere of Jupiter in anticipation of the success of the first Pioneer missions scheduled to arrive at Jupiter in December 1973 and December 1974. Both Pioneers 10 and 11 are expected to return photometric data and occultation data on the Jovian upper atmosphere and ionosphere. The occultation data will provide information on the electron density distribution with height in the Jovian atmosphere. This information will be used to determine the composition of the upper atmosphere by modeling the ionosphere in order to reproduce the observed electron density distribution. These calculations will supplement the photometric data for obtaining a hydrogen–helium ratio in the upper atmosphere—a quantity of extreme value for constructing models of the planetary interior and of the evolution of the solar nebula.

Several studies of the Jovian ionosphere have been made by various authors (Rishbeth, 1959; Zabriskie, 1960; Shimizu, 1964, 1971; Gross and Rasool, 1964; Hunten, 1969; Prasad and Capone, 1971; Tanaka and Hirao, 1973; McElroy, 1973), and it is evident that several factors are important in these models: (1) an accurate model of the neutral upper atmosphere, (2) accurate absorption and ionization cross sections, (3) an accurate description of the ion-neutral reactions and ion loss processes, (4) an accurate description of photoelectron collision processes, and (5) a good description of the atmospheric dynamics. This review will be confined mainly to a description of ion chemistry—the production, reaction, and loss processes of ions—although processes involving fast photoelectrons will also be briefly described. Ion chemistry and photoelectron processes are fairly readily studied in Earth-based laboratories, while accurate models for the neutral atmosphere and atmospheric dynamics must await some *in situ* measurements.

The chemistry of the lower atmosphere of Jupiter is vastly complicated by the presence of methane, ammonia, and perhaps water and hydrogen sulfide as well. The ionosphere, however, is formed in the tenuous upper regions of the atmosphere; far above the heights at which ammonia, water, and hydrogen sulfide are condensed. Methane is present in significant amounts only near the lower boundary of the ionosphere (McElroy, 1973; Prinn, 1970). This reduces the problem considerably to essentially a description of ionization processes in hydrogen–helium mixtures, but does not necessarily mean that the chemistry is simple. This review will show in fact that the ion chemistry assumed by previous ionospheric models is incomplete and that the rate constants used were, for the most part, incorrect. These deficiencies were painfully recognized by most of the modelers, and hopefully this review will bring to light more recent data concerning ionization processes in hydrogen–helium mixtures which will serve for the construction of more accurate models of the Jovian ionosphere. Much work remains to be done, both in the laboratory and *in situ*, and those areas in which more laboratory data are required will be emphasized.

II. Production of Ions by Photoionization

The neutral upper atmosphere of Jupiter consists essentially of molecular hydrogen and atomic hydrogen and helium—the relative abundances of which are not well known. Primary ionization by solar ultraviolet radiation in this medium produces H_2^+ ions from H_2, He^+ ions from He, and H^+ ions from H and H_2. The laboratory measurements and theoretical calculations of photoabsorption and ionization cross sections for H_2, He, and H have been adequately reviewed by Hudson (1971), Hudson and Kieffer (1971), and Marr (1967), and will not be covered here.

Except for the most recent work of McElroy (1973), the dissociative photoionization of hydrogen has not been recognized as a potential source of protons in the Jovian ionosphere. The only source of protons used in previous models is photoionization of hydrogen atoms in the upper atmosphere. While the branching ratio H^+/H_2^+ for photoionization of H_2 is quite small, the neutral density ratio H_2/H is fairly large at heights where the peak ionization rates occur, so that this process may nevertheless be an important source of protons. Figure 1 shows the branching ratio H^+/H_2 as a function of ionizing photon wavelength measured by several experimenters. The values from Samson (1972) are only approximate, and the agreement between the results of Monahan *et al.* (1974) and of Browning and Fryar (1973) is not very good at 584 Å. In spite of the present shortcomings in the experimental data, it seems certain that this process should be considered as a source of protons in the Jovian ionosphere (Monahan *et al.*, 1974).

At short photoionization wavelengths between the threshold at 685 Å and about 420 Å, protons are produced by ionization of H_2 above the dissociation limit of the $^2\Sigma_g^+$ ionic molecular ground state, process (II):

$$hv + H_2 \longrightarrow [H_2^+(^2\Sigma_g^+)]^v + e^- \qquad (I)$$
$$\longrightarrow [H_2^+(^2\Sigma_g^+)]^* + e^-$$
$$\longrightarrow H^+ + H \qquad (II)$$

These protons are produced with little or no excess kinetic energy. At wavelengths shorter than about 420 Å, a second threshold is observed for the formation of protons by ionization of H_2 into the repulsive $^2\Sigma_u^+$ excited electronic state

$$hv + H_2 \longrightarrow H_2^+(^2\Sigma_u^+) + e^-$$
$$\longrightarrow H^+ + H \qquad (III)$$

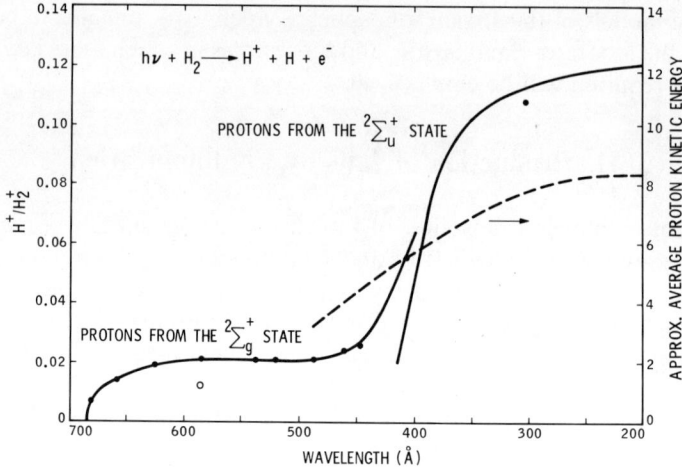

Fig. 1. Branching ratio for the production of protons by dissociative photoionization of hydrogen vs. photon wavelength (left-hand ordinate). Filled dots are the data of Browning and Fryar (1973), the open circle is the datum of Monahan et al. (1974), and the solid curve starting at 420 Å is the approximate branching ratio data from Samson (1972). Dashed curve is the approximate average kinetic energy of that fraction of protons produced from the $^2\Sigma_u^+$ state (Dunn and Kieffer, 1963; Van Brunt and Kieffer, 1970) given vs. photon wavelength (right-hand ordinate).

The protons and hydrogen atoms produced by process (III) have a great deal of excess kinetic energy. The approximate average excess kinetic energy of that fraction of protons produced by ionization of H_2 into the $^2\Sigma_u^+$ state, taken from the work of Kieffer (Dunn and Kieffer, 1963; Van Brunt and Kieffer, 1970), is shown as a function of ionizing wavelength on the right-hand ordinate of Fig. 1. The amount of excess kinetic energy in the H^+ ion is very important for the subsequent ion–molecule reactions involving protons.

The amount of excess internal energy in H_2^+ ions produced by photoionization of hydrogen is also very important for the ion–molecule reactions involving H_2^+. These ions are produced in vibrationally excited states by photoionization into the $^2\Sigma_g^+$ ground state [process (I)] (Chupka et al., 1968). At wavelengths between 804 and about 725 Å, the partitioning between vibrational states in the H_2^+ ions is a strong function of wavelength. The thresholds for the major vibrationally excited H_2^+ ions produced from H_2 appear in this wavelength range, and considerable autoionization structure is observed. Figure 2 gives approximate values for the fraction of H_2^+ ions produced in vibrational states v, $F_v(\lambda)$, vs. photoionization wavelength calculated from the low resolution spectra of Chupka et al. (1968). In making this calculation it was assumed that the ions arising from autoionizing transitions are produced exclusively in the most highly vibrationally excited

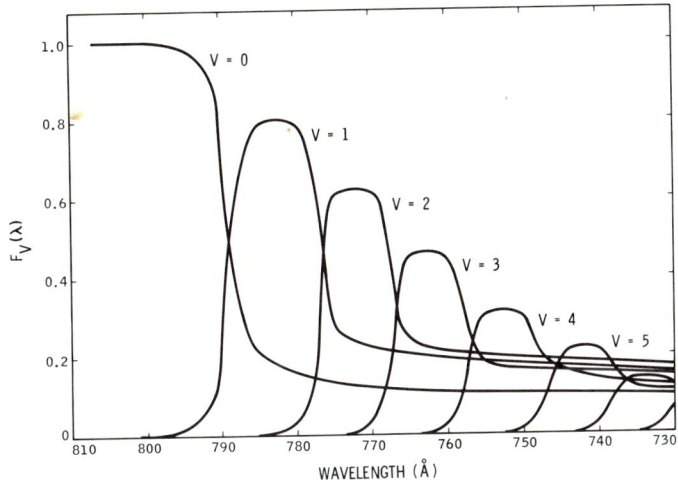

FIG. 2. Approximate distribution of H_2^+ ions produced in vibrational state v by photoionization of H_2 vs. photon wavelength. Calculated from the low resolution data in Chupka et al. (1968).

state possible at that wavelength, and that the continuum underlying the autoionization lines consists of ions in a vibrational distribution predicted by the Franck–Condon principle. At wavelengths less than 725 Å, autoionization structure is not observed and the distribution of vibrational energies in the H_2^+ ions is most likely the Franck–Condon distribution given in Table I (Dunn, 1966).

TABLE I

Franck–Condon Factors for the Ionization of Hydrogen[a]

v	F_v
0	0.09
1	0.16
2	0.17
3	0.15
4	0.12
5	0.09
6	0.06
7	0.04
8	0.03

[a] From Dunn (1966).

III. Ion–Neutral Reactions

A. H_2^+

The H_2^+ ion is the major ion produced by solar photoionization of the Jovian ionosphere. The production rate for H_2^+ ions at heights near the ionization peak is more than ten and one hundred times the production rates for protons and He^+ ions, respectively (Gross and Rasool, 1964; Hunten, 1969; Prasad and Capone, 1971; Shimizu, 1971; Tanaka and Hirao, 1973; McElroy, 1973). The H_2^+ ion disappears rapidly in hydrogen–helium mixtures by ion–neutral reaction. Figure 3 is taken from the work of Theard and Huntress (1974), and shows the temporal behavior of a swarm of ions in a mixture of hydrogen and helium at a total pressure corresponding to the neutral particle density, 3.6×10^{11} cm^{-3}, near the ionization peak in the Jovian upper atmosphere. The relative abundance of helium is exaggerated

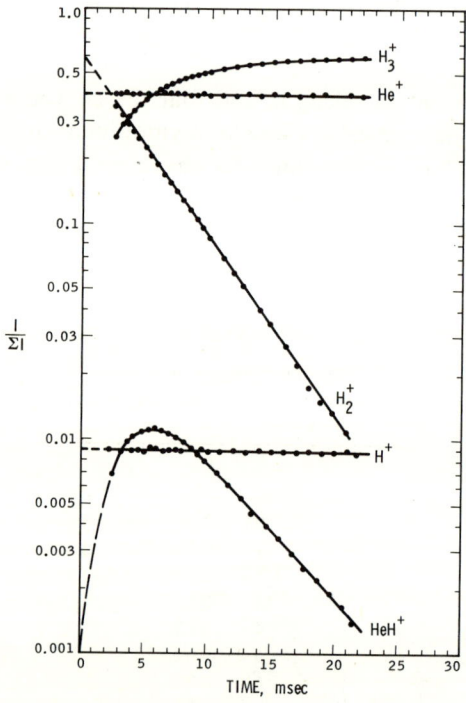

FIG. 3. Temporal behavior of ions following a short (1 msec) burst of 25 eV ionizing electrons in a hydrogen–helium mixture at a low total pressure comparable to that expected in the Jovian ionosphere. $p[H_2] = 2.6 \times 10^{-6}$ Torr; $p[He] = 8.5 \times 10^{-6}$ Torr. Ion cyclotron resonance data from Theard and Huntress (1974).

in these experiments in order to produce sufficient HeH^+ ions for detection. The ions are formed at time $t = 0$ in these experiments by electron impact at 50 eV, which produces a Franck–Condon distribution of vibrationally excited states in the H_2^+ ion corresponding to photoionization of H_2 at wavelengths below about 725 Å. The decay of the H_2^+ ions with time in Fig. 3 is due primarily to reaction in the neutral H_2 molecules to produce H_3^+ ions. The He^+ ions do not react rapidly with H_2, so that the production of HeH^+ ions is due to reaction of H_2^+ ions with He. The HeH^+ ions decay by reaction with H_2 to produce H_3^+ ions. The H_3^+ ion is the terminal ion formed by ion–molecule processes in hydrogen–helium mixtures.

Ion–neutral reactions in hydrogen–helium mixtures have been the subject of a large number of both experimental and theoretical studies (reviewed in Theard and Huntress, 1974). The reactions which have been shown to occur for H_2^+ ions in these mixtures are

$$H_2^+(v) + H_2 \xrightarrow{k_{1a}^v} H_3^+(v') + H \tag{1a}$$

$$\xrightarrow{k_{1b}^v} H_2^+(v' < v) + H_2(v - v') \tag{1b}$$

$$+ He \xrightarrow{k_{2a}^v} HeH^+ + H \tag{2a}$$

$$\xrightarrow{k_{2b}^v} H_2^+(v' < v) + He \tag{2b}$$

A charge transfer reaction with H atoms is also possible

$$H_2^+(v) + H \xrightarrow{k_3^v} H^+ + H_2(v') \tag{3}$$

Experiments by Chupka and co-workers (1968; Chupka and Russell, 1968) have shown that the rate constants k_{1a} and k_{2a} are dependent on the quantum number v. The change in the rate constant k_{1a} with quantum number v and photoionizing wavelength, determined from the measurements of Chupka and co-workers (1968), is shown in Fig. 4. The right-hand ordinate in Fig. 4 gives the absolute rate constant for reaction (1a) from the work of Theard and Huntress. The vertical lines show the thresholds for the appearance of vibrationally excited ions, and the data points show the relative rate constant measured by Chupka and co-workers for autoionizing lines in the H_2^+ photoionization spectrum at the wavelength of the line (Chupka et al., 1968; Chupka and Russell, 1968). The dotted step function shows the average of the data points taken in each vibrational manifold relative to the average of the data points for $v = 0$. The solid line corresponds to the measured rate constant as a function of ionizing wavelength. This function is slightly different than that given by the points since the autoionization lines overlie a continuum due to direct ionization which gives a distribution of vibrational states from $v = 0$ to $v = v_{max}$ at the wavelength of the photon.

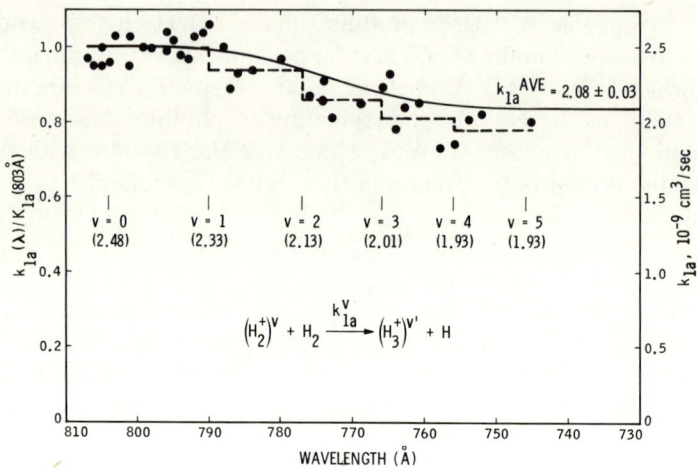

FIG. 4. Relative and absolute rate constants for reaction (1a) vs. ionizing photon wavelength and vibrational quantum number v in the H_2^+ ion. See text.

The numbers in parenthesis in Fig. 4 are the absolute rate constants k_{1a}^v for H_2^+ ions in the indicated vibrational state as determined from the absolute scale on the right-hand ordinate. The absolute scale is established from measurement of the rate constant k_{1a}^{ave} for a Franck–Condon distribution of vibrational states in H_2^+ ions formed at very high ionization energies ($\lambda < 725$ Å).

The rate constant k_{1a}^v decreases slightly with increasing v, becoming constant for $v \geq 4$. It may not be a bad approximation in model ionospheres to simply use the average rate constant k_{1a}^{ave} independent of wavelength since the change in k_{1a}^v amounts to only $\sim 20\%$ between $v = 0$ and $v = \infty$. It is important to note that the actual value for the rate constant of reaction (1a) is fully four times the incorrect value (Giese and Maier, 1963) used in all previous model ionospheres.

The rate constant for reaction (2a) is also very strongly dependent on vibrational energy in the H_2^+ ion. This reaction does not occur for H_2^+ ions in the ground vibrational state at ion temperatures appropriate to the Jovian ionosphere. The threshold for the reaction occurs at $v = 3$. Figure 5 shows the rate constant k_{2a}^v for reaction (2a) as a function of the wavelength of the ionizing photon as determined from the data of Chupka and Russell (1968). The data points in Fig. 5 show a large amount of scatter, which is for the most part real structure. The data points that fall above the average curve correspond to autoionizing lines in the H_2^+ photoionization spectrum which produce ions predominantly in the maximum vibrational state at that wavelength. The data points below the average curve generally correspond

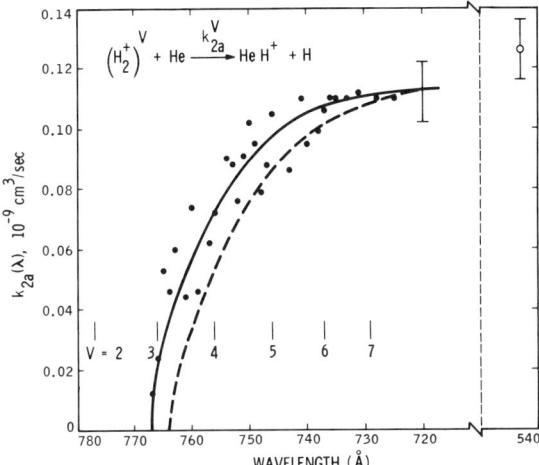

FIG. 5. Absolute rate constant for reaction (2a) vs. ionizing photon wavelength. Filled dots, solid curve, and dashed curve are taken from the data of Chupka and Russell (1968) (see text). Open circle is taken from Theard and Huntress (1974) for electron impact ionization corresponding to a photon wavelength of 543 Å.

to prominent valleys in the H_2^+ photoionization spectrum where direct ionization dominates and a distribution of ions in vibrational states from $v = 0$ to $v = v_{\max}$ at that wavelength is produced. Since the rate constant for reaction (2a) increases dramatically with increasing v, the rate constant k_{2a} at autoionizing wavelengths will be larger than at wavelengths where direct ionization dominates.

The data in Fig. 5 were obtained for H_2^+ ions with a total complement of approximately 3 kcal/mole initial rotational and kinetic energy. Since the rate constant k_{2a} is extremely sensitive to the amount of internal energy in the H_2^+ ion, the average rate constant for this reaction is probably less than that given by the solid-line curve in Fig. 5 at the lower rotational and kinetic temperatures appropriate to ions in the Jovian ionosphere. The dotted curve in Fig. 5 shows a theoretical correction (Chupka and Russell, 1968) to the data which gives the rate constant k_{2a} as a function of wavelength for ions with no initial rotational and translational energy. Because of the strong dependence of k_{2a} on excess internal energy in the H_2^+ ion, it may not be a good approximation in model calculations of HeH^+ ion densities to use a wavelength independent k_{2a}^{ave} for a Franck–Condon distribution of initial vibrational energies in the H_2^+ ion.

Theard and Huntress have shown that collisional deactivation of vibrationally excited H_2^+ ions occurs in nonreactive collisions between these ions and helium atoms [reaction (2b)]. The rate constant for this process is large and approximately equal to the rate constant for the reactive channel; $k_{2b}^v \sim k_{2a}^v$. The main effect of reaction (2b) is to reduce the production rate of HeH^+ ions via reaction (2a) for large He/H_2 ratios. An additional process which may reduce the production rate of HeH^+ ions for low He/H_2 ratios is the collisional deactivation of vibrationally excited H_2 ions in nonreactive collisions with H_2 [reaction (1b)]. It is clear from Fig. 4 that a significant fraction of the collisions between vibrationally excited H_2^+ ions and H_2 are unreactive. For H_2^+ ions in vibrationally excited states $v \geq 4$, the maximum rate constant for reaction (1b) is 5.5×10^{-10} cm^3/sec. Dunn (1964) has identified reaction (1b) and has placed a lower limit of approximately 2.5×10^{-10} cm^3/sec on the value for k_{1b}^{ave}. Reactions (1b) and (2b) may not significantly affect the calculated electron densities in model ionospheres, or the densities of the major ions in the system (H^+, H_3^+), but calculations of HeH^+ densities will probably be too high without including these reactions.

The charge transfer reaction of H_2^+ ions with H atoms [reaction (3)] will be important only at extreme heights in the Jovian ionosphere where the density of H atoms becomes comparable to the density of H_2 molecules. Reaction (3) has not been investigated in the laboratory, and consequently the rate constant is unknown. In order to proceed at a rapid rate, thermal energy charge transfer reactions generally require (1) matching of energy levels between the reactants and products, and (2) favorable Franck–Condon transition probabilities—in this case between the matching vibrational states in the H_2^+ ion and H_2 neutral (Bowers and Elleman, 1972; Laudenslager and Bowers, 1972). Because of the extreme sensitivity of thermal energy charge transfer reactions to the proper conditions of energy level matching and transition probabilities, it is very difficult to calculate an accurate theoretical rate constant for reaction (3). Without an experimental measurement of k_3, all that can be said is that k_3 may have a value between 0 and the maximum theoretically predicted collision rate constant, 2.6×10^{-9} cm^3/sec. Reaction (3) could be a significant source of protons in the extreme ionosphere.

B. He^+

The helium ion has not been observed to react with hydrogen, even though several exothermic reaction channels are available.

$$He^+ + H_2 \longrightarrow H_2^+ + He \qquad (4a)$$
$$\longrightarrow H^+ + H + He \qquad (4b)$$
$$\longrightarrow HeH^+ + H \qquad (4c)$$

The sum of the rate constants for these reactions is less than 10^{-13} cm^3/sec

(Fehsenfeld et al., 1966). A possible exothermic reaction with hydrogen atoms exists

$$He^+ + H \longrightarrow H^+ + He \tag{5}$$

but this reaction has not been examined experimentally. Reaction (5) has been overlooked in all model ionosphere calculations, but it is almost certain that the rate constant for this reaction is extremely small. Reaction (5) is a charge transfer process in which the excess energy of the reaction, 11 eV, would have to be transferred to internal excitation of the proton or helium atom. Since the proton has no excited states and the first excited state of helium lies near 20 eV, this reaction is highly improbable according to the usual rules which govern thermal energy charge transfer reactions. Transfer of the excess energy into translational modes is not favorable and has not been observed in nondissociative thermal energy charge transfer reactions.

The low reactivity of He^+ ions with H_2 and H raises problems with the calculation of the loss rate for He^+ ions in Jovian ionospheric models. Hunten has shown that a total rate constant as small as 2×10^{-17} cm^3/sec for k_4 would be significant. The current experimental upper limit is four orders of magnitude greater than this number, and it is at present beyond experimental technology to measure a rate constant much less than 10^{-13} cm^3/sec for gas-phase ion–molecule reactions. Alternative loss processes for He^+ ions are discussed in Section IV.

C. H^+

Thermal energy protons have no exothermic reaction channels available for collisions with hydrogen and helium. The ion processes controlling the density of protons in the Jovian ionosphere will therefore include one or several of the mechanisms discussed in Section IV. The calculation of loss rates can be made with more confidence in the case of protons since, unlike helium ions, there are no possible bimolecular ion–neutral reactions with immeasurably slow rate constants which can control the loss rate.

Kinetically excited protons produced by the dissociative photoionization of H_2 at wavelengths less than about 420 Å have sufficient excess energy to react with H_2 via the endothermic process

$$H^+ + H_2 \longrightarrow H_2^+ + H \tag{6a}$$

The cross section for this reaction as a function of proton kinetic energy has been measured in ion beam experiments (Holliday et al., 1971; Maier, 1971). The rate constant k_{6a} calculated from these measurements is given in Fig. 6 as a function of proton kinetic energy. The theoretical rate constant for collisions between protons and hydrogen molecules is 2.8×10^{-9} cm^3/sec.

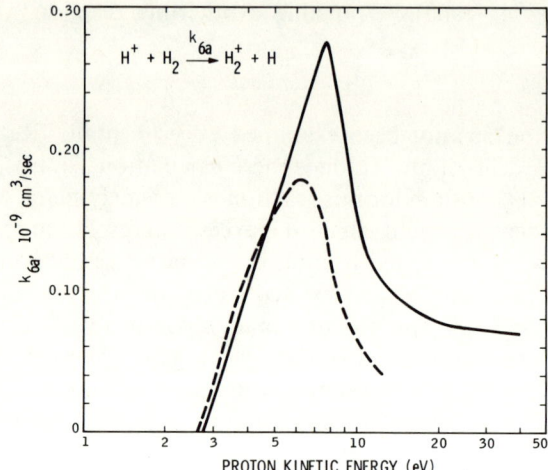

FIG. 6. Rate constant for reaction (6a) vs. proton kinetic energy determined from the ion beam data of Maier (1971) (solid curve), and of Holliday et al. (1971) (dashed curve).

If the experimental data in Fig. 6 are correct, then a large fraction of the collisions between kinetically excited protons and hydrogen molecules may not be reactive. In this case, nonreactive collisions can result in thermalization of energetic protons by inelastic processes:

$$H^+ + H_2 \longrightarrow H_2^* + H^+ \tag{6b}$$

$$\longrightarrow H + H + H^+ \tag{6c}$$

or by elastic recoil in collisions with H_2 and He. The rate constants and energy loss per collision for these processes have not yet been measured.

Thermalization of kinetically excited protons can also occur via symmetric charge transfer with hydrogen atoms

$$H^+ + H \longrightarrow H + H^+ \tag{7}$$

The theoretical (Dalgarno, 1958; Dewangen, 1973) and experimentally measured (Fite et al., 1962; Belyaev et al., 1967) rate constants for this reaction are given in Fig. 7 as a function of the kinetic energy of the proton. Experimental data for the rate constant of reaction (7) exist only for protons with very high kinetic energies (> 10 eV). For kinetically excited protons, the charge transfer reaction (7) will leave the resulting hydrogen atoms with most of the kinetic energy of the original protons. The velocity of an 8-eV hydrogen atom is about 40 km/sec. The escape velocity of a hydrogen atom

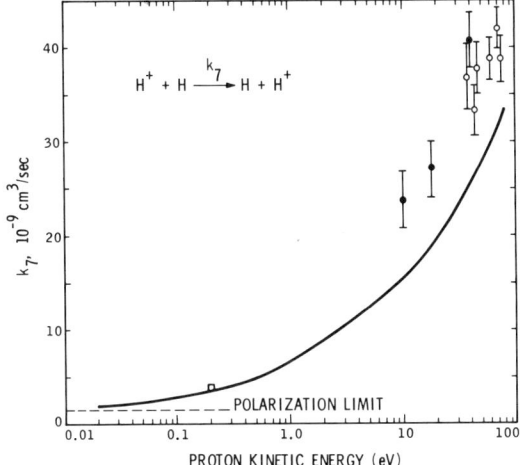

FIG. 7. Rate constant for reaction (7) as a function of proton kinetic energy. Filled circles, data of Belyaev et al. (1967), open circles, data of Fite et al. (1962), open square, calculation of Dalgarno (1958), solid curve, calculation of Dewangen (1973), and the dashed line, polarization limit at low energies.

is 60 km/sec at Jupiter and 36 km/sec at Saturn. However, since the production rate of energetic hydrogen atoms from kinetically excited protons by this mechanism is likely to be quite small, this process may not result in significant nonthermal loss of hydrogen atoms from the atmospheres of these planets. Symmetric charge transfer between H_2^+ ions and H_2 molecules does not occur at thermal kinetic energies (Clow and Futrell, 1972), and symmetric charge transfer between He^+ ions and He is probably not an important ion cooling mechanism since He^+ ions are not excited when produced by photoionization at wavelengths $\lambda > 200$ Å.

D. HeH^+

Previously, ion–electron recombination has been presumed to be the principal loss mechanism for HeH^+ ions in the Jovian ionosphere (Gross and Rasool, 1964; Hunten, 1969; Prasad and Capone, 1971; Shimizu, 1971; McElroy, 1973; Tanaka and Hirao, 1973). The exothermic reaction

$$HeH^+ + H_2 \longrightarrow H_3^+ + He \qquad (8)$$

has been overlooked in all ionospheric models to date, perhaps because the rate constant for reaction (8) has only recently been measured (Theard and Huntress, 1974). The value for k_8 is large, 1.83×10^{-9} cm^3/sec, and if a nominal value for the dissociative recombination rate of HeH^+ ions with

TABLE II

BIMOLECULAR REACTIONS OF THE MAJOR REACTIVE IONS

Reaction		Rate constant[a]	Ref.	Comments
$H_2^+(v) + H_2$	$\xrightarrow{k_{1a}^{avc}}$ $H_3^+(v') + H$	2.08	Theard and Huntress (1974)	Value for $\lambda < 725$ Å. See Fig. 4 for k_{1a} vs. λ
	$\xrightarrow{k_{1b}^{avc}}$ $H_2^+ + H_2(v')$	~0.4	Theard and Huntress (1974)	Inferred value. Experimental lower limit of 0.25×10^{-9} from Dunn (1964)
+ He	$\xrightarrow{k_{2a}^{avc}}$ $HeH^+ + H$	0.13	Theard and Huntress (1974)	Value for $\lambda < 725$ Å. See Fig. 5 for k_{2a} vs. λ
	$\xrightarrow{k_{2b}^{v}}$ $H_2^+ + \mathbf{He}$	$k_{2b}^{v} \simeq k_{2a}^{v}$		
+ H	$\xrightarrow{k_3^{v}}$ $H^+ + H_2(v')$	0–2.6	Theard and Huntress (1974)	Not measured. Difficult to predict theoretically. Maximum theoretical rate constant is 2.6×10^{-9}
$He^+ + H_2$	$\xrightarrow{k_{4a}}$ $H_2^+ + He$			
	$\xrightarrow{k_{4b}}$ $H^+ + H + He$	<0.0001	Fehsenfeld et al. (1966)	No reaction observed. Experimental upper limit
	$\xrightarrow{k_{4c}}$ $HeH^+ + H$			
+ H	$\xrightarrow{k_5}$ $H^+ + He$	0	See text	Simple rules for thermal energy charge transfer predict zero rate constant
$HeH^+ + H_2$	$\xrightarrow{k_8}$ $H_3^+ + He$	1.83	Theard and Huntress (1974)	
+ H	$\xrightarrow{k_9}$ $H_2^+ + He$	0.9	Rutherford and Vroom (1973)	Experimental rate may be low (see text). Unimportant except at extreme heights

[a] In units of 10^{-9} cm^3/sec.

electrons of 10^{-8} cm^3/sec (Gross and Rasool, 1964; Hunten, 1969; Prasad and Capone, 1971; Shimizu, 1971; McElroy, 1973; Tanaka and Hirao, 1973) is assumed, then using either Hunten's (1969) model or the Gross and Rasool (1964) model for the Jovian ionosphere it is clear that reaction (8) will completely dominate over recombination with electrons as the major loss mechanism for HeH$^+$ ions.

The HeH$^+$ ion does not react with He atoms, and the reaction with H atoms

$$\text{HeH}^+ + \text{H} \longrightarrow \text{H}_2^+ + \text{He} \qquad (9)$$

could only be an important loss mechanism for HeH$^+$ ions at extreme heights in the Jovian ionosphere. The beam experiments of Rutherford and Vroom (1973) give a rate constant of $\sim 0.9 \times 10^{-9}$ cm^3/sec for reaction (9), although the Langevin rate constant is 2.3×10^{-9} cm^3/sec. It is quite difficult to accurately measure absolute cross sections (and therefore rate constants) in beam experiments, especially when the neutral reactant is not a stable species. Rate constants calculated from absolute cross sections as measured in beam experiments usually give low values—as is the case for the rate constant k_{1a} calculated from the data of Giese and Maier. Exothermic proton transfer reactions generally exhibit rate constants close to the Langevin value, so that to use a rate constant of 2.3×10^{-9} cm^3/sec for reaction (9) may not be inappropriate.

IV. Terminal-Ion Loss Processes

A. Ion–Electron Recombination

The terminal ions produced in the Jovian ionosphere are H$^+$, He$^+$, and H$_3^+$. For the polyatomic H$_3^+$ ion, the principal loss mechanism is dissociative ion–electron recombination.

$$\text{H}_3^+ + \text{e}^- \xrightarrow{\alpha} (\text{H} + \text{H}) + \text{H} \qquad (10)$$

The rate constant, or recombination coefficient α, is large for this process. The most recent values for $\alpha(\text{H}_3^+)$ are given in Fig. 8 from Leu et al. (1973). It is not known whether the products of reaction (10) are H$_2(v)$ + H or 3H. The total energy released by reaction (10) is more than sufficient for dissociation into three hydrogen atoms. Reaction (10) proceeds through an excited neutral H$_3$ intermediate molecule (Bardsley and Biondi, 1970). Immediate dissociation of this unstable molecule into three hydrogen atoms is plausible. Knowledge of the product distribution for reaction (10) is of some importance for ionospheric chemistry since the photoionization of hydrogen,

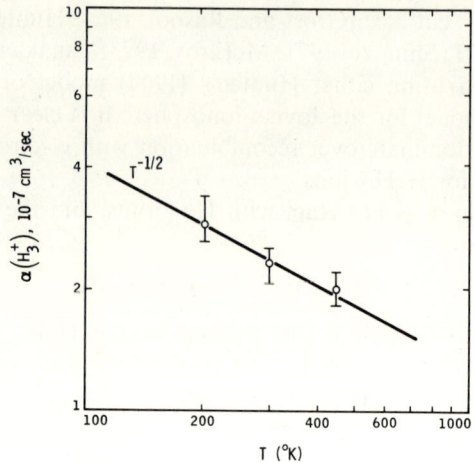

FIG. 8. Dissociative recombination coefficient for H_3^+ ions as a function of ion temperature (electron and ion temperatures equal) from Leu et al., (1973).

followed by recombination of H_3^+ ions, is a major source of hydrogen atoms in the upper atmosphere of Jupiter (Gross and Rasool, 1964; Hunten, 1969; Prasad and Capone, 1971; Shimizu, 1971; McElroy, 1973; Tanaka and Hirao, 1973). It has been previously assumed in ionospheric models that the products are $H_2 + H$, in which case the photoionization of H_2 results in the formation of two hydrogen atoms via reactions (1a) and (10). If three hydrogen atoms are produced in reaction (10), then the overall process results in the formation of four hydrogen atoms.

Recombination of the atomic ions H^+ and He^+ with electrons must occur via the radiative processes

$$H^+ + e^- \longrightarrow H + h\nu \tag{11}$$

$$He^+ + e^- \longrightarrow He + h\nu \tag{12}$$

The electron densities are too low in the Jovian ionosphere, less than 10^6 cm^{-3}, for three-body collisional recombination to occur. The radiative recombination coefficients for atomic ions[1] are quite small compared to the dissociative recombination coefficients of polyatomic ions (Table III), so that H^+ and He^+ ions are not rapidly removed from the Jovian ionosphere by this mechanism. If the loss rates for H^+ and He^+ are small, then these become the dominant ions in the Jovian ionosphere, and the number density of electrons is kept fairly high.

[1] Recombination coefficients quoted from Hunten's (1969) extrapolation of the theoretical values of Bates and Dalgarno (1962) to 150°K.

TABLE III

Ion-Electron Recombination Coefficients for Terminal Ions

Reaction	Recombination coefficient (cm^3/sec)	Ref.	Comments
$H_3^+ + e^- \longrightarrow (H + H) + H$	3.4×10^{-7}	Leu et al. (1973)	150°K under the conditions that the temperature of the electrons and ions are all equal. See Fig. 8
$H^+ + e^- \longrightarrow H + h\nu$	6.25×10^{-12}	Hunten (1969)	150°K
$He^+ + e^- \longrightarrow He + h\nu$	6.25×10^{-12}	Hunten (1969)	150°K

Because of the low recombination coefficients for H^+ and He^+ ions, any process which might convert these atomic ions into polyatomic ions would significantly increase the total ion loss rate and greatly influence the calculated electron density. It is for this reason that a rate constant as small as 10^{-17} cm^3/sec for reaction (4) is still important as a possible loss mechanism for He^+ ions. Several processes including radiative association, three-body ion–neutral reactions, bimolecular reaction with CH_4 at the lower boundary of the ionosphere, and reaction with vibrationally excited H_2 molecules have been considered which might increase the loss rates for H^+ and He^+ ions.

B. Radiative Association

Radiative association reactions of the type

$$H^+ + H_2 \longrightarrow H_3^+ + h\nu \qquad (13)$$

have very slow rates and have never been observed in the laboratory. Several workers (Hunten, 1969; Prasad and Capone, 1971; Tanaka and Hirao, 1973) have nevertheless suggested that these reactions can control the loss rate of atomic ions if the rate constants are on the order of 10^{-13}–10^{-17} cm^3/sec. However, McElroy (1973) has recently questioned whether the rate constant for reaction (13) can be this large. A closer examination is made of these reactions here, and it is shown that the rate constants for radiative association in this case are likely to be much less than 10^{-19} cm^3/sec. It is also shown that larger rate constants are possible, but that even in this case radiative association reactions may not be significant loss mechanisms compared to radiative ion–electron recombination and three-body reactions of atomic ions.

Radiative association proceeds through an intermediate excited ion which necessarily contains all of the excess energy of the association reaction, i.e.,

$$H^+ + H_2 \xrightarrow{k_a} [H_3^+]^* \xrightarrow{k_r} H_3^+ + h\nu$$

The overall rate at which the total reaction proceeds depends critically on the radiative lifetime of this excited intermediate ion. The probability for stabilization of the intermediate ion by radiative emission of excess energy is given by the ratio of the time during which the ion and neutral are in intimate contact and the time required for emission to take place. The time during which the ion and neutral are in intimate contact is defined by the time it takes the ion–neutral pair to traverse the potential surface for the collision. For "hard-sphere" collisions, this time is on the order of one vibrational period τ_v. Typically, $\tau_v \lesssim 10^{-13}$ sec so that

$$\eta \lesssim (\tau_v/\tau_r)$$
$$\simeq 10^{-13}/\tau_r \qquad (A)$$

where η is the probability of reaction (13) occurring per collision, and $\tau_r = k_r^{-1}$ is the radiative lifetime of the excited intermediate ion.

Radiative lifetimes for optically allowed transitions of electronically excited states are in the range 10^{-9}–10^{-6} sec, corresponding to a collisional efficiency of $\eta = 10^{-4}$–10^{-7}. This is in agreement with a value of $\sim 10^{-6}$ predicted by a more sophisticated theoretical treatment of this problem (Carrington, 1972). Since ion–neutral collision rate constants are on the order of 10^{-9} cm^3/sec, reaction rate constants of 10^{-13}–10^{-16} cm^3/sec are predicted for radiative association using this simple theory. This analysis is most likely the basis for the numbers used in previous ionospheric models.

The fault in using large rate constants (10^{-13}–10^{16} cm^3/sec) for radiative association reactions such as (13) is that the intermediate ions for the possible reaction partners in the Jovian ionosphere are not formed in electronically excited states. In most cases, the ground state reactants do not have sufficient energy to reach the first electronically excited states of the intermediate ions, so that the intermediates can only be formed in vibrationally excited states of the ground electronic state. In this case, the rate constants for radiative association are probably much less than 10^{-13}–10^{-16} cm^3/sec.

The appropriate reactions, and the amount of excess energy in the intermediate ions formed, are

$$H^+ + H_2 \longrightarrow [H_3^+]^*, \quad E^* = 4.47 \text{ eV} \qquad (14)$$
$$+ \text{He} \longrightarrow [\text{HeH}^+]^*, \quad E^* = 1.99 \text{ eV} \qquad (15)$$
$$+ H \longrightarrow [H_2^+]^*, \quad E^* = 2.69 \text{ eV} \qquad (16)$$
$$\text{He}^+ + \text{He} \longrightarrow [\text{He}_2^+]^*, \quad E^* = 2.31 \text{ eV} \qquad (17)$$
$$+ H_2 \longrightarrow [\text{HeH}_2^+]^*, \quad E^* = 9.30 \text{ eV} \qquad (18)$$
$$+ H \longrightarrow [\text{HeH}^+]^*, \quad E^* = 12.97 \text{ eV}. \qquad (19)$$

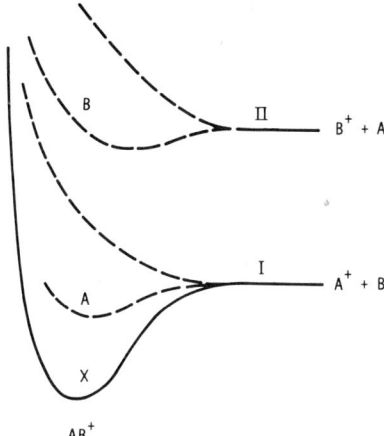

FIG. 9. Schematic representation of potential energy curves involved in the collisions of hypothetical ion–neutral pairs $A^+ + B$ and $B^+ + A$. Solid curve is the ground state of the AB^+ ion, dashed curves are various families of excited states of the AB^+ ion. Radiative association must proceed by a transition from an excited state of the AB^+ ion to a bound state. See text.

A schematic diagram of the potential energy surfaces for these reactions is given in Fig. 9. Reactions (14)–(17) proceed along the potential surface for the dissociation of the intermediate ion into the lowest energy asymptotic products given by the set of curves labeled I in Fig. 9. In this case the excitation energy of the excited intermediate ion is simply the dissociation energy of the ground state ion. For the ions H_2^+, He_2^+, HeH^+, and H_3^+, bound electronically excited states of the type labeled A in Fig. 9 do not exist (see, for example, H_2^+: Sharp, 1971; He_2^+: Gupta and Matsen, 1967; Browne, 1966; HeH^+: Michels, 1966; H_3^+: Geller and King, 1974), so that no electronic transitions are available. The repulsive states are of no consequence since the total energy at 150°K is insufficient for the reactant pair to ride up high enough on the repulsive curve in order for transitions to bound states to take place. For reactions (14)–(17), only vibrational excitation in the intermediate ion is permitted.

The radiative decay time τ_r for vibrationally excited ions is much longer than for electronically excited ions, on the order of 10^{-3} sec or greater, so that for the radiative stabilization of the intermediate ions in reactions (14)–(17), $\eta \leq 10^{-10}$ and $k \leq 10^{-19}$ cm³/sec. This is a very small rate constant, and much less than the number 10^{-17} cm³/sec given by Hunten as the minimum rate constant for effective competition with electron–ion radiative recombination.

Reactions (18) and (19) proceed along the potential surface for dissociation of the intermediate into higher energy asymptotic products, given by the set of curves labeled II in Fig. 9. For the HeH$^+$ ion, bound electronically excited states of the type labeled B in Fig. 9 do exist (Michels, 1966). These states have a very small dissociation energy into He$^+$ and H, and the equilibrium internuclear separation is very large compared to the ground state. The potential energy curves for these bound excited states are strongly repulsive in the region of internuclear distances equivalent to the separation of the nuclei in ground vibronic states. In this situation, the probability for an electronic transition to the ground state is small and the radiative lifetime of the excited state is long, whether or not the transition is allowed by selection rules. For reaction (19), electronic transitions from the first excited states to the ground state of the intermediate ion are likely to be slow. Sando, Cohen, and Dalgarno (1971), however, have calculated a theoretical rate constant for radiative association of $\sim 2 \times 10^{-15}$ cm^3/sec for reaction (19)

$$\text{He}^+ + \text{H} \longrightarrow \text{HeH}^+ + h\nu \qquad (19\text{a})$$
$$\longrightarrow \text{H}^+ + \text{He} + h\nu \qquad (19\text{b})$$

and give $k_{19a}/k_{19b} \cong 1.7$.

Reaction (18) is essentially equivalent to reaction (4). Several dissociation modes are available to the excited intermediate complex, which yield the products given in reactions (4a)–(4c). Because of the exothermicity of these dissociation pathways, it is much more likely that dissociation will occur before radiation. In this case, the important quantity is not the radiative association rate constant, but the rate constant for reaction (4).

From the preceding discussion, it is concluded that the rate constants for the radiative association reactions of H$^+$ and He$^+$ ions in the Jovian ionosphere are less than 10^{-19} cm^3/sec [except perhaps for reaction (19)]. There is a circumstance, however, in which these rate constants can be considerably larger. This arises because of the possibility that for intermediate ions formed in ion–neutral collisions, the lifetimes of the intermediate ions toward dissociation, τ_v, can be considerably greater than that given by the "hard-sphere" collision time. This possibility is discussed in the following section where the relationships between the three-body reaction, the radiative association reaction, and the dissociative and radiative lifetimes of the intermediate ions are examined. The rate constants for both radiative and collisional de-excitation of the intermediate ions in reactions (14)–(19) are critically dependent on the dissociative lifetime of the intermediate ion.

C. Three-Body Processes

If the pressure is sufficiently high, collisions with a third body can remove the excess energy in the intermediate ions shown in reactions (14)–(19) and a stable molecular product ion is formed. This process may lead to significant loss of H^+ and He^+ ions in the lower ionosphere where the number density of molecules is the greatest. The three-body reactions

$$H^+ + H_2 + H_2 \longrightarrow H_3^+ + H_2 \tag{20}$$

and

$$He^+ + He + He \longrightarrow He_2^+ + He \tag{21}$$

have been identified and their rate constants measured (Miller et al., 1968; Niles and Robertson, 1965) (Table IV). Dalgarno (1972) first recognized the importance of reaction (20) as a major loss process for protons in the Jovian ionosphere, and this reaction has been used in recent model ionospheres by McElroy (1973) and Tanaka and Hirao (1973).

The three-body reaction can be broken down into a two-step mechanism. First, the formation of an excited intermediate ion

$$A^+ + B \underset{k_d}{\overset{k_a}{\rightleftarrows}} [AB+]^*, \tag{B}$$

where k_a is the rate constant for the association reaction and k_d is the unimolecular rate constant for the dissociation of the intermediate ion, followed by stabilization of the excited intermediate ion by collision with a third body

$$[AB+]^* + C \xrightarrow{k_s} AB^+ + C^*, \tag{C}$$

where k_s is the rate constant for the stabilization reaction. Solution of the kinetic equations for this reaction mechanism shows that the rate constant for the three-body reaction, defined by the loss equation

$$d[A^+]/dt = -k_{(3)}[A+][B][C], \tag{D}$$

can be written in terms of the rate constants for the sequential bimolecular processes (B) and (C) according to the relation

$$\begin{aligned} k_{(3)} &= k_s k_a / k_d \\ &= k_s k_a \tau_d \end{aligned} \tag{E}$$

where $\tau_d = k_d^{-1}$ is the lifetime toward dissociation of the excited intermediate ion. The lifetime τ_d is the same as the lifetime τ_v in the preceding section. The stabilization rate constant k_s is given by the product of the bimolecular collision rate constant k_c and a factor γ which gives the probability of stabilization per collision.

$$k_{(3)} = \gamma k_c k_a \tau_d \tag{F}$$

TABLE IV

THREE-BODY REACTIONS OF TERMINAL IONS WITH H_2 AND He

Reaction		Rate constant	Ref.	Comment
$H^+ + H_2 + H_2$	$\xrightarrow{k_{20}}$ $H_3^+ + H_2$	3.2×10^{-29} cm^6/sec	Miller et al. (1968)	Measured at 300°K. Temperature dependence undetermined
$H^+ + H_2 + He$	$\rightarrow H_3^+ + He$	Unknown		Most probable product is H_3^+
$H^+ + He + He$	$\rightarrow HeH^+ + H_2$	Unknown		Unimportant for low He/H_2 values
	$\rightarrow HeH^+ + He$			
$He^+ + H_2 + H_2$	$\xrightarrow{k_{22a}} H_3^+ + H + He$	Unknown		Assumed to be at least as fast as reaction (21). Most probable product is H_3^+
	$\rightarrow H_2^+ + H_2 + He$			
	$\rightarrow H^+ + H + H_2 + He$			
	$\rightarrow HeH_2^+ + H_2$			
	$\rightarrow HeH^+ + H + H_2$			
$He^+ + H_2 + He$	$\rightarrow H_2^+ + 2He$	Unknown		
	$\rightarrow H^+ + H + 2He$			
	$\rightarrow HeH_2^+ + He$			
	$\rightarrow HeH^+ + H + He$			
	$\rightarrow He_2H^+ + H$			
$He^+ + He + He$	$\xrightarrow{k_{21}} He_2^+ + He$	$3.19 \times 10^{-29}T^{-1}$, °K cm^6/sec	Niles and Robertson (1965)	Probably unimportant for low He/H_2 values
$H_3^+ + H_2 + H_2$	$\xrightarrow{k_{34}} H_5^+ + H_2$	$< 0.1\, k_{20}$	Saporoschenko (1965)	

If we assume that $\gamma = 1$, $\tau_d = 10^{-13}$ sec, and use $k_c = k_a = 10^{-9}$ cm^3/sec for typical bimolecular ion–molecule collision rate constants, then three-body reaction rate constants on the order of $k_{(3)} = 10^{-31}$ cm^6/sec are predicted. This is in excellent agreement with the measured value for reaction (21) at 300°K. Since the helium atom is a very poor sink for carrying away excess energy in single collisions, this agreement between theoretical and experimental rate constants indicates that three-body ion–molecule reactions are quite efficient and that a rate constant of 10^{-31} cm^3/sec is probably a good lower limit for any such hypothetical reaction. On the other hand, the rate constant for reaction (20) is quite large; $k_{20} = 3.2 \times 10^{-29}$ cm^6/sec at 300°K. This rate constant is some 50 times greater than is predicted by Eq. (F) using the hard-sphere collision time. The theoretical maximum rate constants for bimolecular collisions between H^+ and H_2, and between H_3^+ and H_2, are 2.8×10^{-9} and 2.1×10^{-9} cm^3/sec, respectively. Assuming again that $\gamma = 1$ in Eq. (F), the lifetime τ_d of the $[H_3^+]^*$ complex in reaction (20) must be approximately 5×10^{-12} sec. Of course, if $\gamma < 1$, then the lifetime τ_d must be even larger. Lifetimes even several orders of magnitude larger than 10^{-12} sec are not unusual for the intermediate ions in bimolecular ion–molecule collisions. Unlike neutral–neutral collisions, the ion-induced dipole potential between the ion and neutral collision partners leads to orbiting collisions and consequently to the formation of long-lived collision complexes in which weakly bound, excited intermediate ions are formed.

If the dissociative lifetimes for the excited intermediate ions in reactions (14)–(19) are considerably greater than one vibrational period—or the "hard sphere" collision time τ_v—then the rate constants for radiative association are significantly increased. The relative importance of radiative to collisional de-excitation of the excited intermediate ions depends on the neutral density, the bimolecular collision rate constant, and the radiative lifetime of the excited intermediate ion:

$$\frac{[d/dt]_r}{[d/dt]_{(3)}} = \frac{k_c \eta [H_2]}{\gamma k_c k_a \tau_d [H_2]^2}$$

$$\cong (k_c \tau_r [H_2])^{-1} \tag{G}$$

where $\gamma \simeq 1$ and $k_a \simeq k_c$, and where $[d/dt]_r$ and $[d/dt]_{(3)}$ are the loss rates for radiative association and three-body reaction, respectively. Since $k_c \simeq 10^{-9}$ cm^3/sec and $\tau_r \geq 10^{-3}$ sec, the three-body reaction will dominate over radiative association for neutral densities greater than about 10^{12}/cm^3 (or less if $\tau_r > 10^{-3}$ sec). The rate constant for radiative association can be approximated from the three-body rate constant by using relation (G) to obtain

$$k_r \cong k_{(3)}/k_c \tau_r$$
$$\cong 10^9 k_{(3)}/\tau_r \qquad \text{(H)}$$

In addition to simple de-excitation of the excited intermediate complex in reactions (14)–(19), collisions with a third body can lead to chemical rearrangement involving the third body to yield a new stable product ion. This kind of reaction considerably increases the number of possible products for three-body reactions. For example, the loss of He$^+$ ions by three-body reaction in the Jovian atmosphere probably occurs mainly by reaction with two hydrogen molecules, not by reaction (21).

$$\text{He}^+ + \text{H}_2 \longrightarrow [\text{HeH}_2^+]^* \xrightarrow{\text{H}_2} \text{HeH}_2^+ + \text{H}_2 \qquad \text{(22a)}$$
$$\longrightarrow \text{HeH}^+ + \text{H} + \text{H}_2 \qquad \text{(22b)}$$
$$\longrightarrow \text{H}_2^+ + \text{H}_2 + \text{He} \qquad \text{(22c)}$$
$$\longrightarrow \text{H}^+ + \text{H} + \text{H}_2 + \text{H}_2 \qquad \text{(22d)}$$
$$\longrightarrow \text{H}_3^+ + \text{H} + \text{He} \qquad \text{(22e)}$$

Reactions (22a)–(22d) all involve processes in which the third body acts either to stabilize the intermediate ion or to promote the dissociation of the intermediate ion into products other than the original reactants. Reaction (22e), however, is entirely different in that the third body has chemically participated in the rearrangement of the intermediate ion. In this particular case, reaction (22e) is most likely the preferred reaction pathway since the H_3^+ ion is the terminal ion formed in any further reaction among the product ions. Another way of envisioning reaction (22e) is that the affinity of the H_2 molecule for a proton is much greater than that of the hypothetical and unstable HeH molecule, so that in any encounter between HeH_2^+ ions and H_2 molecules, the H_2 molecule will grab off the proton, resulting in the formation of an H_3^+ ion. Since proton transfer reactions are among the most facile of ion–molecule reactions and generally dominate over most other available processes at thermal energies, reaction (22e) is predicted to be the major channel for the three-body reaction between He$^+$ ions and two hydrogen molecules.

The exothermic three-body reactions of H^+, He^+, and H_3^+ ions that are possible in He–H_2 mixtures are given in Table IV. Except for reactions (20) and (21), these reactions have not been studied, and their product distributions and rate constants are not known. Three-body reactions involving hydrogen atoms are not given in Table IV since the density of H atoms relative to He and H_2 is low at the altitudes in the Jovian ionosphere where three-body reactions are expected to be important. The reactions with two hydrogen molecules [reactions (20) and (22)] are expected to be the dominant three-body reactions for two reasons: (i) their rates are probably larger than three-body reactions involving He atoms because of the increased efficiency of the hydrogen molecule as a sink for distributing the excess internal energy in the intermediate ions and as a third body to carry away the excess energy, and (ii) the He/H_2 ratio in the Jovian ionosphere is expected to be less than unity.

Unlike the case for bimolecular ion–molecule reactions, the rate constants for three-body ion–molecule reactions are temperature dependent. This arises because the lifetime toward dissociation τ_d of the intermediate ion depends strongly on its internal energy, and hence on the temperature of the reactant ions. In this case, the rate constant *increases* with decreasing temperature. Reaction (21) has been shown to follow a T^{-1} dependence. This is probably typical for most three-body ion–molecule reactions.

D. Reactions with Methane

If methane molecules reach to sufficiently high altitudes in the upper atmosphere of Jupiter, then loss of H^+ and He^+ ions can occur by bimolecular reaction with CH_4. The rate constant and product distribution for the reaction of He^+ with CH_4 are given in Table V.

$$He^+ + CH_4 \longrightarrow CH_2^+ + H_2 + He \quad (23a)$$
$$\longrightarrow CH^+ + H_2 + H + He \quad (23b)$$
$$\longrightarrow CH_3^+ + H + He \quad (23c)$$
$$\longrightarrow CH_4^+ + He \quad (23d)$$

The reaction of H^+ ions with CH_4 may proceed either by charge transfer to give the CH_4^+ ion, or by H^- abstraction from CH_4 to give the CH_3^+ ion and an H_2 molecule.

$$H^+ + CH_4 \longrightarrow CH_4^+ + H \quad (24a)$$
$$\longrightarrow CH_3^+ + H_2 \quad (24b)$$

Unfortunately, there is no experimental determination of the product distribution or rate constant for reaction (24) at thermal kinetic energies. The rate constant for the charge transfer reaction is probably close to the theoretical value of 4.3×10^{-9} cm^3/sec. This is expected since the experimental rate constant for the charge transfer reaction between Kr^+ and CH_4 is so close to the theoretical value (Laudenslager et al., 1974). The difference in ionization potential between Kr and CH_4 ($\Delta IP = 1.3$ eV) is only slightly greater than the difference in ionization potential between H and CH_4 ($\Delta IP = 0.9$ eV).

TABLE V

Bimolecular Reactions of Terminal Ions with Methane

Reaction		Rate constant[a]	Ref.
$He^+ + CH_4$	$\longrightarrow CH_2^+ + H_2 + He$	0.93	Huntress et al. (1974a)
	$\longrightarrow CH^+ + H_2 + H + He$	0.24	Huntress et al. (1974a)
	$\longrightarrow CH_3^+ + H + He$	0.06	Huntress et al. (1974a)
	$\longrightarrow CH_4^+ + He$	0.04	Huntress et al. (1974a)
$H^+ + CH_4$	$\longrightarrow CH_4^+ + H$	(4.3)	Not measured,
	$\longrightarrow CH_3^+ + H_2$		see text
$H_3^+ + CH_4$	$\longrightarrow CH_5^+ + H_2$	2.4	Huntress et al. (1974b)

[a] In units of 10^{-9} cm^3/sec.

The CH_4^+, CH_2^+, and CH^+ product ions formed in reactions (23) and (24) all react with H_2 by hydrogen atom abstraction (Huntress and Kim, 1974).

$$CH_4^+ + H_2 \longrightarrow CH_5^+ + H \quad (25)$$

$$CH_2^+ + H_2 \longrightarrow CH_3^+ + H \quad (26)$$

$$CH^+ + H_2 \longrightarrow CH_2^+ + H \quad (27)$$

The CH_3^+ and CH_5^+ ions ultimately produced do not react with H_2, He, or H, with the possible exception of the thermoneutral reaction.

$$CH_5^+ + H \longrightarrow CH_4^+ + H_2 \quad (28)$$

The CH_3^+ ions can react further with CH_4 (Huntress and Pinizzotto, 1973)

$$CH_3^+ + CH_4 \longrightarrow C_2H_5^+ + H_2 \quad (29)$$

or possibly by three-body reaction with H_2

$$CH_3^+ + H_2 + H_2 \longrightarrow CH_5^+ + H_2 \quad (30)$$

The CH_5^+ and $C_2H_5^+$ ions do not react further with H_2 or CH_4. Both $C_2H_5^+$ ions and CH_4^+ ions may possibly react with H atoms

$$C_2H_5^+ + H \longrightarrow C_2H_4^+ + H_2 \quad (31)$$

$$CH_4^+ + H \longrightarrow CH_3^+ + H_2 \quad (32)$$

Unfortunately, there is no experimental information on ion–neutral reactions with hydrogen atoms. Neither have any of the possible three-body reactions of these hydrocarbon ions with H_2 been studied.

TABLE VI

BIMOLECULAR REACTIONS OF HYDROCARBON IONS

Reaction		Rate constant[a]	Ref.
$CH_4^+ + H_2$	$\longrightarrow CH_5^+ + H$	0.041	Huntress and Kim (1974)
$+ H$	$\longrightarrow CH_3^+ + H_2$	Not measured	
$+ CH_4$	$\longrightarrow CH_5^+ + CH_3$	1.11	Huntress and Pinizzotto (1973)
$CH_3^+ + CH_4$	$\longrightarrow C_2H_5^+ + H_2$	0.89	Huntress and Pinizzotto (1973)
$CH_2^+ + H_2$	$\longrightarrow CH_3^+ + H$	0.72	Huntress and Kim (1974)
$+ CH_4$	$\longrightarrow C_2H_5^+ + H$	0.25	Huntress et al. (1974a)
	$\longrightarrow C_2H_4^+ + H_2$	0.47	Huntress et al. (1974a)
	$\longrightarrow C_2H_3^+ + H + H_2$	0.24	Huntress et al. (1974a)
	$\longrightarrow C_2H_2^+ + 2H_2$	0.14	Huntress et al. (1974a)
$CH^+ + H_2$	$\longrightarrow CH_2^+ + H$	1.01	Huntress and Kim (1974)
$+ CH_4$	$\longrightarrow C_2H_4^+ + H$	0.07	Huntress et al. (1974a)
	$\longrightarrow C_2H_3^+ + H_2$	0.98	Huntress et al. (1974a)
	$\longrightarrow C_2H_2^+ + H_2 + H$	0.14	Huntress et al. (1974a)
$CH_5^+ + H$	$\longrightarrow CH_4^+ + H_2$	Not measured	
$C_2H_5^+ + H$	$\longrightarrow C_2H_4^+ + H_2$	Not measured	

[a] In units of 10^{-9} cm^3/sec.

The production and reaction of hydrocarbon ions at the lower boundaries of the Jovian ionosphere is an intriguing question and should be more thoroughly investigated than the brief mention given here. The importance of these reactions vitally depends on transport effects and on the position of the turbopause in the upper atmosphere (Section V). McElroy (1973) has even considered that significant concentrations of CH_3, C_2H_2, C_2H_4, and C_2H_6 may be present in the lower ionosphere, so that reactions of protons with these species may also be important.

E. REACTIONS WITH EXCITED NEUTRAL SPECIES

Reactions with vibrationally excited H_2 molecules may provide a mechanism for overcoming the endothermicity of reaction (6a) for thermal protons and for promoting an increase in the rate constant for reaction (4). Reaction (6a) is exothermic for hydrogen molecules in vibrationally excited states $v = 4$ and greater. These reactions are potentially important loss processes for H^+ ions (McElroy, 1973) and He^+ ions. Unfortunately, reactions with vibrationally excited H_2 have not been experimentally investigated, and the distribution of vibrationally excited H_2 in the Jovian ionosphere is a difficult function to calculate at best. Detailed data on the absolute cross sections and excitation functions for the production of vibrationally excited H_2 by a variety of processes are required.

Processes which could contribute to the formation of vibrationally excited H_2 include collisional excitation by fast photoelectrons; recombination of electrons with H_3^+ ions via reaction (10); and energy transfer in reactions such as (1b), (3), and three-body reactions. McElroy has suggested that a fair fraction of the ultraviolet energy absorbed in the Jovian upper atmosphere may appear as vibrational excitation of the H_2 molecule and that in spite of the low kinetic temperatures in the ionosphere (100–150°K), vibrational temperatures greater than 700° may be possible.

V. Ion Loss Rates

A. H^+ AND He^+

Figures 10 and 11 show the loss rates for H^+ and He^+ ions calculated for the various ion loss mechanisms discussed in the previous sections. The atmospheric models of Hunten (1969) and of Gross and Rasool (1964) have been used. The altitude scale refers to the height above the cloud tops at the 2.5-atm. level. Methane densities are taken from Prinn (1970). The electron densities can vary considerably with height, depending on the model. Between 200 and 450 km, both Hunten and Gross and Rasool give electron densities between 10^5 and 10^6 per cubic centimeter. Shimizu (1971) has shown how transport effects can lower this value to 10^4 per cubic centimeter.

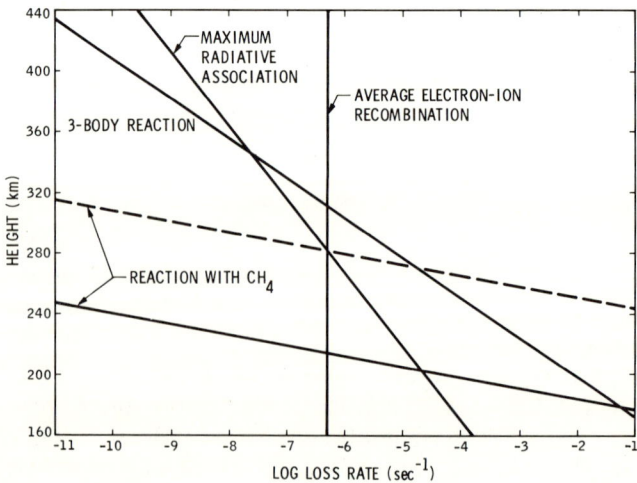

FIG. 10. Ion loss rates vs. height calculated for protons in the Jovian ionosphere. The loss rate for reaction with CH_4 is calculated using both the Gross and Rasool turbopause (solid line) and the Hunten turbopause (dashed line).

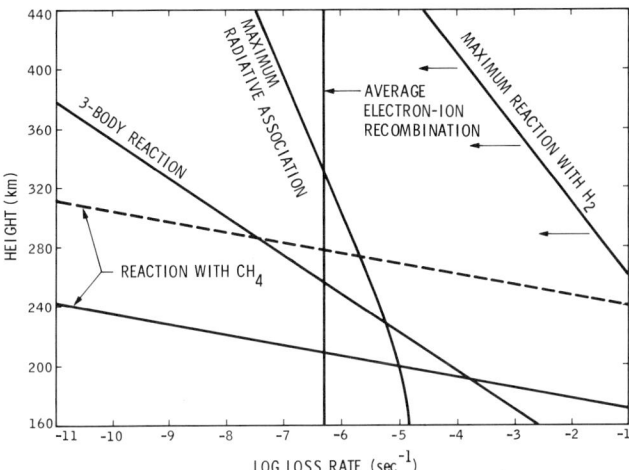

FIG. 11. Ion loss rates vs. height calculated for He^+ ions in the Jovian ionosphere. The loss rate for reaction with CH_4 is calculated using both the Gross and Rasool turbopause (solid line) and the Hunten turbopause (dashed line).

For comparison purposes, an average value of 10^5 per cubic centimeter, constant with height, is used for the calculation of the ion–electron radiative recombination rates in Figs. 10 and 11.

The loss rates calculated for three-body reactions assume reaction with two hydrogen molecules [reactions (20) and (22)]. Although reaction (22) involves an intermediate ion with a lower dissociation energy than reaction (20) does, it is assumed that this reaction is at least as fast as reaction (21) for the reasons discussed in Section IV. A rate constant of 2×10^{-31} cm^6/sec is therefore used as a lower limit for reaction (22) at 150°K.

The loss rates in Fig. 10 for the radiative association of H^+ ions are calculated using relation (H) by assuming that $\tau_r = 10^{-2}$ sec for the radiative lifetime of the $(H_3^+)^*$ intermediate ion. Reactions of vibrationally excited $(H_3^+)^*$ ions have been observed experimentally (Huntress and Bowers, 1973, 1974), the results indicating a radiative lifetime for these ions considerably in excess of 10^{-3} sec. The radiative lifetimes of the intermediate ions in reactions (14)–(17) are likely to be quite long, possibly even longer than 10^{-2} sec, because of the very small change in dipole moment involved in the vibrational transitions. The loss rates for radiative association given in Fig. 10 are therefore maximum loss rates, and the real loss rates due to this process are likely to be much less. The results in Fig. 10 show that the electron–ion radiative recombination process is clearly the dominant loss mechanism for H^+ ions at altitudes greater than ~ 300 km in these models. Below these altitudes, three-body reactions with H_2 and bimolecular reaction with CH_4

will dominate. The radiative association process is not likely to be an important loss mechanism for protons and can probably be ignored in model ionosphere calculations.

The loss rates in Fig. 11 for the radiative association of He^+ ions are calculated using Dalgarno's value for association with H atoms; $k_{19} = 2 \times 10^{-15}$ cm^3/sec. The loss rates for radiative association with He atoms are probably insignificant compared to reaction (19) because of the small rate constant $k_{17} \leq 2 \times 10^{-19}$ cm^3/sec calculated for this reaction from relation (H) and k_{21}. The charge transfer reaction (without radiation) between He^+ ions and H atoms is also ignored in Fig. 11, since the rate constant k_5 is likely to be much less than the theoretical value for the association reaction (19b).

The loss rates shown in Fig. 11 for the bimolecular reaction of He^+ ions with H_2 are calculated using the experimental upper limit for the rate constant of reaction (4), $k_4 = 10^{-13}$ cm^3/sec. The rate constant for this reaction could easily be zero or less than the value 10^{-17} cm^3/sec that Fig. 11 shows would be significant. Until the rate constant k_4 can be accurately measured, or the upper limit lowered by more than four orders of magnitude, it is not likely that the overall loss rate for He^+ ions in the ionosphere can be calculated with any confidence. Reaction (4) has not been observed, so that the practice of using the experimental upper limit of 10^{-13} cm^3/sec for k_4 in ionospheric models is not justified, and the loss rate of He^+ ions can be orders of magnitude smaller than previously assumed. This would have a considerable effect on the calculated ion densities in these models. McElroy has also shown that reaction (4b) becomes an important source of protons in the ionosphere even for a rate constant as small as 10^{-20} cm^3/sec. The rate constant, mechanism, and products of the bimolecular and three-body reactions of He^+ ions with H_2 remain the principal problems in the radiation chemistry of $He-H_2$ mixtures.

The loss rates for reactions with methane depend critically on the location of the turbopause. The dashed lines in Figs. 10 and 11 are drawn for the location of the turbopause as given by Hunten, and the solid lines are for the location of the turbopause as given by Gross and Rasool. In the absence of significant transport effects (the Gross and Rasool model), these reactions are of only minor importance. However, for eddy diffusion coefficients as large as $K = 5 \times 10^6$ cm^2/sec (the Hunten model), these reactions are extremely important and will be a major loss mechanism for H^+ and He^+ ions in the ionosphere.

The loss rates for H^+ and He^+ ions due to reaction with vibrationally excited H_2 molecules are not calculated because of the lack of any experimental information on these processes or on the distribution of vibrationally excited H_2 molecules in the Jovian ionosphere. The vibrational energy level separations in the H_2 molecule are quite large (1.0 eV between

$v = 0$ and $v = 4$) so that very high temperatures are required to populate vibrationally excited states. If it is assumed that k_{6a} is on the order of 10^{-9} cm^3/sec for H$_2$ molecules in the vibrationally excited $v = 4$ state, then equilibrium vibrational temperatures in excess of 1000°K are required if this reaction is to compete with radiative ion–electron recombination at 320 km in Fig. 10. Correspondingly higher vibrational temperatures, or large departures from vibrational equilibrium, would be required at higher altitudes or for lower rate constants. These temperatures seem excessive in view of the present knowledge of the Jovian ionosphere (McElroy, 1973). Reactions with metastable electronically excited neutrals may also be possible, but large metastable densities, $\sim 10^2$/cm^3, would be required for effective competition with ion–electron recombination above 320 km. In view of the discussion given in Section VI,A, the density of metastables is likely to be several orders of magnitude less than 10^2/cm^3.

B. H_3^+

Because of the large rate constant for the dissociative recombination of polyatomic ions with electrons, the loss rate of H_3^+ ions is dominated by this process. In the presence of significant transport effects, the reaction (Huntress et al., 1974b).

$$H_3^+ + CH_4 \longrightarrow CH_5^+ + H_2 \qquad (33)$$

can become a major loss process for H_3^+ ions. For the Hunten turbopause, reaction (33) becomes significant at an altitude of 250 km. The three-body reaction

$$H_3^+ + H_2 + H_2 \longrightarrow H_5^+ + H_2 \qquad (34)$$

has been identified experimentally (Saporoschenko, 1965), but the rate constant is considerably smaller than for reaction (20); in which case this process may not be a significant loss process for H_3^+ ions compared to the bimolecular reaction with CH$_4$.

The H_3^+ ions are produced by reaction (1a) in a distribution of vibrationally excited states. This distribution has not yet been well characterized, but it has been shown that for a Franck–Condon distribution of initial vibrational energies in the H_2^+ reactant ions, H_3^+ product ions with as much as 1.9 eV excess internal vibrational energy are produced (Huntress and Bowers, 1973a,b). This excess vibrational energy is removed in subsequent collisions with H$_2$ molecules by transfer of vibrational energy from the ion to vibrational or translational energy in the neutral

$$H_3^+(v) + H_2 \longrightarrow H_3^+ + H_2(v') \qquad (35)$$

The rate constant for reaction (35) is $k_{35} \sim 3 \times 10^{-10}$ cm^3/sec (Huntress and Bowers, 1973a,b).

In the Jovian atmosphere, the vibrationally excited H_3^+ ions are most likely completely deactivated to the ground state ions by collisions with H_2 before further reactions can take place. The rate constant for dissociative recombination of H_3^+ ions with electrons is probably fairly sensitive to excess internal energy in the H_3^+ ion. However, the experimental rate constants given in Fig. 8 are most likely the rate constants for reaction of ground state ions since the values were measured in a high pressure experiment. The rate constant for the reaction of H_3^+ ions with methane is independent of vibrational energy in H_3^+, but the product distribution is not (Huntress et al., 1974b). For excited H_3^+ ions, the reaction

$$(H_3^+)^v + CH_4 \longrightarrow CH_3^+ + 2H_2 \tag{36}$$

is observed.

The conversion of the internal energy of H_2^+ and H_3^+ ions into vibrational and kinetic energy of neutrals via reactions (35), (1b), and (2b) provide a mechanism for thermal heating of the neutral ionosphere. Several other mechanisms for heating of the neutral ionosphere include the transfer of the excess kinetic energy of protons via reactions (6) and (7), the conversion of the internal energy of the intermediate complexes in reactions (14)–(19) via collisions with a third body, and the conversion of the excess energy of the dissociative recombination reaction (10). Superelastic collisions of electrons with vibrationally excited H_3^+ ions

$$e^- + H_3^+(v) \longrightarrow H_3^+ + e^- \tag{37}$$

can provide a mechanism for heating of thermal electrons in the ionosphere.

C. H_2^+, He_2^+, He_2H^+, HeH^+, AND HeH_2^+

For the reactive ions, H_2^+ and HeH^+, the dominant loss mechanism is bimolecular ion–molecule reaction with H_2—except perhaps at extreme altitudes. These reactions proceed at rapid rates to give the H_3^+ ion (Table II). Any He_2^+, He_2H^+, and HeH_2^+ ions formed by three-body reactions or in chemi-ionization reactions (Section VI) will also react rapidly with H_2. The reactions of these exotic ions with H_2 and H are given in Table VII. Since the most likely mode of formation of He_2^+, He_2H^+, and HeH_2^+ ions is by three-body reactions in the lower ionosphere, reactions with H_2 are probably more important than reactions with H atoms. The product distribution of the reaction of He_2^+ ions with H_2 is not well known, but the rate constant has been measured (Adams et al., 1970; Veatch and Oskam, 1973). The He_2H^+ and HeH_2^+ ions can react with H_2 only by proton transfer to form H_3^+ ions. Although the latter two reactions have not been examined experimentally, the rate constants for these facile proton transfer reactions is expected to be close to the theoretical rate constants given in Table VII.

TABLE VII

REACTIONS OF EXOTIC IONS OF HELIUM

Reaction		Rate constant[a]	Comments
$He_2^+ + H_2$	$\rightarrow H^+ + H + 2He$	⎫	Product distribution unknown. Veatch and Oskam (1973) claims both H^+ and H_2^+ are produced. Adams et al. (1970) give He_2H^+ as possible major product
	$\rightarrow H_2^+ + 2He$	⎬ 0.53; (Adams et al., 1970)	
	$\rightarrow HeH^+ + H + He$	⎪	
	$\rightarrow HeH_2^+ + He$	⎪	
	$\rightarrow He_2H^+ + H$	⎭	
$+ H$	$\rightarrow H^+ + 2He$	Not measured	Unimportant except at extreme heights
	$\rightarrow HeH^+ + He$		
$He_2H^+ + H_2$	$\rightarrow H_3^+ + 2He$	(1.80)	Expected rate constant (theoretical)
	$\rightarrow H_2^+ + He$		Most probable product is H_2^+.
$+ H$	$\rightarrow HeH_2^+ + He$	Not measured	Unimportant except at extreme heights
$HeH_2^+ + H_2$	$\rightarrow H_3^+ + H + He$	(1.88)	Expected rate constant (theoretical)
	$\rightarrow H_3^+ + He$		Most probable product is H_3^+.
$+ H$	$\rightarrow H^+ + H_2 + He$	Not measured	Unimportant except at extreme heights
	$\rightarrow HeH^+ + H_2$		

[a] In units of 10^{-9} cm^3/sec.

VI. Ionization Processes of Photoelectrons

A. Production of Metastable Excited Species and Secondary Ionization

The electrons resulting from photoionization of molecules in the ionosphere can have considerable kinetic energy and are capable of causing secondary ionization either by direct electron impact ionization or through the production of metastable electronically excited species via the reactions

$$e^- + He \longrightarrow He(2^3S), He(2^1S) + e^- \quad (38)$$
$$+ H \longrightarrow H(2S) + e^- \quad (39)$$
$$+ H_2 \longrightarrow H_2^* + e^- \quad (40)$$
$$\longrightarrow H(2S) + H + e^- \quad (40b)$$

The excitation energies of the metastable $He(2^1S)$, $He(2^3S)$, and $H(2S)$ states are 20.6 eV, 19.8 eV, and 10.2 eV, respectively (Muschlitz, 1968). Various metastable excited states of hydrogen exist with excitation energies greater than 11.8 eV (Olmsted et al., 1965). The lifetimes of the $He(2^1S)$, $He(2^3S)$, and $H(2S)$ metastable species are 0.019 sec (Drake et al., 1969), 7900 sec (Drake, 1971; Moos and Woodworth, 1973) and 0.12 sec (Muschlitz, 1968), respectively. These long-lived, highly excited neutral species can, in turn, produce ions by the chemi-ionization (or Penning ionization) processes (Smith and Muschlitz, 1960; Sholette and Muschlitz, 1962; Hotop and Niehaus, 1968; Bolden et al., 1970; Dunning and Smith, 1970; Schmeltekopf and Fehsenfeld, 1970; Shaw et al., 1971).

$$He(2^1S), He(2^3S) + H_2 \longrightarrow H_2^+ + He + e^- \quad (41a)$$
$$\longrightarrow HeH^+ + (H + e^-) \quad (41b)$$
$$\longrightarrow HeH_2^+ + e^- \quad (41c)$$
$$\longrightarrow H^+ + He + (H + e^-) \quad (41d)$$
$$+ He \longrightarrow He_2^+ + e^- \quad (42)$$
$$+ H \longrightarrow H^+ + He + e^- \quad (43a)$$
$$\longrightarrow HeH^+ + e^- \quad (43b)$$
$$H(2S) + H_2 \longrightarrow H_3^+ + e^- \quad (44)$$
$$H_2^* + H_2 \longrightarrow H_3^+ + (H + e^-) \quad (45)$$

Reactions (41)–(44) have been identified and their rate constants measured (Table VIII).

TABLE VIII

Chemi-Ionization Reactions of Metastable Excited Neutrals

Reaction	Rate constant[a]	Product distribution	Comments
$He(2^3S) + H_2$	0.032 (Schmeltekopf and Fehsenfeld, 1970; Sholette and Muschlitz, 1962; Bolden et al., 1970)	H_2^+ : 0.88 ⎫ Hotop and HeH^+ : 0.02 ⎬ Niehaus, HeH_2^+ : 0.10 ⎭ 1968	Veatch and Oskam (1973) claim H^+ as a product ion, does not observe HeH_2^+, and gives a total rate constant of 0.052
$He(2^1S) + H_2$	0.021	Not measured	Total rate constant calculated from $k(2^1S)/k(2^3S)$ given in Sholette and Muschlitz (1962) and Dunning and Smith (1970). Value given for $k(2^1S)$ in Schmeltekopf and Fehsenfeld (1970) probably in error; see Dunning and Smith (1970). Product distribution probably the same as for $He(2^3S)$
$He(2^3S) + He$	0.86	He_2^+	Smith and Muschlitz (1960)
$He(2^1S) + He$	0.65	He_2^+	Smith and Muschlitz (1960)
$He(2^3S) + H$	0.62	H^+ major product HeH^+ minor product	Shaw et al. (1971); $HeH^+/H^+ \sim 1/10$
$H(2S) + H_2$	~ 0.03	H_3^+	Chupka et al. (1968). Nonreactive deactivation rate constant may be much larger; see text
+ He	~ 2.8	(Deactivation)	Extrapolated to thermal energies from the data of Kass and Williams (1971)

[a] In units of 10^{-9} cm^3/sec. Only known reactions are given in this table. All rate constants are quoted at $\sim 300°K$.

Metastable species can also be formed as the products of ion–electron recombination reactions, such as (10), (11), (12), (56), and (57). Unfortunately, neither the product distributions nor the product excitation energies have been measured for these reactions. The energy level separation between the 2S and 2P levels in the hydrogen atom is extremely small (Lamb and Retherford, 1950) (4.09×10^{-6} eV), so that metastable H(2S) atoms are readily quenched in collisions with neutrals by conversion to the short-lived H(2P) state.

$$H(2S) \xrightarrow{(H_2, He, H)} H(2P) \rightsquigarrow H(1S) + h\nu \qquad (46)$$

Reaction (46) has been examined for collisions with helium atoms (Kass and Williams, 1971), and has shown to have an extremely large rate constant (Table VIII). Quenching of the metastable He(2^1S) atoms probably occurs primarily in reactive encounters [reactions (41)–(43)], since the energy level separation between the 2^1S and 2^3S levels in helium is large (0.8 eV). Helium (2^1S) metastables are rapidly converted to He(2^3S) metastables by superelastic collisions with electrons.

$$e^- + He(2^1S) \longrightarrow He(2^3S) + e^- \qquad (47)$$

The rate constant for this reaction is $\sim 3 \times 10^{-7}$ cm^3/sec for slow electrons (Phelps, 1955).

The relative importance of these secondary ionization processes depends on two factors: (*i*) the fraction of photoelectrons with sufficient kinetic energy to cause direct ionization and formation of metastable species, and (*ii*) the fraction of these energetic electrons which survive competing thermalization processes. An estimate of the importance of these processes may be obtained by briefly examining order-of-magnitude estimates for the production of energetic electrons and for the cross sections of energy-loss processes of energetic electrons.

The He II line at 303 Å can be used as an example of a strong line in the solar extreme UV spectrum. The kinetic energy of the electrons resulting from ionization of hydrogen at this wavelength will be about 25 eV. These electrons are sufficiently energetic to ionize H_2 molecules and to form the metastable states of He, H_2, and H. However, these electrons will also lose their energy by collisions with neutrals, thermal electrons, and ions in the ionosphere. For 25 eV electrons, the relative cross sections for these processes are approximately

$$\sigma_{e-c} \simeq 10^{-14} \text{ cm}^2$$

for elastic scattering with thermal electrons (Dalgarno, 1969) and

$$\sigma_s \simeq 10^{-15} \text{ cm}^2$$
$$\sigma_i \simeq 10^{-16} \text{ cm}^2$$
$$\sigma_e \simeq 10^{-16} \text{ cm}^2$$
$$\sigma_d \simeq 10^{-16} \text{ cm}^2$$
$$\sigma_m \simeq 10^{-18} \text{ cm}^2$$

for collisions with H_2; where σ_s, σ_i, σ_e, σ_d, and σ_m are the cross sections for elastic scattering (Trajmar et al., 1970), ionization (Rapp and Englander-Golden, 1965), excitation (Trajmar et al., 1970), dissociation (Corrigan, 1965), and metastable production (Clampitt and Newton, 1969) (both H* and H_2^*), respectively. Although the cross section for elastic scattering in collisions with electrons is quite large, these collisions are not very competitive with scattering processes in collisions with neutrals because of the low number density of electrons compared to neutrals in the ionosphere: $[e^-]/[H_2] \lesssim 10^{-4}$ at 450 km. Elastic collisions with neutrals are not effective in removing the energy of the electron because of the very small fraction of energy lost per collision in these events (Dalgarno, 1969), $\Delta E \simeq 5 \times 10^{-4} E_0$. Ionization of H_2 is effective in removing about 15 eV of the energy of the electron. Dissociation of H_2 also removes from 4.5 to 18 eV of the energy of the electron and can lead to emission in the Lyman lines of H (McGowan et al., 1969; Vroom and de Heer, 1969). The excitation of short-lived electronically excited states can also remove a large fraction of the total kinetic energy and yield various emission features (Stone and Zipf, 1972).

The process of thermalization of 25 eV electrons is therefore initiated by collisions with H_2 which lead to ionization, excitation, and dissociation of H_2. Metastable excited species, H_2^* and H*, are formed in less than 1% of these collisions. Collisions of 25-eV electrons with helium are not effective in removing the energy of the electron since dissociative modes are not available and since the cross section for ionization (Rapp and Englander-Golden, 1965) and excitation (Brongersma et al., 1972) is small at this energy. The cross section for ionization and excitation of He does increase significantly at higher electron energies. The cross section at 25 eV for excitation of the metastable 2^3S and 2^1S states of He is about $\sigma_m = 10^{-17}$ cm^2 (Brongersma et al., 1972). This is ten times larger than σ_m for H_2, but it is likely that the number density of helium atoms in the Jovian atmosphere is about 10 times fewer than hydrogen molecules. In this case, collisions with H_2 will be the major energy loss mechanism for 25-eV photoelectrons, and less than 1% of the total number of collisions will result in the formation of helium metastables.

Generalizing the above arguments to include wavelengths other than 303 Å capable of producing photoelectrons of sufficient energy to form metastable species, the number of metastables produced per ionizing photon is likely to be less than 10^{-3}. This means that reactions (41)–(45) are responsible only for the production of minor, if somewhat exotic, ions in the ionosphere. On the other hand, secondary ions produced by direct impact of energetic photoelectrons

$$e^- + H_2 \longrightarrow H_2^+ + 2e^- \tag{48}$$

$$+ He \longrightarrow He^+ + 2e^- \tag{49}$$

can amount to a significant fraction of those produced by primary photoionization. A more accurate estimate of the importance of these ionization processes of energetic photoelectrons will require a much more detailed calculation than that given here.

B. Electron Impact Excitation and Dissociation of Ions

The dissociation of H_2^+ ions into H^+ ions and H atoms by the impact of energetic electrons has been studied in the laboratory.

$$e^- + H_2^+ \longrightarrow H^+ + H + e^- \tag{50}$$

This process has a large rate constant even at low electron energies (Dunn and Van Zyl, 1967) (7×10^{-9} cm^3/sec at 10 eV). This particular process is not important in the Jovian ionosphere, however, because of the fast ion–molecule reaction (1a) and the low relative densities of electrons to neutrals. For the nonreactive H_3^+ molecular ion, the rate constant for dissociation by electron impact is unknown. Even if the rate constant for H_3^+ were as large as that for H_2^+, however, this process would not compete with ion–electron recombination as a major loss mechanism for H_3^+ ions.

C. Production of Negative Ions

A layer may possibly be formed at the very lower boundary of the Jovian ionosphere where the principal negative charge carriers are negative ions rather than electrons, similar to the D layer in the Earth's ionosphere. Negative ions are probably not as readily produced in the Jovian atmosphere as in the terrestrial atmosphere, however, because of the very low electron attachment cross section for hydrogen as compared to oxygen. No H_2^- ions have been observed by electron attachment processes in hydrogen. The cross section for the production of the H^- ion by dissociative electron attachment

$$e^- + H_2 \longrightarrow H^- + H \tag{51}$$

is small and highly dependent on electron energy (Rapp et al., 1965; Asundi and Schultz, 1967) (Fig. 12). Hydride ion formation by the pair process

$$hv, e^- + H_2 \longrightarrow H^- + H^+ \tag{52}$$

has not been observed in molecular hydrogen by photon impact, but it is suspected to be responsible for hydride ion formation by electron impact at energies greater than 17.2 eV (Rapp et al., 1965). The radiative attachment process in atomic hydrogen has a negligibly small cross section

$$e^- + H \longrightarrow H^- + hv \tag{53}$$

and can probably be ignored as a source for hydride ions (Gross and Rasool, 1964).

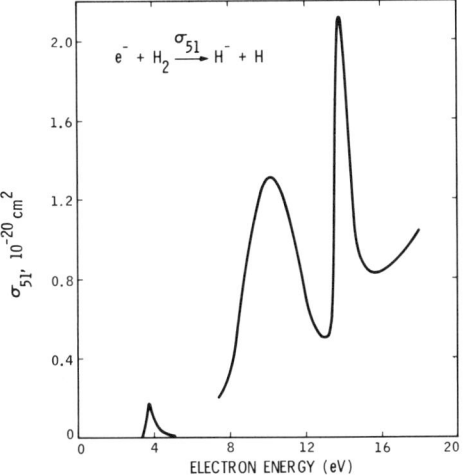

FIG. 12. Cross section for the formation of hydride ions in hydrogen vs. electron impact energy. The data at 4 eV is from Asundi and Schultz (1967), and the data from 8 to 18 eV is from Rapp et al. (1965).

Loss mechanisms for the hydride ion include the inverse of reactions (51)–(53): photodetachment, associative detachment, and ion–ion recombination

$$hv + H^- \longrightarrow H + e^- \tag{54}$$

$$H^- + H \longrightarrow H_2 + e^- \tag{55}$$

$$+ H^+ \longrightarrow H^* + H \tag{56}$$

$$+ H_3^+ \longrightarrow H_2 + H_2. \tag{57}$$

The hydride ion does not react with H_2 or CH_4. The rate constants for reactions (55) and (56) have been measured experimentally (Schmeltekopf et *al.*, 1967; Janev and Taňcic, 1972; Moseley *et al.*, 1970) and are given in Table IX. The cross section for photodetachment as a function of photon energy is given by Branscomb (1962) and shows a broad peak over the range 2000–16,000 Å with a peak cross section of $\sigma = 4 \times 10^{-17}$ cm^2 near 8000 Å.

TABLE IX

REACTIONS OF HYDRIDE IONS

Reaction	Rate constant[a]	Ref.	Comments
$H^- + H \longrightarrow H_2 + e^-$	~1.3	Schmeltekopf *et al.* (1967)	300°K
$+ H^+ \longrightarrow H(n = 2, 3) + H(1s)$	126	Moseley *et al.* (1970)	Extrapolated to 150°K from the data of Moseley *et al.* (1970). Excited hydrogen atoms produced
$+ H_3^+ \longrightarrow H_2 + H_2$	Not measured		

[a] In units of 10^{-9} cm^3/sec.

VII. Additional Photon Impact Processes

A. Formation and Ionization of Metastables by Photon Impact

Metastable electronically excited hydrogen atoms can be formed by the photodissociation of hydrogen between 800 and 845 Å (Chupka *et al.*, 1968). These excited H atoms can produce H_3^+ ions via reaction (44). Metastable excited H_2 molecules may also be produced by photon impact in this wavelength region (Chupka *et al.*, 1968). Chupka and co-workers have shown, however, that the contribution to the total H_3^+ ion intensity produced by reactions (44) and (45) over the wavelength range 725–850 Å is negligible compared to the production of H_2^+ ions (and subsequently H_3^+ ions) by photon impact.

Metastable species can also be destroyed by photoionization. The cross section for the ionization of metastable He(2^1S) and He(2^3S) atoms has been measured by Stebbings *et al.* (1973) and has been shown to be in the range $\sigma = 0.5$–1.0×10^{-17} cm^2 over the wavelength range 2400–3100 Å. The loss rate of helium metastables due to photoionization is approximately 100 times smaller than that due to reaction with H_2 at altitudes near 300 km.

B. Photodissociation of Ions

The cross section for the photodissociation of H_2^+ ions into H^+ ions and H atoms has been measured by Dunn (1964) and by J. A. Burt (private communication), giving an average value of $\sigma = 2.6 \times 10^{-19}$ cm^2 for wavelengths between 2000 and 10,000 Å. This particular process has little significance because of the fast ion–molecule reaction (1a). For the nonreactive H$_3^+$ ion,

$$hv + H_3^+ \longrightarrow H^+ + H_2, \tag{58}$$

Dunn reports that the photodissociation cross section is less than 10^{-20} cm^2 over this wavelength range. Burt, however, gives a cross section for (58) of 4×10^{-18} cm^2. Even for a cross section as large as reported by Burt, reaction (58) is probably not a major loss mechanism for H$_3^+$ ions. The average solar flux over the wavelength range 2000–10,000 Å at Jupiter is approximately 4×10^{11} photons/cm^{2}sec Å (from the solar flux data of Hinteregger (1970) with appropriate scaling to Jupiter), so that the total flux over this bandwidth is about $F = 3 \times 10^{15}$ photons/cm^2 sec. The loss rate for H$_3^+$ ions due to reaction (58) is therefore given by $\sigma F \simeq 10^{-3}$/sec. This is about thirty times smaller than the H$_3^+$ loss rate due to ion–electron recombination (for an average electron density of 10^5 per cubic centimeter).

C. Photoionization of Methane

Ionospheric models to date have ignored the ionization of methane, since methane does not reach into the region of the ionosphere where the peak ionization rates occur. The relative abundance of methane is rapidly diminished above the turbopause by diffusive separation. Significant amounts of methane may, however, encroach upon the lower reaches of the ionosphere—especially if the turbopause is located at a high altitude and transport effects are large. Using Hunten's position for the turbopause, Prinn has shown that sufficient ultraviolet flux reaches to lower altitudes for CH$_4$ to be photoionized to CH$_4^+$, CH$_3^+$, and CH$_2^+$ ions. The subsequent chemistry of these ions is discussed in Section IV,D. Again, the question of the production and importance of hydrocarbon ions in the lower ionosphere rests critically on the existence of significant transport effects.

VIII. Concluding Remarks

We have given here a review of the radiation chemistry of hydrogen–helium mixtures, reported in as much detail as this chemistry is presently known. Processes involving methane in those mixtures have also been included. All known ionization processes occurring in the closed system consisting of hydrogen, helium, methane, and energetic photons have been considered as they may apply to the chemistry of the Jovian ionosphere.

Many of these processes appear to be of only minor importance, such as formation and chemi-ionization reactions of metastable excited species, electron and photon impact dissociation of molecular ions, and negative-ion formation, but are included for completeness since they probably do in fact occur and because they may possibly turn out to be important as our knowledge of the Jovian ionosphere and of molecular physics improves.

Our review has uncovered new data and new reactions which have been overlooked in previous ionospheric models and which may have considerable effects on the calculations of ion and electron densities in the ionosphere: the production of thermal and hot protons via dissociative photoionization of hydrogen, the production of H_2^+ and H_3^+ ions in vibrationally excited states, the collisional deactivation of these excited ions and the dependence of reaction rates on ionizing photon wavelength, the reaction of HeH^+ ions with H_2 and H, reactions of kinetically excited protons, and the possible production of three hydrogen atoms by the dissociative ion–electron recombination of H_3^+ ions. The rate constant for the principal ion–molecule reaction occurring in the ionosphere has been shown to be four times faster than previously assumed, and it has been shown that the radiative association reactions used in some models are probably not important atomic ion loss mechanisms. Three-body reactions and reactions with methane have been considered in some detail. Also, the large effect of transport phenomena on the possible importance of ionization processes involving methane and hydrocarbons has been emphasized, and the fallacy of using a "large" rate constant for the reaction of He^+ ions with H_2 has been pointed out.

In the course of this examination of the radiation chemistry of the Jovian ionosphere, it has become clear that there are many gaps in our knowledge of this chemistry. In order to accurately model the Jovian ionosphere, new data are required and additional processes must be examined. A "shopping list" that the Jovian aeronomer might give to the laboratory kineticist would include the following:

1. Rate constant and product distribution for the reaction $He^+ + H_2$, measured to an upper limit of 10^{-17} cm^3/sec. This reaction is extremely important as a loss mechanism for He^+ ions and also for the production of protons and hydrocarbon ions.

2. Product and energy distribution in the reaction $e^- + H_3^+$; are the products $H_2(v) + H$ or $3H$. This information is extremely important for calculations of H atom concentrations and of the energy distribution in the neutral ionosphere.

3. Rate constants, product distributions, and temperature dependence for the three-body reactions of H^+ and He^+ in mixtures of H_2 and He. These are important atomic ion loss mechanisms.

4. Rate constants and product distributions for the reaction of protons with simple hydrocarbons. These reactions are important as a loss mechanism for protons and as a production mechanism for hydrocarbon ions.

5. Rate constants and product distributions for the reactions of H^+ and He^+ ions with vibrationally excited H_2 molecules. These reactions may possibly be important loss mechanisms for H^+ and He^+ ions.

6. Better data on the reaction of kinetically excited protons with H_2 and H. The absolute rate constants vs. kinetic energy determined by beam experiments for the reaction $H^+ + H_2$ can easily be off by almost an order of magnitude. Although the theoretical calculations for the reaction $H^+ + H$ are probably fairly good, this needs to be confirmed at low relative velocities. These reactions are important for the thermalization processes of hot protons, the loss of hot protons, and the heating of ions, electrons, and neutrals in the ionosphere.

7. Rate constants for the reactions of ions with neutral hydrogen atoms. The reactions of H_2^+ and HeH^+ with H may be important loss mechanisms for these ions at extreme heights. Although the rate constant for the charge transfer reaction between He^+ and H is theoretically very small, experimental verification would increase confidence in this prediction. The reactions of hydrocarbon ions with H may be important for hydrocarbon ion chemistry in the lower ionosphere.

8. Temperature dependence of bimolecular ion–molecule reactions. While these reactions are theoretically predicted to have no temperature dependence, and while experiment tends to confirm this prediction in most cases for nonpolar gases, there are a few exceptions and this assumption should be tested experimentally for the major reactions involved in controlling the ion chemistry in the ionosphere.

9. More detailed information on photoelectron impact processes for the ionization, dissociation, and excitation of the neutral components in the ionosphere. These data are required for including in model ionosphere calculations the processes of secondary ionization, chemi-ionization by metastable neutral excited species, reactions of ions with excited neutral species and formation of negative ions.

10. Three-body associations of electrons in hydrogen–helium–hydrocarbon mixtures. These reactions may be important for negative ion formation. Processes involving negative ions in these mixtures should in general be more vigorously studied. There is a dearth of information on negative ion formation and reaction in these mixtures.

11. More complete data on chemi-ionization reactions and reactions of the exotic ions He_2^+, He_2H^+, and HeH_2^+ in H_2–He–CH_4 mixtures. These reactions may be important in the lower atmosphere during electrical discharges in clouds, if not important in the ionosphere as well.

This is a fairly extensive wish list and presents an interesting challenge to the experimental molecular physicist. Challenges to the aeronomer have also been presented in this review, including the necessity for characterizing transport effects in the Jovian upper atmosphere, locating the turbopause in the mesosphere and determining the temperature profile and composition of neutral atmosphere. Together, these problems promise an adventure in scientific discoveries concerning the major planets of the solar system—an adventure which has only just begun.

ACKNOWLEDGMENT

I would like to thank Professor A. Dalgarno and Dr. Donald M. Hunten for reading the manuscript and providing several helpful suggestions.

REFERENCES

Adams, N. G., Bohme, D. K., and Ferguson, E. E. (1970). *J. Chem. Phys.* **52**, 5101.
Asundi, R. K., and Schultz, G. J. (1967). *Phys. Rev.* **158**, 25.
Bardsley, J. N., and Biondi, M. A. (1970). *Advan. At. Mol. Phys.* **6**, 1.
Bates, D. R., and Dalgarno, A. (1962). In "Atomic and Molecular Processes" (D. R. Bates, ed.), p. 245. Academic Press, New York.
Belyaev, V. A., Brezhnev, B. G., and Erastov, E. M. (1967). *Sov. Phys.—JETP* **25**, 777.
Bolden, R. C., Hemsworth, R. S., Shaw, M. J., and Twiddy, N. D. (1970). *Proc. Phys. Soc., London (At. Mol. Phys.)* [2] **3**, 61.
Bowers, M. T., and Elleman, D. D. (1972). *Chem. Phys. Lett.* **16**, 486.
Branscomb, L. M. (1962). In "Atomic and Molecular Processes" (D. R. Bates, ed.), p. 113. Academic Press, New York.
Brongersma, H. H., Knoop, F. W., and Backx, C. (1972). *Chem. Phys. Lett.* **13**, 16.
Browne, J. C. (1966). *J. Chem. Phys.* **45**, 2707.
Browning, R., and Fryar, J. (1973). *Proc. Phys. Soc., London (At. Mol. Phys.)* [2] **6**, 364.
Carrington, T. (1972). *J. Chem. Phys.* **57**, 2033.
Chupka, W. A., and Russell, M. E. (1968). *J. Chem. Phys.* **49**, 5426.
Chupka, W. A., Russell, M. E., and Refaey, K. (1968). *J. Chem. Phys.* **48**, 1518.
Clampitt, R., and Newton, A. S. (1969). *J. Chem. Phys.* **50**, 1997.
Clow, R. P., and Futrell, J. H. (1972). *Int. J. Mass Spectrom. Ion Phys.* **8**, 119.
Corrigan, S. J. B. (1965). *J. Chem. Phys.* **43**, 4381.
Dalgarno, A. (1958). *Phil. Trans. Roy. Soc. London, Ser. A* **250**, 426.
Dalgarno, A. (1969). *Can. J. Chem.* **47**, 1723.
Dalgarno, A. (1972). In "Physics of Electronic and Atomic Collisions" (T. R. Gover and F. J. de Heer, eds.), p. 381. North-Holland Publ., Amsterdam.
Dewangen, D. P. (1973). *Proc. Phys. Soc., London (At. Mol. Phys.)* [2] **6**, L20.
Drake, G. F. W. (1971). *Phys. Rev. A* **3**, 908.
Drake, G. F. W., Victor, G. A., and Dalgarno, A. (1969). *Phys. Rev.* **180**, 25.
Dunn, G. H. (1964). In "Atomic Collision Processes" (M. R. C. McDowell, ed.), p. 997. North-Holland Publ., Amsterdam.

Dunn, G. H. (1966). *J. Chem. Phys.* **44**, 2592.
Dunn, G. H., and Kieffer, L. J. (1963). *Phys. Rev.* **132**, 2109.
Dunn, G. H., and Van Zyl, B. (1967). *Phys. Rev.* **154**, 40.
Dunning, F. B., and Smith, A. C. H. (1970). *Proc. Phys. Soc., London (At. Mol. Phys.)* [2] **3**, 60.
Fehsenfeld, F. C., Schmeltekopf, A. L., Goldan, P. D., Schiff, H. I., and Ferguson, E. E. (1966). *J. Chem. Phys.* **44**, 4087.
Fite, W. L., Smith, A. C. H., and Stebbings, R. F. (1962). *Proc. Roy. Soc., Ser. A* **268**, 527.
Geller, M., and King, J., Jr. (1974). To be published.
Giese, C. G., and Maier, W. B., II. (1963). *J. Chem. Phys.* **39**, 739.
Gross, S. H., and Rasool, S. I. (1964). *Icarus* **3**, 311.
Gupta, B. K., and Matsen, F. A. (1967). *J. Chem. Phys.* **47**, 4860.
Hinteregger, H. E. (1970). *Ann. Geophys.* **26**, 547.
Holliday, M. G., Muckerman, J. T., and Friedman, L. (1971). *J. Chem. Phys.* **54**, 1058.
Hotop, H., and Niehaus, A. (1968). *Z. Phys.* **215**, 395.
Hudson, R. D. (1971). *Rev. Geophys.* **9**, 305.
Hudson, R. D., and Kieffer, L. J. (1971). *At. Data* **2**, 205.
Hunten, D. M. (1969). *J. Atmos. Sci.* **26**, 826.
Huntress, W. T., Jr., and Bowers, M. T. (1973a). *Int. J. Mass Spectrom. Ion Phys.* **12**, 1.
Huntress, W. T., Jr., and Bowers, M. T. (1973b). *Int. J. Mass Spectrom. Ion Phys.* **12**, 357.
Huntress, W. T., Jr., and Kim, J. K. (1974). To be published.
Huntress, W. T., Jr., and Pinizzotto, R. F., Jr. (1973). *J. Chem. Phys.* **59**, 4742.
Huntress, W. T., Jr., Laudenslager, J. B., and Pinizzotto, R. F., Jr. (1974a). *Int. J. Mass Spectrom. Ion Phys.* **13**, 331.
Huntress, W. T., Jr., Theard, L. P., and Kim, J. K. (1974b). To be published.
Janev, R. K., and Tančic, A. R. (1972). *Proc. Phys. Soc., London (At. Mol. Phys.)* [2] **5**, L250.
Kass, R. S., and Williams, W. L. (1971). *Phys. Rev. Lett.* **27**, 473.
Lamb, W. E., Jr., and Retherford, R. C. (1950). *Phys. Rev.* **79**, 549.
Laudenslager, J. B., and Bowers, M. T. (1972). *Abstr., 25th Annu. Gaseous Electron. Conf.* (unpublished).
Laudenslager, J. B., Huntress, W. T., Jr., and Bowers, M. T. (1974). To be published.
Leu, M. T., Biondi, M. A., and Johnson, R. (1973). *Phys. Rev. A* **8**, 413.
McElroy, M. B. (1973). *Space Sci. Rev.* **14**, 460.
McGowan, J. W., Williams, J. F., and Vroom, D. A. (1969). *Chem. Phys. Lett.* **3**, 614.
Maier, W. B., II. (1971). *J. Chem. Phys.* **54**, 2732.
Marr, G. V. (1967). "Photoionization Processes in Gases." Academic Press, New York.
Michels, H. H. (1966). *J. Chem. Phys.* **44**, 3834.
Miller, T. M., Moseley, J. T., Martin, D. W., and McDaniel, E. W. (1968). *Phys. Rev.* **173**, 115.
Monahan, K. M., Huntress, W. T., Jr., Lane, A. L., Ajello, J. M., Burke, T. E., LeBreton, P., and Williamson, A. (1974). *Planet. Space Sci.* **22**, 143.
Moos, H. W., and Woodworth, J. R. (1973). *Phys. Rev. Lett.* **30**, 775.
Moseley, J., Aberth, W., and Peterson, J. (1970). *Phys. Rev. Lett.* **24**, 435.
Muschlitz, E. E., Jr., (1968). *Science* **159**, 599.
Niles, F. E., and Robertson, W. W. (1965). *J. Chem. Phys.* **42**, 3277.
Olmstead, J., III, Newton, A. S., and Street, K., Jr. (1965). *J. Chem. Phys.* **42**, 2321.
Phelps, A. V. (1955). *Phys. Rev.* **99**, 1307.
Prasad, S. S., and Capone, L. A. (1971). *Icarus* **15**, 45.
Prinn, R. G. (1970). *Icarus* **13**, 424.
Rapp, D., and Englander-Golden, P. (1965). *J. Chem. Phys.* **43**, 1480.
Rapp, D., Briglia D. D., and Sharp, T. E. (1965). *Phys. Rev. Lett.* **14**, 533.
Rishbeth, H. (1959). *Aust. J. Phys.* **12**, 466.
Rutherford, J. A., and Vroom, D. A. (1973). *J. Chem. Phys.* **58**, 4076.

Samson, J. A. R. (1972). *Chem. Phys. Lett.* **12**, 625.
Sando, K. M., Cohen, J., and Dalgarno, A. (1971). *In* "VII ICPEAC, Abstracts of Papers" (L. M. Branscomb *et al.*, eds.), Vol. 2, p. 973. North-Holland Publ., Amsterdam.
Saporoschenko, M. (1965). *J. Chem. Phys.* **42**, 2760.
Schmeltekopf, A. L., and Fehsenfeld, F. C. (1970). *J. Chem. Phys.* **53**, 3173.
Schmeltekopf, A. L., Fehsenfeld, F. C., and Ferguson, E. E. (1967). *Astrophys. J.* **148**, L155.
Sharp, T. E. (1971). *At. Data* **2**, 119.
Shaw, M. J., Bolden, R. C., Hemsworth, R. S., and Twiddy, N. D. (1971). *Chem. Phys. Lett.* **8**, 148.
Shimizu, M. (1964). *Progr. Theor. Phys.* **32**, 977.
Shimizu, M. (1971). *Icarus* **14**, 273.
Sholette, W. P., and Muschlitz, E. E., Jr. (1962). *J. Chem. Phys.* **36**, 3368.
Smith, G. M. and Muschlitz, E. E., Jr. (1960). *J. Chem. Phys.* **33**, 1819.
Stebbings, R. F., Dunning, F. B., Tittel, F. K., and Rundel, R. D. (1973). *Phys. Rev. Lett.* **18**, 815.
Stone, E. J., and Zipf, E. C. (1972). *J. Chem. Phys.* **56**, 4646.
Tanaka, T., and Hirao, K. K. (1973). *Planet. Space Sci.* **21**, 751.
Theard, L. P., and Huntress, W. T., Jr. (1974). *J. Chem. Phys.* **60**, 2840.
Trajmar, S., Truhlar, D. G., and Rice, J. K. (1970). *J. Chem. Phys.* **52**, 4502.
Van Brunt, R. J., and Kieffer, L. J. (1970). *Phys. Rev. A* **2**, 1293.
Veatch, G. E., and Oskam, H. J. (1973). *Phys. Rev. A* **8**, 389.
Vroom, D. A., and de Heer, F. J. (1969). *J. Chem.Phys.* **50**, 580.
Zabriskie, F. R. (1960). Dissertation, Princeton University, Princeton, New Jersey.

SUBJECT INDEX

A

Abbe-type comparator, 232
Absorption spectroscopy, dye lasers in, 179–182
see also Dye lasers
Acetone, spectrum of, 135
Alkali metal atoms
  electron impact, excitation by, 71–74
  proton impact, ionization by, 86
Argon
  electron impact, ionization by, 104–105
  proton impact, ionization by, 112
Aromatic hydrocarbons, autoionization in, 163–164
Atom-atom collisions, 113–124
  electron loss and ionization in, 119–124
  excitation in, 114–119
Atomic beam
  magnetic deflection of, 210–212
  recoil deflection of, 212–213
Atomic spectroscopy
  see also Spectroscopy
  atomic-beam deflection in, 210–213
  dye lasers in, 173–217
  nonlinear coherent resonant phenomena in, 215–217
Atomic structure theory, relativistic effects in, 1–50
Atoms
  aufbau process in, 148, 154
  excitation by electron and proton impact, 54–87
  ionization by electron and proton impact, 87–112
  photoionization of, 194–196
  relativistic effect in many-electron, 1–50
  two-step ionization of, 194–195
Aufbau process, in atoms, 148–149, 154
Autoionization, 159–164
  in aromatic hydrocarbons, 163–164
  and satellite decay, 253
  vibrational, 164

B

Beam foil source, 231–232

Bethe-Salpeter wave equation, 6
Born approximation, in collisions,
  alkali metal atoms, electron impact, excitation by, 71–74
  argon, electron impact, ionization by, 104–105
  argon, proton impact, ionization by, 112
  atom-atom collisions, 113–124
  calcium, electron impact, excitation by, 78
  cross sections, relation between those for electron and proton impact, 55–56
  differential cross sections, 57, 88–89, 95, 109, 110
  electron loss, 119–124
  and generalized oscillator strength, 57
  heavy rare gases, electron impact, excitation by, 75–76
    LS coupling in, 75
  helium, electron impact, excitation by, 62–71
  helium, proton impact, ionization by, 107–110
  hydrogen, electron impact, excitation by, 57–62
  hydrogen, electron impact, ionization by, 89–91
  hydrogen, proton impact, ionization by, 106–107
  impact parameter treatment, 55
  ion-atom collisions, 113–124
  lithium, proton impact, ionization by, 97–99
  magnesium, electron impact, excitation by, 78
  magnesium, electron impact, ionization by, 104
  mercury, electron impact, excitation by, 77
  neon, proton impact, ionization by, 111–112
  oxygen, electron impact, excitation by, 78
  oxygen, electron impact, ionization by, 100–101
  velocity-length results, comparison, 70, 84, 85, 93
  wave treatment, 54, 88–89

Subject Index

Breit equation, 7
Breit Hamiltonian
  derivation of, 5
  errors in, 10
  one-photon exchange processes in, 6
Breit interaction, 5–7, 18
  Pauli limit and, 16
Breit perturbation, 16

C

Caesium
  electron impact, excitation by, 73
  proton impact, excitation by, 78
Calcium, electron impact, excitation by, 78
Calcium-like spectra, 256
Central field, effective operator for, 26–27
Central field Hamiltonian, 11–13
Collisions, fast, *see* Born approximation
Correlation effects, relativistic, 49–50
Cross section, photoionization, in photoelectron energy, 157–158
Coulomb central field, effective operator form, 27
Coulomb interaction and Dirac electrons, 6, 9–10
Coulomb wave functions, use in Born approximation, 92

D

Darwin term, in perturbation Hamiltonian, 15
Diatomic molecules, photoionization of, 166
Dirac electron, in central field, 7–10
Dirac equation, 29
Dirac Hamiltonian, 6
Dirac-Hartree-Fock program, 49
Dirac wave equation, 3–5
Doppler absorption profile, 199
Doppler broadening, 206, 210
Doppler-free spectroscopy, 197, 209
Dye lasers
  in absorption spectroscopy, 179–182
  applications of, 177–217
  in atomic spectroscopy, 173–217
  in collisional quenching of resonance radiation, 193
  fluorescence line narrowing in, 205
  in fluorescence spectroscopy, 182–194, 205–209
  in heterodyne spectroscopy, 214–215
  high spectral densities of, 177–197
  level crossing and optical radiofrequency double resonance experiments with, 191–193
  lifetimes measurements with, 182–185
  in optical nutation experiment, 216
  output of, 176–177
  photoionization and photodetachment with, 194–196
  power and energy of, 176–177
  properties of, 174–177
  pulsed fixed-frequency lasers in, 175
  pulse properties of, 177
  quantum-beat and modulation experiments with, 185–191
  radio-frequency and optical spectroscopy, 215–216
  in saturation spectroscopy, 198–204
  spectral width of, 176
  in transient nutation experiment, 216–217
  tunable, 197–217
  tuning range of, 175
  in two-photon processes, 196–197

E

Eclipse solar ultraviolet spectra, 281–283
Effective operator (relativistic)
  for central field, 26
  for Coulomb central field, 27
  for Coulomb interaction, 27–29
  general concepts of, 21–22
  for H_p and H_B, 26–30
  hyperfine structure in, 32–36
  involving fields, 30–40
  in many-electron relativistic effects, 21–40
  nonrelativistic limits in, 25–26
  one-electron, 22–23
  Pauli limit and, 39
  transition probabilities in, 36–40
  two-electron, 24
  Zeeman effect and, 30–32
Electronic correlation effects, in relativistic scheme, 49
Electronic energy, vibrational energy conversion to, 164

SUBJECT INDEX

Electron impact excitation, 54–87 see Born approximation in collisions
ion dissociation and, 332
Electron impact ionization, 89–105 see Born approximation in collisions
Electron loss, in collisions, 119–124
Electron-molecule interactions, 167–168
ESCA (electron spectroscopy for chemical analysis), 134
Ethyl trifluoroacetate, spectrum of, 135
Excitation transfer, with dye lasers, 193–194
Excited neutral species, in Jovian ionosphere, 321–322

F

First-order perturbation theory, relativistic effects, 36
Fluorescence line narrowing, dye lasers and, 205
Fluorescence spectroscopy
 classic absorption and, 205–209
 dye lasers in, 182–194
Foldy-Wouthuysen transformation, 11, 42
Forbidden lines, in spectra classification, 240, 246
Franck-Condon distribution in H_2^+, 299
Frequency offset locked laser spectroscopy, 214

G

Glass lasers, 175
Grazing incidence solar spectrum, 262, 279

H

H_B effective operator form for, 29–30
Halogen derivatives of methane, photoelectron spectra of, 143–144
Hartree-Fock-Slater relativistic approximation, 48
Hartree-Fock functions or techniques
 and Born approximation, 92, 98–99
 multiconfigurational, 237–239
 relativistic, 44–47
 in spectra classification, 236–238, 249, 253
Hartree-X method, 236–237

Heavy ions, spectra of, 257–258
Heavy rare gases, electron impact, excitation by, 75–76
Helium
 electron impact, excitation by, 62–71
 electron impact, ionization by, 92–97
 exotic ions of, 326–327
 generalized oscillator strengths, 65–68, 97
 proton impact, excitation by, 80–86
 proton impact, ionization by, 107–110
Helium-hydrogen ion, in Jovian ionosphere 307–309
Helium ion, in Jovian ionosphere, 304–305
Helium-like spectra, 250–253
Heterodyne spectroscopy, dye lasers in, 214–215
Highly ionized atoms
 in solar spectra, 262–283
 spectra classification for, 223–285
Highly ionized systems, line classifications 239–261
Hydrids (isoelectronic with inert gases) photoelectron spectra for, 139–142
Hydrogen
 in Jovian ionosphere, 304–306, 333–334
 molecular ion, Franck-Condon distribution in, 299
 photoionization of molecular, 298
Hydrogen atomic
 electron impact, excitation by, 59–60
 electron impact, ionization by, 89–91
 proton impact, excitation by, 78–80
 proton impact, ionization by, 106–107
Hydrogen-helium mixtures, radiation chemistry of in Jovian ionosphere, 295–338
Hyperfine structure, effective operator and, 32–36

I

Inert gas
 electron impact, excitation by, 62–71, 75–77
 electron impact, ionization by, 91–97, 101–105
 partitioning of three protons from nucleus of, 141
 proton impact, excitation by, 80–86
 proton impact, ionization by, 107–112

Interelectronic interaction, Breit formulation of, 2
Ion, photodissociation, 334–335
Ion-atom collisions, 113–124
　electron loss and ionization in, 119–124
　excitation in, 114–119
Ion dissociation, electron impact excitation and, 332
Ionization
　see Born approximation in collisions
　secondary, 328–330
Ionization stages, separation of, 234–235
Ion-neutral reactions, in Jovian ionosphere, 300–309

J

Jahn-Teller splitting, 142
Jovian ionosphere
　ion-electron recombination in, 309–311
　ion loss rates in, 322–327
　ion-neutral reactions in, 300–309
　methane photoionization in, 335
　negative ion production in, 332–334
　radiative association in, 311–314
　rate constants required, 336–337
　terminal-ion loss processes in, 309–322
　three-body processes in, 314–319
Jupiter
　see also Jovian ionosphere
　exploration of, 295–296
　lower atmosphere of, 296
　upper atmosphere of, 297

L

Landé-g factor, relativistic, 31
Laser frequency, in Doppler-free spectroscopy, 197–198
Laser-produced plasmas, highly ionized spectra of, 228–230
Lasers, dye, see Dye lasers
Level-crossing experiments, with dye lasers, 191
Lifetimes measurement, with dye lasers, 182–185
Light sources, in spectral analysis, 225–232
Line classification, published data on, 258–261

Line separation, source density and, 235–236
Lithium
　electron impact, excitation by, 72
　electron impact, ionization by, 97–99
　proton impact, excitation by, 86
Lorentz invariant theory, 6

M

Magnesium
　electron impact, excitation by, 78
　electron impact, ionization by, 104
Magnetic dipole moment, 33
Many-electron atom, relativistic effects, 1–50
　effective operators in, 21–40
　interaction with external field, 13–14
　nonrelativistic limits in, 10–21
MCHF (multiconfigurational Hartree-Fock method), 237–239
Mercury, electron impact, excitation by, 77
Metastable atoms
　destruction in fast atom-atom collisions, 124
　in Jovian ionosphere, 329
Metastable systems in Jovian ionosphere, 328–332
Methane, halogen derivatives of photoelectron spectroscopy, 143–144
Methane, in Jovian atmosphere, 296
Methane, in Jovian ionosphere,
　photoionization of, 335
　reactions, 319–320, 324
Michelson interferometer, 233
Modulation experiments, with dye lasers, 185–193
Molecules
　multiple bonded diatomic, 144–149
　triatomic, 151–154
Multiple bonded systems, photoelectron spectra and structure of, 144–151

N

Negative ions, in Jovian ionosphere, 332–334
Neon
　electron impact, ionization by, 101–103
　proton impact, ionization by, 111–112

SUBJECT INDEX

Nitrogen, atomic, electron impact, excitation by, 99–100
Nonrelativistic Hamiltonian, 20–21
Nonrelativistic limits
  higher-order terms in, 17–19
  in many-electron atom, 10–21, 25–26
Nucleus, size and mass of, 48–49

O

One-electron effective operators, 22–23
Optical nutation experiment, dye lasers in, 216
Optical resonances, 209–214
Orbital approximation, photoionization process in, 157
Oscillator strength, generalized, 57, 89
  of helium, 65–68, 97
Oxygen, atomic
  electron impact, excitation by, 78
  electron impact, ionization by, 100–101
Oxygen, molecular
  absorption energy loss spectrum of, 160
  autoionization in, 162
  potential energy curves of, 161

P

Parametric method, in relativistic radial wave functions, 42–43
Partial cross sections, in photoionization of diatomic molecules, 166
Pauli Hamiltonian, 12
Pauli limit
  effective operator and, 39
  in higher-order terms, 17–18
  of two-body interactions, 15–16
  Zeeman effect and, 32
Pauli spin matrix, 9
Perturbation Hamiltonian, 15–16
Perturbations, relativistic plus nonrelativistic, 40–42
Perturbation theory, relativistic effects, 36–40
  second-order, 19
Photodetachment, with dye lasers, 194–196
Photoelectron energy, photoionization cross section in, 157–158

Photoelectrons
  angular distribution of, 164–166
  ionization processes of, 328–334
Photoelectron spectra
  intensities of, 155–156
  of halogen derivatives of methane, 143–144
  of multiple bonded diatomic molecules, 144–149
  orbital assignment of bands in, 154
  of triatomic molecules, 151–154
Photoelectron spectroscopy, 131–169
  angular distribution of photoelectrons in, 164–166
  electron-molecule interactions in, 167–168
  success of, 169
  technical development of, 133–134
  transient species in, 168
  ultraviolet, 136–155
  X-ray, 134–136
Photoionization
  with dye lasers, 194–196, 214
  kinetic energy from, 328
  in orbital approximation, 157
  production of ions by, 297–299
Photoionization cross section, 155
Photoionizing bands, Rydberg bands and, 162
Photon impact processes, 334–335
Pioneer mission, to Jupiter, 296
Plasma machine, in spectral analysis, 226–228
Plasmas, laser-produced, 228–230
Polyatomic molecules, spectra and structure of, 149–151
Potassium
  electron impact, excitation by, 73
  proton impact, excitation by, 86
Proton impact excitation, 54–87, see Born approximation in collisions
Proton impact ionization, 105–112, see Born approximation in collisions
Pseudopotential method, 44–46
p-type orbitals, photoelectron band energies of, 157–158

Q

Quantum-beat experiments, with dye lasers, 185–191

Quantum numbers
  in highly ionized systems, 249–250
  in spectra classification, 254–255

R

Rabi magnetic resonance method, 210–211
Radial wave functions, relativistic, 42–50
  hydrogenic, 43–44
  pseudopotential method in, 44–46
  self-consistent field calculations, in, 46–68
Radiative association, in Jovian ionosphere, 311–314
Radio-frequency spectrometry, vs. optical, 215–216
Raman scattering, dye lasers and, 196
Recombination
  dielectronic and satellites, 253
  in Jovian ionosphere, 309–311
Relativistic effects
  basic concepts in, 3–10
  Breit interaction in, 5–7
  correlation, 49–50
  Dirac electron in central field in, 7–10
  Dirac wave equation in, 3–5
  Hamiltonian, 21
Relativistic radial wave functions, 42–50
  configuration mixing of correlates in, 49–50
  parametric method in, 42–43
  size and mass of nucleus in, 48–49
Resonance radiation
  collisional quenching of, 193
  impact excitation of, 71–74, 86
Ritz combination principle, 240
Rockets, solar spectra from, 224, 243, 246, 248, 262–265, 281–283
Rubidium, electron impact, excitation by, 73
Ruby lasers, 175, 228
Rydberg bands, photoionizing bands and, 162
Rydberg states, 132

S

Satellites, orbiting,
  projects for, 257, 284–285
  solar spectra from, 224, 248–252, 262–265, 281, 283
Satellites, spectra classification for, 250–257
Saturation spectroscopy, dye lasers in, 198–204
Schrödinger equation
  many-electron, 10
  one-electron, 8
Secondary ionization processes, in Jovian ionosphere, 328–330
Slater energy parameters, 237, 241
Slater-Condon theory, 236
Sodium
  electron impact, excitation by, 72
  electron impact, ionization by, 103–104
  proton impact, excitation by, 86
Sodium azide, spectrum of, 135
Solar flare, spectrum of, 248, 250–252
Solar ionization, and Jovian ionosphere, 300–309
Solar spectra
  between 1 and 25 Å, 250–251, 283
  between 20 and 2000 Å, 262–281
  eclipse, 281–283
  emission lines of highly ionized atoms in, 262–283
Source density, line separation and, 235–236
Spectra classification for highly ionized atoms, 223–285 See also Spectral analysis, light sources
  calculations for, 236–238
  first long period in, 255–257
  of heavy elements, 257–258
  ionization stages in, 234–235
  line classification of highly ionized systems in, 239–261
  line separation vs. source density in, 235–236
  measuring and experimental techniques in, 232–236
  published data in, 258–261
  quantum number in, 254–255
Spectral analysis, light sources, 225–232
  beam foil, 231–232
  high voltage vacuum spark, 230–231
  laser produced plasmas, 228–230
  low inductance high capacity spark, 231
  theta pinch and other plasma sources, 226–228
Spectral widths, with dye lasers, 178

SUBJECT INDEX

Spectrograms, time-resolved, 234
Spectroscopy, absorption, *see* Absorption (*adj.*)
  see Dye lasers
Spin-orbit splitting, 241
Sun, soft X-ray spectrum of, 248, 250–252
  see also Solar (*adj*).

T

Terminal-ion loss processes, in Jovian ionosphere, 309–322
Terminal ions in Jovian ionosphere
  and excited neutral species, 321–322
  reaction of with methane, 320–321
  three-body reactions of, 316
Three-body processes, in Jovian ionosphere 314–319
Time-resolved spectrograms, recording of, 234
Toroidal devices, 226
Transition probabilities, effective operator in, 36–40
Transient nutation experiment, dye lasers in, 216
Triatomic molecules, spectra and structure of, 151–154
Tunable dye lasers
  see also Dye lasers
  applications of, 197–217
  Doppler absorption profile and, 198–199
  fluorescence line narrowing with, 205
Two-electron operators, relativistic, 24
Two-photon processes, dye lasers in, 196–197

U

Ultraviolet photoelectron spectroscopy, 136–155
  general features of, 136–139
  photoelectron spectra of hydrids, isoelectronic with inert gases, 139–142
  physical aspects of, 155–169
Ultraviolet spectra
  eclipse solar, 281–283
  solar vacuum, 262

V

Vibrational energy, conversion of to electronic energy, 164
Vibrationally excited ions, in Jovian ionosphere, 313

W

Wigner-Eckhart theorem, 22, 31, 34

X

Xenon lasers, 175
X-ray photoelectron spectroscopy, 134–136

Z

Zeeman effect, 30–32

ECS
PC 3567
£16.80

SCIENCE